I0885410

The IoT and the Next Revolutions Automating the World

Dinesh Goyal
Poornima Institute of Engineering & Technology, India

S. Balamurugan
QUANTS Investment Strategy & Consultancy Services, India

Sheng-Lung Peng
National Dong Hwa University, Taiwan

Dharm Singh Jat
Namibia University of Science and Technology, Namibia

A volume in the Advances in Web Technologies and Engineering (AWTE) Book Series

IGI Global
DISSEMINATOR OF KNOWLEDGE

Published in the United States of America by
IGI Global
Engineering Science Reference (an imprint of IGI Global)
701 E. Chocolate Avenue
Hershey PA, USA 17033
Tel: 717-533-8845
Fax: 717-533-8661
E-mail: cust@igi-global.com
Web site: http://www.igi-global.com

Library of Congress Cataloging-in-Publication Data

Names: Goyal, Dinesh, 1976- editor.
Title: The IoT and the next revolutions automating the world / Dinesh Goyal, Shanmugam Balamurugan, Sheng-Lung Peng, and Dharm Singh Jat, editors.
Description: Hershey, PA : Engineering Science Reference, 2019. | Includes bibliographical references.
Identifiers: LCCN 2018060055| ISBN 9781522592464 (hardcover) | ISBN 9781522592471 (softcover) | ISBN 9781522592488 (ebook)
Subjects: LCSH: Internet of things. | Automatic control.
Classification: LCC TK5105.8857 .I677 2019 | DDC 384.3/3--dc23 LC record available at https://lccn.loc.gov/2018060055

This book is published in the IGI Global book series Advances in Web Technologies and Engineering (AWTE) (ISSN: 2328-2762; eISSN: 2328-2754)

British Cataloguing in Publication Data
A Cataloguing in Publication record for this book is available from the British Library.

All work contributed to this book is new, previously-unpublished material. The views expressed in this book are those of the authors, but not necessarily of the publisher.

For electronic access to this publication, please contact: eresources@igi-global.com.

Advances in Web Technologies and Engineering (AWTE) Book Series

Ghazi I. Alkhatib
The Hashemite University, Jordan
David C. Rine
George Mason University, USA

ISSN:2328-2762
EISSN:2328-2754

Mission

The **Advances in Web Technologies and Engineering (AWTE) Book Series** aims to provide a platform for research in the area of Information Technology (IT) concepts, tools, methodologies, and ethnography, in the contexts of global communication systems and Web engineered applications. Organizations are continuously overwhelmed by a variety of new information technologies, many are Web based. These new technologies are capitalizing on the widespread use of network and communication technologies for seamless integration of various issues in information and knowledge sharing within and among organizations. This emphasis on integrated approaches is unique to this book series and dictates cross platform and multidisciplinary strategy to research and practice.

The **Advances in Web Technologies and Engineering (AWTE) Book Series** seeks to create a stage where comprehensive publications are distributed for the objective of bettering and expanding the field of web systems, knowledge capture, and communication technologies. The series will provide researchers and practitioners with solutions for improving how technology is utilized for the purpose of a growing awareness of the importance of web applications and engineering.

Coverage

- Software agent-based applications
- IT readiness and technology transfer studies
- Strategies for linking business needs and IT
- Web systems engineering design
- Quality of service and service level agreement issues among integrated systems
- IT education and training
- Security, integrity, privacy, and policy issues
- Data analytics for business and government organizations
- Web Systems Architectures, Including Distributed, Grid Computer, and Communication Systems Processing
- Web user interfaces design, development, and usability engineering studies

IGI Global is currently accepting manuscripts for publication within this series. To submit a proposal for a volume in this series, please contact our Acquisition Editors at Acquisitions@igi-global.com or visit: http://www.igi-global.com/publish/.

The Advances in Web Technologies and Engineering (AWTE) Book Series (ISSN 2328-2762) is published by IGI Global, 701 E. Chocolate Avenue, Hershey, PA 17033-1240, USA, www.igi-global.com. This series is composed of titles available for purchase individually; each title is edited to be contextually exclusive from any other title within the series. For pricing and ordering information please visit http://www.igi-global.com/book-series/advances-web-technologies-engineering/37158. Postmaster: Send all address changes to above address.

Titles in this Series

For a list of additional titles in this series, please visit: www.igi-global.com/book-series

Integrating and Streamlining Event-Driven IoT Services
Yang Zhang (Beijing University of Posts and Telecommunications, China) and Yanmeng Guo (Chinese Academy of Sciencs, China)
Engineering Science Reference • ©2019 • 309pp • H/C (ISBN: 9781522576228) • US $205.00

Semantic Web Science and Real-World Applications
Miltiadis D. Lytras (The American College of Greece, Greece) Naif Aljohani (King Abdulaziz University, Saudi Arabia) Ernesto Damiani (Khalifa University, UAE) and Kwok Tai Chui (City University of Hong Kong, ong Kong)
Information Science Reference • ©2019 • 399pp • H/C (ISBN: 9781522571865) • US $195.00

Innovative Solutions and Applications of Web Services Technology
Liang-Jie Zhang (Kingdee International Software Group Co., Ltd., China) and Yishuang Ning (Tsinghua University, China)
Engineering Science Reference • ©2019 • 316pp • H/C (ISBN: 9781522572688) • US $215.00

Dynamic Knowledge Representation in Scientific Domains
Cyril Pshenichny (ITMO University, Russia) Paolo Diviacco (Istituto Nazionale di Oceanografia e di Geofisica Sperimentale, Italy) and Dmitry Mouromtsev (ITMO University, Russia)
Engineering Science Reference • ©2018 • 397pp • H/C (ISBN: 9781522552611) • US $205.00

Innovations, Developments, and Applications of Semantic Web and Information Systems
Miltiadis D. Lytras (American College of Greece, Greece) Naif Aljohani (King Abdulaziz University, Saudi Arabia) Ernesto Damiani (University of Milan, Italy) and Kwok Tai Chui (City University of Hong Kong, Hong Kong)
Engineering Science Reference • ©2018 • 473pp • H/C (ISBN: 9781522550426) • US $245.00

Handbook of Research on Biomimicry in Information Retrieval and Knowledge Management
Reda Mohamed Hamou (Dr. Tahar Moulay University of Saida, Algeria)
Engineering Science Reference • ©2018 • 429pp • H/C (ISBN: 9781522530046) • US $325.00

Global Perspectives on Frameworks for Integrated Reporting Emerging Research and Opportunities
Ioana Dragu (Babes-Bolyai University, Romania) Adriana Tiron-Tudor (Babes-Bolyai University, Romania) and Szilveszter Fekete Pali-Pista (Babes-Bolyai University, Romania)
Business Science Reference • ©2018 • 160pp • H/C (ISBN: 9781522527534) • US $135.00

IGI Global
DISSEMINATOR OF KNOWLEDGE

701 East Chocolate Avenue, Hershey, PA 17033, USA
Tel: 717-533-8845 x100 • Fax: 717-533-8661
E-Mail: cust@igi-global.com • www.igi-global.com

Table of Contents

Detailed Table of Contents

In the present scenario, sensors place a major role for implementing smart devices. Internet of Things (IoT) is an advancement of sensors which can communicate with non-communicate things (devices). Many of the developed counties are using smartness in creating and communicating devices using IoT. In India, major challenges focus on how and where to implement smartness. Hence, authors found some different areas like healthcare, education, transport, water, energy, communication, security & safety, citizen services, and so on. All these areas are covered by a smart way using recent technology (IoT) in smart cities concepts. Various technologies like IoT, Big Data, and cloud computing are used for constructing smartness in the form of devices. In this Chapter, authors focus on a smart waste management system using IoT. They provide various smart bin construction technology, advantages, standards and challenges in detail. It is very useful to the reader to understand the various method of waste management in smart cities development using IoT.

The use of Radio Frequency Identification (RFID) in Supply Chain Management (SCM) is one of the promising innovations in recent decades. This chapter first presents an introduction to the concepts and principles of RFID. It then discusses advantages and disadvantages of this technology in a supply chain setting. Application areas of RFID in the context of supply chains are reviewed to demonstrate best practices and related important implementation issues. Different industries (e.g. automotive, transport, retail) are used to emphasizing the benefits of RFID technology. The chapter also highlights operational and strategic implications of adopting RFID-based technological solutions and summarizes available evidence. Finally, a theoretical framework that links RFID key benefits and information attributes used in decision making is proposed. This chapter also provides comprehensive guidance for those considering the implementation of RFID in their supply chains.

Over the last two decades, the evolution of mobile technologies has led to an unprecedented adoption of cellphones in mainstream society. Consumers have moved from having the ability to make telephone calls from anywhere at any time using the traditional cellphones to having an all-access pass to cyberspace using today's smartphones. This portable device has become a reliable support system for the user because of the accessibility and flexibility it offers to maintain the users' daily routine. Authors in this chapter have chosen software optimization techniques to increase battery efficiency because these techniques are more robust. This chapter introduces a novel idea of an automated system for smartphones that prioritize application access based on the owner's usage patterns and daily routine to conserve battery life. This system will serve two purposes: save battery power and improve the smartphone's artificial intelligence.

Radio Frequency Identification (RFID) is a technology that uses radio frequency signals to identify tagged objects. RFID is an important technology used by the Internet of Things (IoT) applications. This technology enables communication between the main devices used in RFID system, the reader, and the tags. The tags share a common communication channel. Therefore, if more than one tag tries to send information at the same time, the reader will be incapable of differentiating these signals in the case of radio signals interference. This phenomenon is known as tag collision problem. The problem of tag collision is one of the major disadvantages for fast tagged-object identification in supply chain management. This chapter describes four different types of binary search algorithms for avoidance of tag collision, and then presents a performance measurement mechanism for RFID application system. Finally, simulation-based experimental results on the performance of these algorithms are presented.

This research focuses on a solution to assist elderly and limited-mobility people. It aims to improve the autonomy, and, consequently, the quality of life of this target audience by automating daily tasks conducted at home, such as turning on the lights and manipulating electronic devices. However, it is important to consider the costs and quality attributes (e.g., usability) related to the design of solutions to automate a specific environment, that may include hardware platforms and physical adaptations. In this context, the authors present in this chapter the software requirements discovery and elicitation of a home automation app considering the real needs of the elderly and limited-mobility people. Additionally, we conduct the requirements specification using the unified modeling language (UML) to improve completeness, along with graphical user interface (GUI) prototypes. Finally, we present a mobile app prototype using the Android and Arduino platforms to illustrate a use scenario of the solution.

Smart Homes and Offices (SHO) are composed of interlinked components with constant data transfer and services targeted at increasing the lifestyle of the people. This chapter describes about the smart components and how SHO are direct implementation of Internet of Things (IOT). The major paradigm in this chapter is appliances supporting smart aspects of SHO, their applications and change in technology in context of smart Homes and Offices. Here we have also discussed the standardization and personalization of gadgets and how it has been increasing our standard of living. Finally, the chapter focuses on privacy preserving mechanisms, its essence over smart cities, strong architecture related to privacy, preserving mechanism, and various approaches available that can retaliate these issues in a smart city environment.

With the increasing prevalence of diabetes mellitus, it has become of utmost importance for wearable technologies for glucose monitoring to be introduced in different contexts. While there is a high number of research on noninvasive techniques for glucose monitoring of diabetes mellitus, there is a shortage of studies discussing the publication trend of wearable technologies that support glucose monitoring. The primary purpose of this chapter was to conduct a Systematic Mapping Study of publication trends relating to wearable technologies for glucose monitoring. This study adopted a Systematic Mapping Study approach in identifying relevant papers for analysis. Articles were identified from relevant databases including IEEE Xplore, ACM Digital Library, ScienceDirect and Scopus. A total of 29 papers met the inclusion criteria. The findings of this study are expected to inform health informatics experts and academics on the current research and publication trends in wearable technologies for glucose monitoring.

With asthma being one of the leading causes of death in different countries, the emphasis on improving the health of asthma patients is important. While the use of smart technologies is a good approach for improving the health of asthma patients, technologies need to be connected in such a way that all components of smart health form an ecosystem. However, the components of such an ecosystem have not been identified in the current literature. The purpose of this chapter was to identify the components of a smart health ecosystem for asthma patients through a systematic literature review. A total of 28 articles met the inclusion criteria. This chapter identified the components of a smart health ecosystem for asthma patients and provided a conceptual framework. The findings of the systematic literature review are expected to inform researchers on the components required for building a smart health ecosystem for asthma patients.

Chapter 9
Role of Smart Wearable in Healthcare: Wearable Internet of Medical Things (WIoMT).................. 133
Jana Shafi, Prince Sattam Bin Abdul Aziz University, Saudi Arabia
Amtul Waheed, Prince Sattam Bin Abdulaziz University, Saudi Arabia

Remote Health monitoring is encouraged to keep an eye on the patient's health status outside the zone of the clinic. These smart wearable gadgets are now integrated with mobile apps to work efficiently as telemedicine and telehealth to incorporate into the Internet of Medical Things. This chapter introduces WIoMT (Wearable Internet of Medical Things) and reviews smart wearable healthcare devices' scientific terms as well as commercial pains. Internet of Medical Things is displayed through a refined background, wearable computing, wearable technology, cloud frameworks, and architecture design. Included are required hardware and software, body sensors, smartphones, the smart medical application, medical location analyzers for data storage, and finally, a diagnosis. Wearable devices are tested under strict observation fitness, vital signs, and smart environment. Wearable devices are now used for a wide scale of monitoring healthcare.

Chapter 10
Design and Development of Internet of Things-Based Wireless Health Monitoring System 156
Neetu Marwah, Manipal University Jaipur, India

In this world with exponentially increasing fitness risks, the authors are proposing an Internet of Things (IoT) based device named Wireless Health Monitoring System (WHMS), to put patients' thinking at ease. We are dwelling in a world which is no longer viable for a doctor to keep a watch over a patient's indispensable parameters all the time. This device is helpful for aged people staying alone at home, people dwelling on hills, pregnant females anywhere, and for busy people who cannot often contact a doctor. A health practitioner, some distance away from the patient, needs to understand his heart rate and body temperature of the physique to begin preliminary treatment. Keeping this as a preluding landmark, the authors are proposing an embedded gadget which can measure the rate of heart beat and body temperature, and keep the statistics on the cloud server for the doctor to determine the next course of action.

Chapter 11
Smart Roads and Parking.. 168
Rachna Jain, Bharati Vidyapeeth's College of Engineering, India
Kartik Nagia, Bharati Vidyapeeth's College of Engineering, India

The past few years, innovation and technology have reached new heights, much beyond laptops, iPads or smart phones. Innovation and technology is progressing with each passing day. The concepts of smart cities, smart homes, and smart vehicles, are being implemented all over in developed countries. Objects here can get connected to the internet, collect data, and channelize it in order to ease our day-to-day life. The current genre of vehicles is equipped with various kinds of sensors, CPUs, and a software system that has communication capabilities. The increase in population has resulted in more vehicles, congested streets, limited parking spaces, and compromised road safety. Research and industry have proposed many technological advancements, and incorporated a few, in vehicles, but improvements for roads have largely gone unexplored. Smart roads, or Smart highways, are the terms used to define roads having IoT-enabled technologies like smart sensors, wireless connectivity, big data, and cloud computing. They use Solar technology to minimize electricity consumption, making the technology energy efficient. Researchers claim that, in the near future, smart vehicles will not be functional without proper smart roads and smart parking systems. Smart roads and parking contribute to making the drive safe, green, and convenient. They provide real-time information to drivers regarding traffic congestion, weather conditions, natural emergencies, e.g. landslide on the mountains, ice on the roads on high terrain, etc. Smart parking systems help drivers with information regarding available parking spaces as well as warnings about incoming traffic. Smart roads are also conceptualised to be equipped with wireless electric-charging systems and electric-charging stations. Electrical energy generated by the vehicles can be used to light streetlights to provide safe navigation to drivers at night. In this chapter, the concept of Smart Roads and Smart Parking systems is elaborated in a comprehensive manner. Various technologies are highlighted which reduce traffic, limit electricity use, significantly increase safety on the roads, and design a way to use parking spaces more effectively without the need to build new roads/parking spaces. The chapter elaborates various technologies that will lay a strong foundation for smart roads such as transforming legacy roads to smart roads, solar road highways, etc. The chapter also elaborates on transforming traditional parking systems into smart parking systems.

Internet of Things (IoT) is a platform that makes a device smart such that every day communication becomes more informative. A Smart Transportation system basically consists of three components which include smart roads, smart vehicles and a smart parking system. Smart roads are used to describe roads that use sensors and IoT technology which makes driving safer and greener. Smart parking system involves an automated system model that can assist the drivers in selecting the suitable parking spot for them. The data that the system collects will be sent for some analysis. It provides real time information to drivers about various aspects of transportation like weather conditions, traffic scenario, road safety, parking space, and many other things. A well-built Smart Transportation system reduces the risk of accidents, improves safety, increases capacity, reduces fuel consumption, and enhances overall comfort and performance for drivers. Our chapter deals with the in-depth discussion of these various aspects of a smart transportation system enabled with IoT technology.

Preface

Smart home, Smart city and Wearable Technologies are the most exponentially growing applications of Internet of Things (IoT). Wearable Technology is considered to be highly ubiquitous and most phenomenal IoT Application. Smart World and Wearable Technology exhibits a high potential to transform our lifestyle. Ranging from healthcare tracking applications to smart watches and smart bands for personal safety, IoT has started to become one of the most indispensable parts of our lives. It is predicted by Forbes that, by the end of 2017, the wearable IoT market will reach $14 billion (Lamkin, 2016). Also, in another study by Business Insider's premium research service, by the end of 2020, the wearable IoT market is expected to grow and reach 162.9 million units (Meola, 2016). Some of the top-notch applications of IoT include smart parking, smart wearable, smart clothing, smart safety, smart farming, smart industry and so on thereby building next-generation smart world.

Immense foresights on today's communication world seem to vest on development of smart world, where intelligent "things" are connected to serve people better. With the exponential growth of Internet of Things (IoT) and its applications, the implementation of smart world becomes much more feasible. Sensor market plays a vital role in applying IoT to build smart world. Applications such as monitoring of parking space to optimally park the vehicles(Smart Parking), detecting the frequency and intensity of traffic and optimally select routes (Traffic Congestion), prediction of trash levels in garbage collection containers (Smart Waste Management), managing the intensity of street lights based on weather condition and sunlight (Smart Lighting), detecting and controlling excess of hazardous gases coming from industries and vehicles (Smart Air Pollution Controlling), detecting the mix of hazardous chemicals of factories with drinking water (Smart Water Pollution Controlling), efficient monitoring and management of power consumption (Smart Grid), artificial intelligence driven retail marketing (Smart Shopping), detecting health abnormalities (Smart Health) and Home Automation mechanisms fall under the category of applying IoT to build smart world (Delgado, Picking, & Grout, 2000; Patil, Bedara, & Pacharne, 2013).

The Internet of Things is associate degree rising topic of technical, social, and economic significance. Client merchandise, durables, cars and trucks, industrial and utility parts, sensors, and different everyday objects square measure being combined with web property and powerful information analytic capabilities that promise to remodel the manner we have a tendency to work, live, and play. Projections for the impact of IoT on the web and economy square measure spectacular, with some anticipating as several as one hundred billion connected IoT devices and a world economic impact of over $11 trillion by 2025. At identical time, however, the web of Things raises vital challenges that might change the manner of realizing its potential edges. News headlines concerning the hacking of Internet-connected devices, police work issues, and privacy fears have already got captured public attention. Technical challenges stay and new policy, legal and development challenges square measure rising. This summary document is meant to assist the web Society community navigate the dialogue close the web of Things in light-weight of the competitive predictions concerning its guarantees and perils. The web of Things engages a broad set of ideas that square measure advanced and tangled from totally different views.

EMERGING ECONOMY AND DEVELOPMENT ISSUES BY MORE AUTOMATED WORLD

The internet of things holds substantial promise for delivering social and monetary benefits to emerging and growing economies. This includes regions along with sustainable agriculture, water excellent and use, healthcare, industrialization, and environmental management, amongst others. As such, IoT holds promise as a tool in accomplishing the United international locations Sustainable improvement dreams. The vast scope of IoT challenges will no longer be precise to industrialized nations. Growing areas also will want to respond to understand the capacity advantages of IoT. Similarly, the unique needs and demanding situations of implementation in much less-advanced areas will need to be addressed, along with infrastructure readiness, market and investment incentives, technical talent requirements, and coverage resources. Whilst the capacity ramifications are tremendous, a number of ability demanding situations may stand within the manner of this imaginative and prescient—specifically in the areas of security; privacy; interoperability and requirements; legal, regulatory, and rights problems; and the inclusion of rising economies. The net of factors includes a complicated and evolving set of technological, social and policy issues across a numerous set of stakeholders. The internet of factors is going on now, and there's a need to deal with its challenges and maximize its blessings even as decreasing its risks (Gubbi, 2013). The net Society cares about IoT because it represents a growing element of the way human beings and institutions are in all likelihood to interact with and contain the internet and network connectivity into their private, social, and economic lives. Answers to maximizing the advantages of IoT at the same time as minimizing the dangers will not be located by means of conducting a polarized debate that pits the guarantees of IoT towards its possible perils. As an alternative, it's going to take informed engagement, talk, and collaboration across number stakeholders to plan the best methods forward.

Artificial intelligence is remodeling each component of our lives, now not least the economy. As a standard-cause generation, AI's packages are doubtlessly infinite. Even as it could be used to automate responsibilities formerly completed by using people, it may also make human labor extra efficient, thereby growing labor demand.

Lamentably, the modern-day trend in industrial AI development is in the direction of increasingly automation, with probably disastrous consequences for society (Memon, 2015). To be sure, automation has been an engine of productiveness growth considering the fact that the beginning of the industrial Revolution, when, starting inside the late eighteenth century, weaving and spinning had been mechanized. However, the tide of automation does not routinely carry all boats. By way of replacing hard work with machines in production duties, automation reduces hard work's share of price brought (and country wide income), contributes to inequality, and might reduce employment and wages.

And but most cutting-edge economies have skilled strong salary and employment boom for the reason that industrial Revolution. As automation has displaced employees in performing sure tasks, other technology have emerged to restore hard work's critical role in the manufacturing technique via growing new obligations wherein human beings have a comparative gain. These technologies have no longer only contributed to productiveness increase; however have also extended employment and wages, generating an extra equitable distribution of resources in the procedure.

Recall agricultural mechanization, which started out in the nineteenth century. At the beginning, the substitution of machines for uncooked labor did lessen the percentage of labor in cost added, displacing a big percentage of the United States team of workers that had previously been hired in farming. But, on the same time, burgeoning new industries wanted employees to carry out novel responsibilities and pursue emerging occupations. Clerical positions increased both in services and manufacturing, in which a finer department of hard work boosted productivity, employment, and salary increase.

UNDERSTANDING THE ISSUES AND REVOLUTIONS AUTOMATING THE WORLD

As the era is advancing, we're witnessing automation in each and every discipline. A totally self sufficient office is going to be the destiny. Human beings regularly desire to have computerized control over diverse electric appliances in office like fan, mild, PC and microwave oven (Karimi & Atkinson, 2013). This undertaking offers a solution which allows in undertaking the challenge effectively. A frequent transfer has been found out the usage of Arduino Uno Atmel Microcontroller, Android application and GSM modem in conjunction with hearth and human sensors the entire of which constitutes office Automation device (Dogo, 2014; Mowad, Fathy, & Hafez, 2014).

The main goal is to design and implement an office Automation device the use of IoT (net of things) this is capable of controlling and automating maximum of the office appliances via an easily potential android application.

It is solely targeted for the aged, physically handicapped and for the ease of controlling the switches without virtually accomplishing for it. The designing of login page in android utility and utilization of human and fireplace sensors whilst designing the hardware keeps the machine safe, distinctly blunders-free and green. The machine has the scope for adjustments, and extra gadgets can be delivered. Also, Adriano that is an open source has made viable to realize the difficult obligations pretty without difficulty due to its enriched libraries.

We are able to make use of voice command generation in offices (Kumar & Solanki, 2016). We're already seeing the popularity of wearable gadgets upward push within the shape of fitness monitors, which might be simply the beginning of the possibilities for wearable era and workplace automation. Perhaps in the future our wearable technology will screen the workplace appliances around us for most efficient power conservation.

This gadget is designed for a purchasing complicated mall but it could be also used in various companies like instructional observe board device or at Railway station, Bus stand and Air-port to display the information and notification. In mall it is also used to control the humidity and temperature of mall thru crucial AC by means of the usage of temperature sensor. In Business Corporation it can be extensively utilized. E-display system can be used to display Emergency message in Hospitals. A few regions in which IoT often used

- **Smart Towns:** To make the town as a smart town to have interaction with the facts exhaust created from your metropolis and community.
    - Tracking of parking areas availability within the town.
    - Monitoring of vibrations and cloth conditions in buildings, bridges and ancient monuments.
    - Detect Android gadgets, iPhone and in well-known any device which goes with Bluetooth interfaces or Wi-Fi (Folea, 2012).

- **Security and Emergencies:**
    - **Perimeter Get Entry to Manipulate:** Detection and manipulate of people in non legal and confined.
    - **Liquid Presence:** Liquid detection in data centers, touchy constructing grounds and warehouses to prevent breakdowns and corrosion.
    - **Radiation Ranges:** In nuclear electricity stations surroundings distributed measurement of radiation stages to generate leakage signals.
    - **Explosive and Unsafe Gases:** Detection of gas leakages and stages in industrial environments, environment of chemical factories and inside mines.
- **Smart Agriculture:**
    - **Wine First-Rate Improving:** Tracking soil moisture and trunk diameter in vineyards to manipulate the quantity of sugar in grapes and grapevine health.
    - **Green Houses**: Manage micro-weather conditions to maximize the manufacturing of fruits and greens and it's first-class.
    - **Golf Courses:** Selective irrigation in dry zones to reduce the water assets required inside the green.
    - **Meteorological Station Network:** Examine of weather conditions in fields to forecast ice formation, rain, drought, and snow or wind changes.
    - **Compost:** control of humidity and temperature degrees in alfalfa, hay, straw, and many others. To prevent fungus and other microbial contaminants.
- **Domestic Automation:** In home by means of the use of the IoT gadget remotely monitor and control our domestic appliances and reduce down in your monthly payments and resource utilization.
    - **Energy and Water Use:** Strength and water supply consumption tracking to acquire advice on a way to shop fee and sources.
    - **Faraway Control Appliances:** Switching on and stale remotely home equipment to avoid injuries and keep power.
    - **Intrusion Detection Systems:** Detection of windows and doors openings and violations to prevent intruders.
    - **Artwork and Items Maintenance:** Monitoring of conditions inside museums and art warehouses.
- **Clinical Subject:**
    - **All Detection:** Help for elderly or disabled humans dwelling impartial.
    - **Medical Refrigerators:** Monitoring and manipulate of situations inside freezers storing medicines, vaccines, and organic factors.
    - **Sportsmen Care:** Vital symptoms monitoring in excessive performance facilities and fields.
    - **Patients Surveillance:** Monitoring of situations of patients inside hospitals and in antique human's domestic.
    - **Ultraviolet Radiation:** Measurement of UV solar rays to warn humans now not to be uncovered in positive hours.
- **Business Control:**
    - **System-to-Gadget Applications:** Device car-diagnosis the hassle and manipulate.
    - **Indoor Air Pleasant:** Monitoring of oxygen stages and toxic gas inner chemical plants to make sure employees and goods protection.
    - **Temperature Monitoring:** Screen the temperature within the enterprise.

    - **Ozone Presence:** In food factories monitoring of ozone levels at some stage in the drying meat procedure.
    - **Vehicle Car Diagnosis:** Records collection from Can Bus to send actual time alarms to emergencies or offer advice to drivers.
- **E-Advertisement Machine:** It may be a version of IOT primarily based E-commercial device for the programs of shopping department stores & different agencies. This proposes model will update the advertisement system in huge shopping complicated like huge bazaar, Reliance fresh etc. Even we can hold the humidity inside the massive purchasing malls with none Human efforts. Also we will use this prototype gadget for the educational company or Railway stations.

The IoT promises to deliver a step alternate in people "first-class of existence and corporations" productiveness (Stankovic, 2014). Via a widely disbursed, domestically smart network of smart devices, the IoT has the ability to enable extensions and enhancements to essential services in transportation, logistics, safety, utilities, schooling, healthcare and other areas, whilst imparting a brand new ecosystem for utility development. A concerted effort is needed to transport the enterprise beyond the early levels of marketplace development closer to maturity, driven via not unusual expertise of the wonderful nature of the possibility. This marketplace has wonderful traits within the regions of provider distribution, enterprise and charging models, abilities required to deliver IoT offerings, and the differing demands those services will region on mobile networks.

Connecting those smart gadgets (nodes) to the net has also started out happening, although at a slower price. The pieces of the technology puzzle are coming together to house the net of factors earlier than the majority assume. Just because the net phenomenon befell not so long ago and caught like a wildfire, the net of things will touch every element of our lives in less than a decade.

ORGANIZATION OF THE BOOK

The book is organized into 18 chapters. A brief description of each of the chapters follows:

Chapter 1 focuses on smart waste management system using IoT. Here a solution is provided for various smart bin construction technologies, advantages, standards and challenges in detail. It is very useful to the reader to understand the various method of waste management in smart cities development using IoT.

Chapter 2 discusses the use of Radio Frequency Identification (RFID) in Supply Chain Management (SCM), one of the promising innovations in recent decades. Here application areas of RFID in the context of supply chains are reviewed to demonstrate best practices and important implementation issues. Different industries (e.g. automotive, transport, retail) are used to emphasizing the benefits of RFID technology. The chapter also highlights operational and strategic implications of adopting RFID-based technological solutions. A conceptual framework that links RFID key benefits and information attributes used in decision making has been proposed.

Chapter 3 illustrates software optimization techniques to increase battery efficiency because these techniques are more robust and proposes a novel idea of an automated system for smart phones that prioritize application access based on the owner's usage patterns and daily routine to conserve battery life. This system will serve two purposes: save battery power and improve the Smartphone's artificial intelligence.

Chapter 4 describes four different types of binary search algorithms for avoidance of tag collision, and then presents a performance measurement mechanism for RFID application system. Finally, simulation-based experimental results on the performance of these algorithms are presented.

Chapter 5 focuses on IoT based solution to assist elderly and limited mobility people. It aims to improve the autonomy, and, consequently, the quality of life of this target audience by automating daily tasks conducted at home, such as turn on the lights and manipulates electronic devices. In this chapter the software requirements discovery and elicitation of a home automation app considering the real needs of the elderly and limited mobility people. It presents a mobile app prototype using the Android and Arduino platforms to illustrate a use scenario of the proposed solution.

Chapter 6 illustrates Smart Homes and Offices (SHO), which are composed of interlinked components with constant data transfer and services targeted at increasing the life style of the people. This chapter describes about the smart components and how SHO are direct implementation of Internet of Things (IOT). The chapter also focuses on privacy preserving mechanisms, its essence over smart cities, strong architecture related to privacy, preserving mechanism and various approaches available that can retaliate these issues in a smart city environment.

Chapter 7 conducts a Systematic Mapping Study of publication trends relating to wearable technologies for glucose monitoring. This study adopts a Systematic Mapping Study approach in identifying relevant papers for analysis. The findings of this study are meant for inform health informatics experts and academics on the current research and publication trends in wearable technologies for glucose monitoring.

Chapter 8 attempts to identify the components of a smart health ecosystem for asthma patients through a systematic literature review and provided a conceptual framework. The findings of the systematic literature review are expected to inform researchers on the components required for building a smart health ecosystem for asthma patients.

Chapter 9 introduces to WIoMT (Wearable Internet of Medical Things) and reviews smart wearable healthcare devices scientific terms as well commercial pains. Internet of Medical Things is displayed through a refined background, wearable computing, wearable technology, cloud frameworks, architecture design, including required hardware and software, body sensors, smart phones, the smart medical application as well medical location analyzers for data storage and finally a diagnosis. Wearable devices are tested under strict observation fitness, vital signs, and smart environment. Wearable devices are now used for a wide scale of monitoring healthcare.

Chapter 10 proposes an embedded gadget which can measure the rate of heart beat and body temperature and keep the statistics on the cloud server for the doctor to take next course of action.

Chapter 11 gives a brief introduction on smart roads and smart parking. Smart roads or Smart highways is the term used for defining roads that use different IOT enabled technology such as sensors, Solar technology and many more to minimize the electric usage. In the near future concept of smart vehicles would not the functional without smart roads and smart parking systems.

Chapter 12 deals with the in-depth discussion of these various aspects of smart transportation system enabled with IoT technology as, a well-built Smart Transportation system reduces the risk of accidents, improve safety, increase capacity, reduce fuel consumption and enhance overall comfort and performance for drivers.

Chapter 13 proposes an idea of automatic irrigation method. Automatic irrigation is done using a soil moisture sensor. The manual method of irrigation is done by using automated process. In this proposed method, apart from moisture sensor, other sensors like PIR sensor, ultrasonic sensor, humidity, temperature sensor and water level sensors are used. This method has additional features like GSM.

Chapter 14 focuses on Smart Water Level Monitoring System for Farmers and provides a smart way to manage water resources on farms in the most cost effective and convenient way for farmers.

Chapter 15 is aimed to discuss critical topics for the implementation of smart agriculture in India. The smart agriculture employs superior technologies such as decision support system, expert system, IoT, GPS, machine learning, robotics and application of connected devices.

Chapter 16 describes how intelligent sockets are advancement in approach to better the features and convenience offered by the existing switchboards, that too using machine learning, System initially classifies various types of analog meters whereas currently, the lbph classifier is trained to detect analog meter with needle and Analog meter with text readings.

Chapter 17 proposes a secure AES algorithm with randomized round keys, which provides better security with negligible overheads and is ideal for use in IoT networks.

Chapter 18 integrates the concept of IoT with roads to make them smart and also talks about using the IoT technologies, with the commencement of smart cities, to reduce the risk of runoff road collisions.

Dinesh Goyal
Poornima Institute of Engineering & Technology, India

S. Balamurugan
QUANTS IS & Consultancy Services, India

Sheng-Lung Peng
National Dong Hwa University, Taiwan

Dharm Singh Jat
Namibia University of Science and Technology, Namibia

REFERENCES

Delgado, A. R., Picking, R., & Grout, V. (2000). Remote-controlled home automation systems with different network technologies. Wrexham, UK: Centre for Applied Internet Research (CAIR), University of Wales.

Dogo, E. M. (2014). *Development of feedback mechanism for microcontroller based SMS electronic strolling message display board*. Academic Press.

Folea, S., Bordencea, D., Hotea, C., & Valean, H. (2012). Smart home automation system using WI-FI low power devices. In *Automation Quality and Testing Robotics (AQTR), 2012 IEEE International Conference on Date of Conference* (pp. 569-574). IEEE. 10.1109/AQTR.2012.6237775

Gubbi, J., Buyya, R., Marusic, S., & Palaniswami, M. (2013). Internet of things (IoT): A vision, architectural elements, and future directions. *Future Generation Computer Systems*, *29*(7), 1645–1660. doi:10.1016/j.future.2013.01.010

Karimi, K., & Atkinson, G. (2013). *What the internet of things (IoT) needs to become a reality* (White Paper). FreeScale and ARM.

Kumar, S., & Solanki, S. S. (2016). *Voice and touch control home automation RAIT-2016*. Academic Press.

Lamkin, P. (2016, February 17). *Wearable tech market to be worth $34 billion by 2020*. Retrieved from https://www.forbes.com/sites/paullamkin/2016/02/17/wearable-tech-market-to-be-worth-34-billion-by-2020/#14e108283cb5

Memon, A. R. (2015). *An electronic information desk system for information dissemination in educational institutions*. Paper presented at the Computing for Sustainable Global Development (INDIACom), 2015 2nd International Conference, New Delhi, India.

Meola, A. (2016, December 19). *Wearable technology and IoT wearable devices*. Retrieved from https://www.businessinsider.com/wearable-technology-iot-devices-2016-8

Mowad, M. A. E., Fathy, A., & Hafez, A. (2014). Smart home automation control system using android application and microcontroller. *International Journal of Scientific & Engineering Research*, *5*(5).

Patil, M., Bedara, A., & Pacharne, V. (2013). *The design and implementation of voice controlled wireless intelligent home automation system based on Zigbee*. Academic Press.

Stankovic, J. (2014). Research directions for the internet of things. *IEEE Internet of Things Journal*, *1*(1), 3–9. doi:10.1109/JIOT.2014.2312291

Chapter 1
Waste Management System for Smart City Using IoT

Golden Julie E.
Anna University Tirunelveli, India

ABSTRACT

In the present scenario, sensors place a major role for implementing smart devices. Internet of Things (IoT) is an advancement of sensors which can communicate with non-communicate things (devices). Many of the developed counties are using smartness in creating and communicating devices using IoT. In India, major challenges focus on how and where to implement smartness. Hence, authors found some different areas like healthcare, education, transport, water, energy, communication, security & safety, citizen services, and so on. All these areas are covered by a smart way using recent technology (IoT) in smart cities concepts. Various technologies like IoT, Big Data, and cloud computing are used for constructing smartness in the form of devices. In this Chapter, authors focus on a smart waste management system using IoT. They provide various smart bin construction technology, advantages, standards and challenges in detail. It is very useful to the reader to understand the various method of waste management in smart cities development using IoT.

INTRODUCTION

In the introduction part, start with overview of smart city services followed by waste monitoring and management.

Nowadays government have started many tasks for improving cleanliness in our nation. Peoples also got somewhat awareness about clean their surroundings as much as possible. Regarding this various positive movement are take over towards smart implementation of waste management. In many of the citied we can found wastes throw outside because of overloaded dustbins. Normally it will create unhygienic for the people and also it crate bad smell.

DOI: 10.4018/978-1-5225-9246-4.ch001

SMART CITY MODEL

Definition of smart city is the ability of a city focus on multiple aspects. One more definition is identifying the special characteristics for development of city. Giffinger et al., (2007) proposed a model of city characterized by six fields, which is derived from smart connection of self-decisive, self-reliant citizens and institutions. The term smart is used for education of its inhabitants. Furthermore, smart city represent the gap between the city government administration and its people. Smart cities are further discuss the use of modern technology, industry, education, participation, technical infrastructure which improve urban traffic.

Six Aspect Operations of Smart City

Giffinger et al., (2007) proposed six factor for smart city characteristics 1. Smart Economy: It mainly focus on productivity flexibility of the labour market includes national and international market. 2. Smart people: It will not focus on the education or qualification of the people belong to the city. It main objective is based on the quality of social interactions regarding integration. 3. Smart Governance contains the aspect of political participation, public and social relationship. Smart Mobility objectives are provided information and communication technology, modern and sustainable transport systems. Smart environment describes pollution, Environment monitoring and protection. Finally, Smart Living contains quality of life as culture, health, safety, housing, tourism, etc.

Figure 1. Characteristics of Smart City (Giffinger et al., 2007)

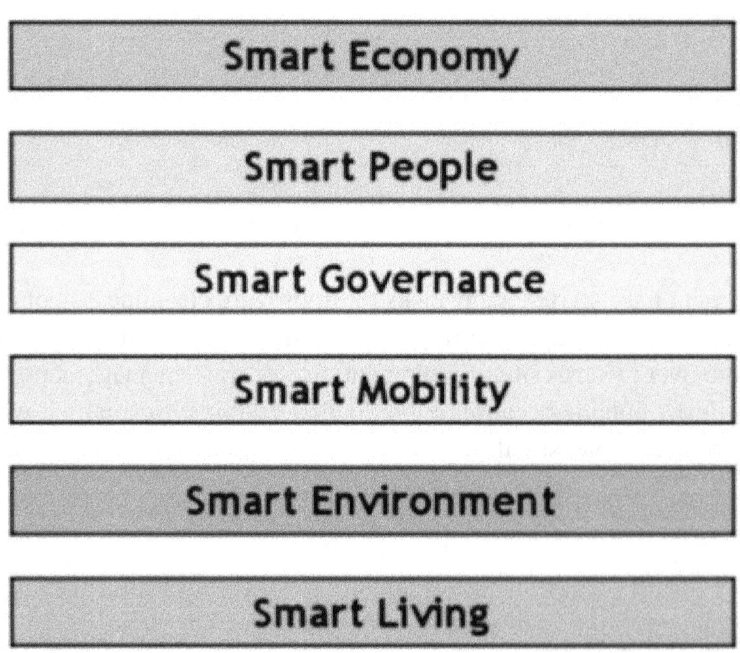

Figure 2. Characteristics and Factors of Smart City(Giffinger et al., 2007)

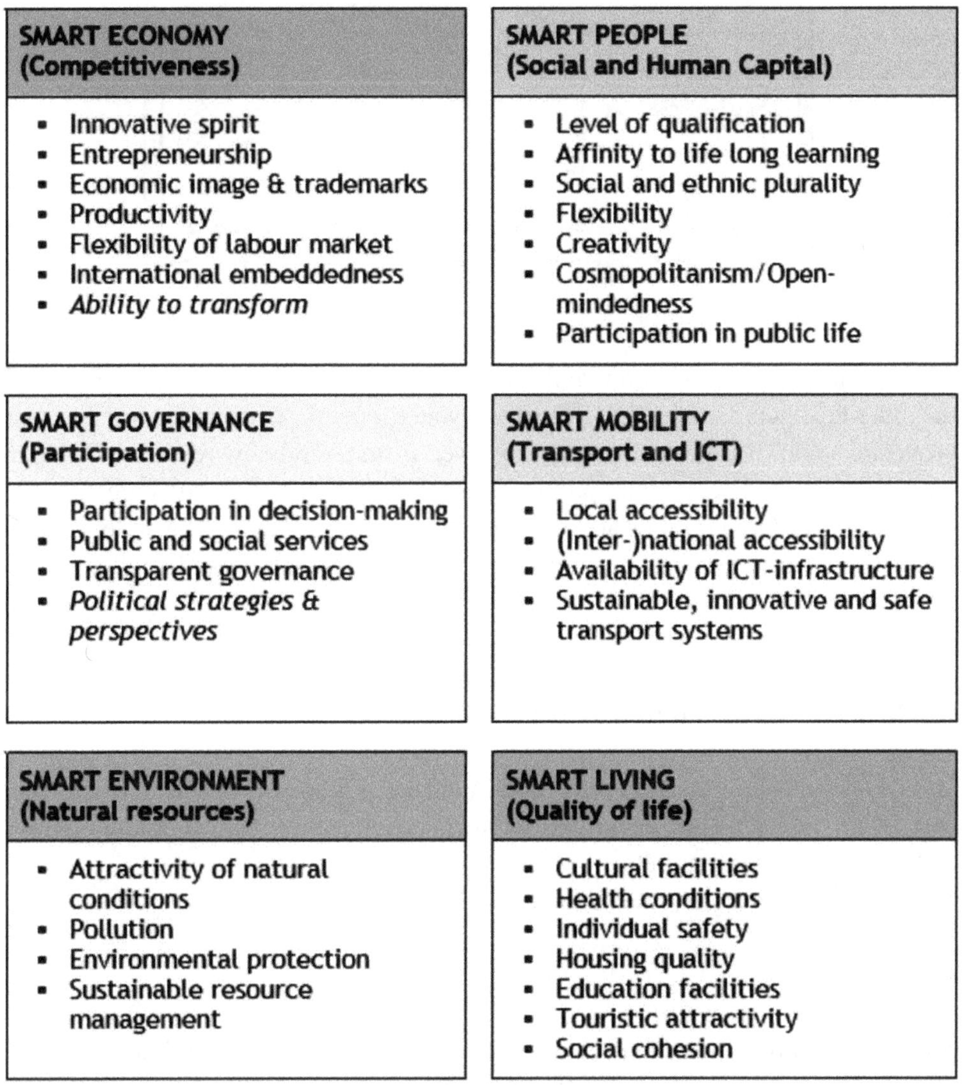

Internet of Things (IoT)

IoT is evaluated from multiple technologies like wireless sensor, embedded system, machine learning, instrument control data analytics and automation. 'Things' refer daily objects or devices that communicated with other devices through internet to monitor and control the objects. Smart IoT belong to Information and Communication Technologies (ICT) applications access through IoT devices. Smart cities are developed under prevention of incidents rather than avoidances of occurrences of after the incidents. E.g, .g, fire detection, crime prevention, floods and climate prediction. Smart city is one of the applications of IoT. It contains three parts data generation, data management, and application handling.

At present population of India is nearly 1.35 billion based on the most recent UN data. Population is growing the double size in just 40 years. Basically population in the city is high compare with villages.

Figure 3. Smart Cities & IoT (Silva et al, 2018)

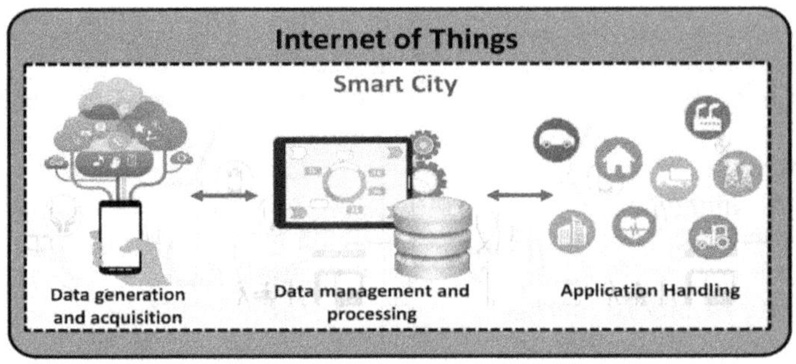

In India major challenges focus on how and where to implement smartness. Hence we found some different areas like healthcare, education, transport, water, energy, communication, security & safety, citizen services and so on. Based on the report received from Grandview research (2018), Smart cities are directly connected to house, people, public places and centre to collect/ share data. AlEnez et al, (2018) propose the goal of smart cities are developed with safe and quality lifestyle with minimal cost. Fig shows the IoT layer architecture of data gathering, data management and application processing in smart city project.

SMART CITIES APPLICATIONS AND CHALLENGES

In smart cities development application enhanced in different fields like home automation, smart parking, Transport, Water management, security, monitoring, pollution monitoring and control, education, health and solid waste management etc.

Here, we Each IoT application is detailed here.

Smart Home

Advance wireless technologies used to control and monitor home held devices like fan, Air conditioners, Washing Machine, Television, LED lights. By using internet connectivity, smart surveillance is used to monitor and control the home even from the workplace.

Challenges in Smart Home

1. Lack of knowledge in the technology and standardization leads to create problem to the automatic devices.
2. Smart Home is having security issues like hacking.
3. Smart Home is inflexibility of interconnected devices and high cost of connectivity.
4. Smart devices are always depends on internet connectivity. If any problem occurs in network connections then the devices are not automated.

Smart Transport and Parking

Day by today our cities is coved with mass number of vehicles on the road. Due to this increasing factor traffic and parking is a major issue. In smart transport and parking slot reduce the traffic using automated system. Parking slots are accurately predicted the available free slots are intimated to the driver through speaker. IoT sensors are used to reduce the traffic based on arrival and departure of vehicles. Vehicular adhoc networks (VANET) provide collusion free traffic in the busy road and also it reduces accidents.

Challenges in smart Transport and Parking: In smart vehicle traffic and parking system contain mobility. Due to dynamic nature system should guide the traffic and taking final decision about parking slot is very time effective.

Smart Health

The significance of Smart system based health monitoring applications is potentially realized in health-care domain compared to any other domains. Remote patient monitoring can reduce the cost of chronic disease management by 10-20% by 2025. The growth of sensors helps to track the patient's medical conditions anytime, anywhere. health monitoring using wearable sensors, elderly assistance, tracking of patients status in 24x7. Solanas et al, (2014) propose Mobile health (m health) it uses mobile phone service to global monitoring of health related issues.

Smart Agriculture

The increasing demand for food, both in terms of quantity and quality, has raised the need for intensification and industrialization of the agricultural sector. The "Internet of Things" (IOT) is a highly promising family of technologies which is capable of providing innovative farming methods towards the precision agriculture (PA). Soil monitoring is one of the major issues in open field agriculture. To accomplish it, various types of wireless sensor nodes (WSNs) have been widely used in soil and climate monitoring deployments in open field agriculture. These wireless sensor nodes interact with physical objects or their environment, communicate with their neighboring nodes and build networks through which data are usually forwarded towards the remote infrastructure

Smart Waste Management

In smart bin concept many dustbins are connected through IoT device to monitor the waste collection box and alert the condition of bin. These types of smart bin are constructed with low cost sensor nodes with embedded system. The main objectives of solid waste management system are to develop our city smarter and improvement in the environment from pollution free city. This help the people to reduce cost, pollutions, increasing green nature by utilize the solid waste for agricultural purpose and finally improve healthy air for breathing.

Figure 4. Smart Cities applications (Talari et al, 2017)

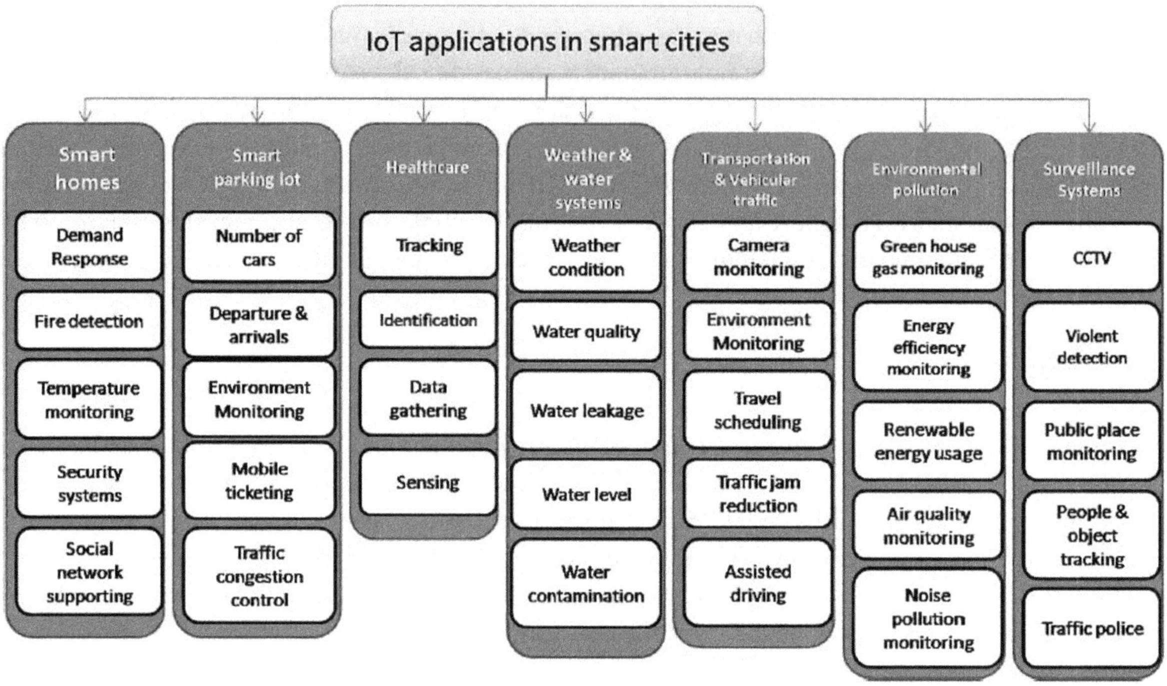

ISSUES IN WASTE COLLECTION AND MANAGEMENT

Kumar et al., (2009) proposed Municipal solid waste management (MSWM),is a one of the important element for developing the cities from health issues. It contains different stages like waste segregation, collection, Transport, Recycle and disposal with minimal impact to the environment.

Waste Segregation

In India we don't have special method to solve the problem of waste segregation. Mostly this segregation takes place after collecting all the wastes under very unsafe and very bad condition. CPCB Report, 2013 says in many occasions during the travel and improper handling the wastes is mixed. Singhal & Pandey, (2001) are discussed the major reason is lack of segregation for proper disposal of waste.

Waste Collection

Usually wastes are collected from home in different ways:

1. House-to-House Waste collectors: People from Municipal visit individual houses and collect it through their vehicle.
2. Group Bins Users: In each area or region they have communal bins. People normally put wastes from their home. As per a set timetable Municipal people will collect it through their vehicle.
3. Self-Delivered Generators: Some people collect their home wastes and straightforwardly to transfer locales or treatment place.

Mostly, house held wastes are community bins. Street sweepings also use this community bins for collecting wastes. Kumar et al., (2009) convey that some municipal authorities paying some amount for waste disposal.

Reuse/Recycle

The collection of waste contain some materials for making new products. Always community bins based waste collection all the wastes are dumped. so its very difficult to find recycle materials. However, rag-Picker usually collects recyclable things like plastic and bottles from community bins. Better use separate community bins for recyclable material collections leads to improve waste management.

Transportations

MSWM practised in India usually uses hand rickshaws, compactors, trucks, tractor and trailers. In trucks are having capacity of some ton and they used without proper covering system. Hand rickshaws are used to collect from home as per timetable. In most causes this type of vehicles also need to maintain their availability as well as their working conditions. If any vehicle got breakdown the entire collection and disposal effectively reduced. Joseph, (2002) says that transfer stations can be found in some metropolitan e.g. Mumbai.

Disposal

In India many of the cities, town and villages uses instinctive disposal of MSW. This type of disposal the collected wastes could not reach the destination and also lost in the cities Peripherals, outskirts, along the road, low lying area, along the drain, green areas, etc. Some of the disposal techniques are open dumping, Land filling, Landfill gas-to-energy plants and Biological treatment of organic waste.

Open Dumping

In MSW are directly disposed on low lying area in routine or dumped in the road side of the outskirts of town. Irrational dumping of waste leads to affect the surface water contamination. During monsoon wastes are flooded in around then city.

Land Filling

It is one of the generally fixed practices in India. Based on CPCB, 2013 report, India uses 59 landfill sites and 376 sites are in implementation stage.1305 sites were identified for further use.

Landfill Gas-To-Energy Plants

Landfill mainly produce two gases methane (CH4) and carbon dioxide (CO2). United Nations Environmental Program (UNEP) shows a study about green house gas emission from landfill. It can be reduced by using appreciate hazards management techniques Like Minimize the waste production, Recycle and reuse etc.

Biological treatment of organic waste: In India has about 50% of organic waste generated as compared to 30% generated by developed countries. In India they have some common biological treatment like Aerobic composting. Vermi- composting. Anaerobic digestion and Thermal treatment.

VARIOUS SOLID WASTE MANAGEMENT SYSTEMS

In this section describe various solid waste management systems; they describe their guidelines on different technical, economical and managerial skills for developing smart waste management. In Many developing counties like India researches are going on for developing smart and low-cost waste management methods.

Wireless Sensor Network Architecture for Solid Waste Management

Longhi et al., (2012) proposed wireless sensor network architecture for solid waste management. In real time many applications like home, industry, environment, underwater, building using wireless sensor networks. A new architecture is developed for improving the performance of on- site waste handling and smart transportations using optimized way. Sensor nodes are used to collect and make use of Data Transfer Nodes (DTN). Garbage bin filling is measured and remotely transfer through DTN. Based on the collected data from sensor further provide Decision Support System (DSS) associate with the structure by a web program. This system can measure level of environmental change occur and also very useful to prevent our environment from damage. In pollution monitoring system contain mobile data acquisition unit integrated with sensors, microcontroller and Global System for Mobile Communications Modem (GSM-Modem), and a Global Positioning System Module (GPS-Module). This unit gather information about level of bin and physical location with time. This collected data can be used for further research analysis for predict approximate time of filling the bin.

Figure 5. WSN architecture

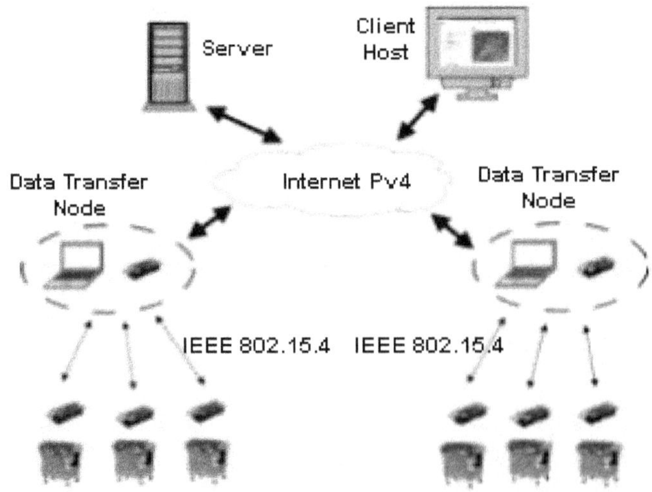

WSN Architecture Contain Three Layer

1. Communication Layer: Quected l M10 GSM/GPRS modules used for long – range communication through DTNS. ARM processor is installed and programmed for utilizing open CPU.
2. Server layer: It is a middle layer of the architecture of WSNs.It contain two arrangement based on TCP/IP and SMS.
3. User interface: In user interface layer also contain to interaction results, first customer having client PC and find DB about all collected data. Second interface is using distributed computing and access through a Web application.

Novel Prototype and Simulation Model for Real Time Solid Waste Bin Monitoring System

Al Mamun, et al., (2014) was proposed wireless sensor network based solid waste management for real time environment. Wireless sensor network architecture is planned with sensor node zigbee and GPRS. Zigbee sensor node is used to measure the status of waste bin and GPRS used for communication. Basically, the system contains three parts such as smart bin, Gateway and Control station. Typically, this architecture is divided in to layers Lower layer, Middle Layer and Upper Layer.

Lower Level Layer

This layer is like physical layer contain sensor nodes and RF communicator. Sensor nodes are divided in to two groups. One group of sensor nodes are place in the underneath of smart bin with accelerometer, a hall effect, an ultrasound, a temperature and a humidity sensor. Another group contain load cell sensor place in the bottom of the bin. The accelerometer sensor used to track the bin cover. Hall Effect sensor monitors the overloading of bin with waste. This sensor information is forwarded to next layer through ZigBee-PRO communication module.

Middle Layer

It processes the data received from lower layer and store it into local database. Then GPRS send a request to the server for connection establishment with control station. After the connection establishment the gateway sends a SMS to the specific cell number with collected data.

Upper Layer

The upper layer contains control station with servers. Control station is responsible for connection establishment with middle layer when send a request. The collected sensor data are further stored in a database for further processing. Using user-friendly web applications shows the nature and status of the bin in remotely.

Figure 6. Architecture of the real time bin monitoring system

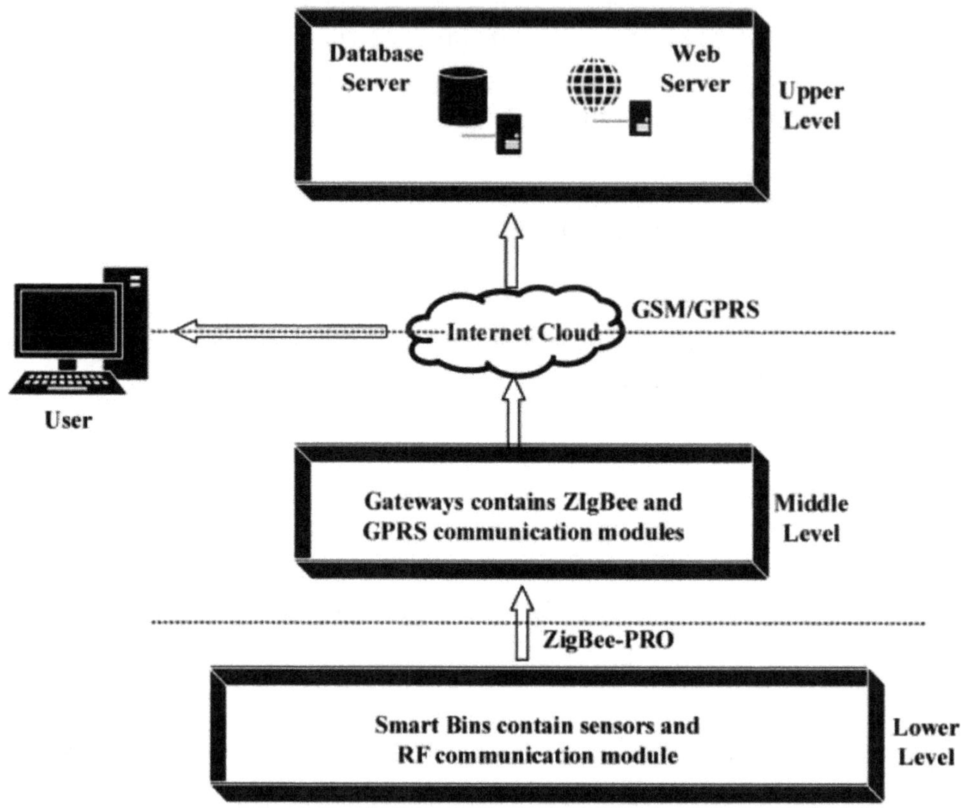

A Smart Waste Management with Self-Describing Objects

Yann Glouc & Paul Couderc,(2013) proposed smart waste management with self-describing objects. The main objectives are reducing waste production, ensuring that wastes are properly dispose and Re-cycling and re-using disposed products. Sorting of wastes are done by Radio Frequency Identification (RFID) tag. In this method wastes are separated by using selective sorting process. Waste management architecture has five step process Waste description, wastes identification, trash bag, collective container and waste collection.

Waste description: In selective sorting method wastes are identified by their digital descriptions about the item using RFID tag.

Waste Identification: User only identify the sorting method for select appropriate dust bin for disposal

IOT BASED WASTE MANAGEMENT SYSTEM

In today modern world all the activities are monitories and controlled by computer through internet. These evaluations take place in the world to connect all the things (even Machines) through internet. The main objective of this IoT is to reduce time, improve accuracy and reduce human errors. It is one of the important tool in every human lifestyle. This automation is mainly implemented into home for security

Figure 7. Waste Flow and Global Architecture of the System

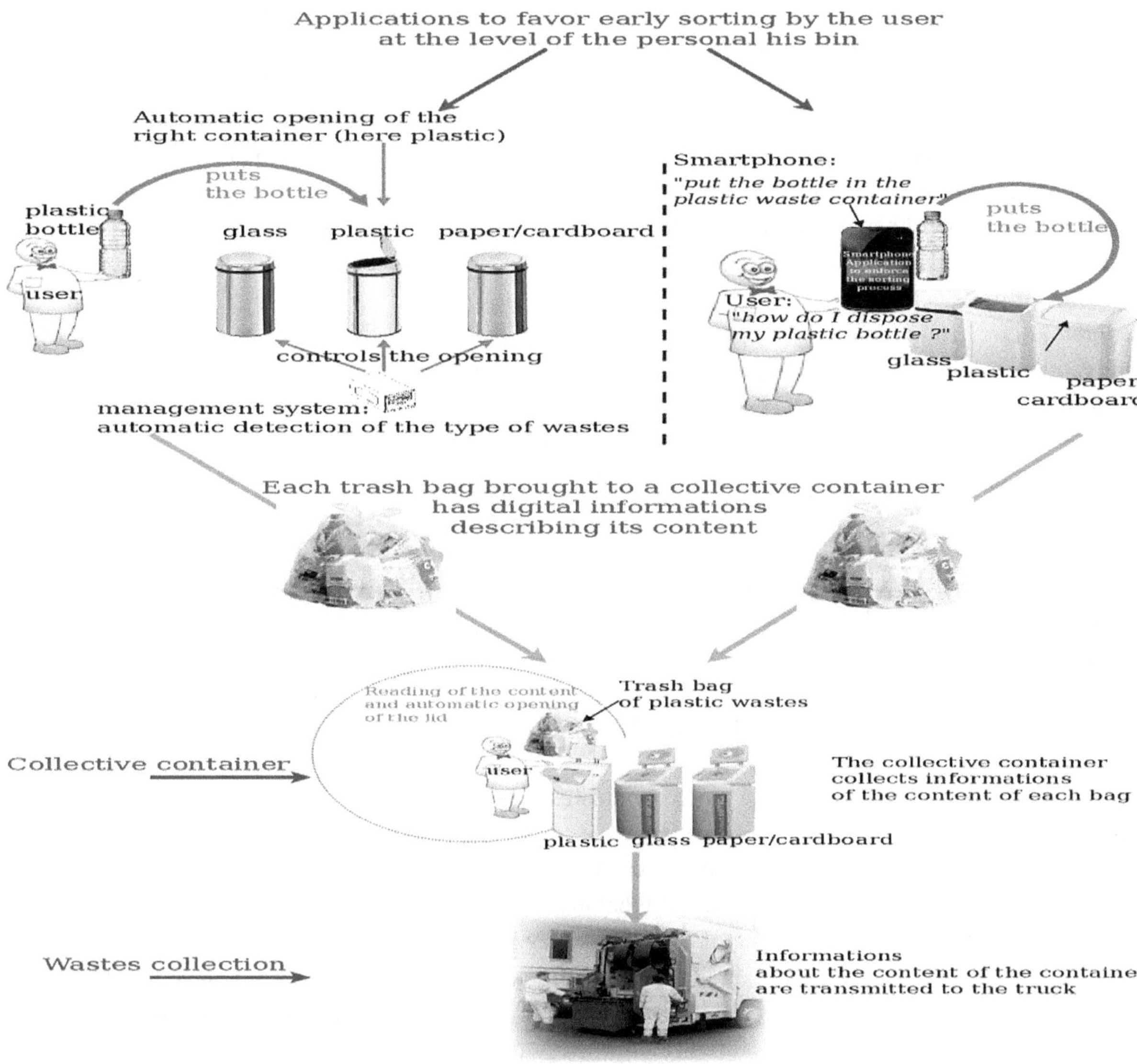

purpose then industries to control huge machines and also monitor the working nature of the machine. Now days these IoT techniques are used in our waste management with various technical changes for improving our friendly eco system.

IoT Based Smart Garbage and Waste Collection Bin

S. S. Navghane et. al (2016) was proposed Smart Garbage focus on the process on sensor to collect data and monitor the environmental changes using remote system. Here they used two sensors one is weight and another is IR sensor. In this IR sensor used to measure various level of garbage present in the dustbin. Weight sensors are used to indicate the threshold level of the dustbin. This will forward the data to the

ARM LPC2148 micro controller and further transmit to the remote system using Wi-Fi modem. At the receiver side mobile will be connected with Wi-Fi modem to view the status of the dustbin.

Somu Dhana Satyamanikanta and M.Narayanan (2017) proposed" Smart Garbage Monitoring System Using Sensors with RFID over Internet of Things" This technique uses RFID card reader with 3 sensors for monitoring garbage collection process. In this technique contains one special sensor name as photo electric sensor which is used to collect the clear representation of the object including Weight sensor and IR sensor. In this method weight sensor is placed under the dustbin to find the weight. This architecture is worked based on the RFID features. Using RFID CARD reader,It will read all the information about the person and what he dropped inside the bin and information about smart bin will be sent to the remote centre.

Indu Anoop et al. 2017 developed an application as "IoT based Smart Waste Management." This application is mainly used to avoid overloaded dustbins. This method of smart waste management contain ultrasonic sensor with arduino. This system is controlled by remote by using Wi-Fi modem. It will reduce human resource and enhance the smartness through notification.

K. Nagalingeswari and Krishna Prasad Satamraju (2017) proposed Efficient Garbage Management System for Smart Cities. In this system *uses Internet of Things architecture.* It ensures seamless robust system used to monitor continuously about the status of the bin. Find low cost collection method using greedy approaches for dustbin collection. It contains LolinNodeMCUIoT board, and Ultrasonic sensor module. This module is designed based on threshold value, once it reaches threshold value it will give alert and trigger the system remotely.

IoT Based Smart Garbage Segregation

M. K. Pushpal et al. (2015) proposed "Microcontroller Based Automatic Waste Segregator" it separates the waste in to three major categories such as metal, dry waste and wet waste. The working principle is containing one simple 8051 microcontroller act as a heart of the system for split the waste. Based on the waste collected from the home the system split that in to three categories. Example safety pin is belonging to metal waste. Paper, class are some of dry waste. Vegetable peels and organic waste generated from home belong to wet waste. By using inductive proximity sensor metals are separated from the waste. In Blower section dry and wet wastes are separated based on the weight. Low weight wastes are sent out by blower.Hight weight wastes are stayed in the belt.

G. Aahash,et al 2018 developed an application "Automatic Waste Segregator using Arduino." This application contains two sensor nodes and one dust bin with two outlet. Here they used metal sensor for detecting metals present in the waste. This sensor works based on the electromagnetic induction for dispatch metals. Based on the detection the corresponding door of the dustbin will open for metal collection. The belt used to move the non-metallic wastes to other opening of the bin. The whole system is controlled by remote using IoT concepts.

ADVANTAGES OF SMART WASTE MANAGEMENT

The main advantages of using smart system are to reduce human resources and increase the performance efficiency.

1. Reduced vehicle fuel consumption and reduce manual monitoring.
2. Monitoring the level of bin. If overloaded sent notification or alert to remote system.
3. Reduce operational cost.
4. IoT based application collects large data and used for analysis or forecast the waste separation., Monitor and control.

CONCLUSION AND FUTURE WORK

The main objective of this chapter focuses on the fundamentals of Smart city services and layers of smart services. Then discuss with IoT usage in smart city projects, followed by issues and challenges in smart city projects. Finally discuss in detail about waste management and monitoring using different techniques Like WSN, IoT, etc. This chapter gives overview of solid wastes and collection methods and improvement with waste segregation.

Based on the collection of information provided in the chapter one can find new technology proposal and review the techniques for improving performance to achieve more efficient and time-consuming waste management and monitoring.

REFERENCES

Al-Maaded, M., & Madi, N. K., Kahraman, R., Hodzic, A., & Ozerkan, N. G. (2012, March). An Overview of Solid Waste Management and Plastic Recycling in Qatar. *Journal of Polymers and the Environment*, *20*(1), 186–194. doi:10.100710924-011-0332-2

Al Mamun, M. A., Hannan, M. A., & Hussain, A. (2014). A Novel Prototype and Simulation Model for Real Time Solid Waste Bin Monitoring System. *Jurnal Kejuruteraan*, *26*, 15–19. doi:10.17576/jkukm-2014-26-02

AlEnezi, A., AlMeraj, Z., & Manuel, P. (2018, April). Challenges of IoT Based Smart-Government Development. In *Green Technologies Conference (GreenTech)*, 2018 (pp. 155-160). IEEE.

Anoop, I., Jain, A., Pathak, S., & Yadav, G. (2017). IOT based Smart Waste Management. *International Journal of Advanced Research in Computer and Communication Engineering, 6*(1).

Giffinger, R., Fertner, C., Kramar, H., Kalasek, R., Pichler-Milanović, N., & Meijers, E. (2007). Smart Cities: Ranking of European Medium-Sized Cities. Vienna, Austria: Centre of Regional Science (SRF), Vienna University Technology.

Glouche, Y., & Couderc, P. (2013). A Smart Waste Management with Self-Describing objects. *The Second International Conference on Smart Systems, Devices and Technologies (SMART'13)*, Rome, Italy.

Grandview Market research report. (2018). *Smart Cities Market Size, Share & Trends Analysis Report By Application (Education, Governance, Buildings, Mobility, Healthcare, Utilities), By Component (Services, Solutions), And Segment Forecasts 2018–2025.* Retrieved from https://www.grandviewresearch.com/industry-analysis/smart-cities-market

Kasliwal Manasi, H., & Suryawanshi Smithkumar, B. (2016). A Novel approach to Garbage Management Using Internet of Things for smart cities. *International Journal of Current Trends in Engineering & Research, 2,* 348-53.

Kumar, S., Bhattacharyya, J. K., Vaidya, A. N., Chakrabarti, T., Devotta, S., & Akolkar, A. B. (2009). Assessment of the status of municipal solid waste management in metro cities, state capitals, class I cities, and class II towns in India: An insight. *Waste Management,* February 2009, Pages 883-895.

Longhi, S., Marzioni, D., Alidori, E., Buò, G. D., Prist, M., Grisostomi, M., & Pirro, M. (2012, May). Solid waste management architecture using wireless sensor network technology. In 2012 5th International Conference on New Technologies, Mobility and Security (NTMS), (pp. 1-5). IEEE. 10.1109/NTMS.2012.6208764

Mahajan, K. (2014, July). Waste Bin Monitoring System Using Integrated Technologies. *International Journal of Innovative Research in Science, Engineering and Technology, 3*(7).

Monika, K. A., Rao, N., Prapulla, S. B., & Shobha, G. (2016). *Smart Dustbin-An Efficient Garbage Monitoring System. International Journal of Engineering Science and Computing*, *6*, 7113–7116.

Nagalingeswari, K., & Satamraju, K. P. (2017). Efficient Garbage Management System for Smart Cities. *International Journal of Engineering Trends and Technology (IJETT), 50*(5).

Naveen, B., Kavya, G. K., Kruthika, S. N., Ranjitha, K. N., & Sahana, C. N. (2018). Automated Waste Segregator Using Arduino. *International Journal of Advance Engineering and Research Development, 5*(05), May -2018.

Navghane, S. S., & Killedar, M. S., & Rohokale, V. M. (2016, May). IoT Based Smart Garbage and Waste Collection Bin. *International Journal of Advanced Research in Electronics and Communication Engineering*, *5*(5), 1577.

Navghane, S. S., Killedar, M. S., & Rohokale, D. V. (2016). *IoT Based Smart Garbage and Waste Collection Bin. International Journal of Advanced Research in Electronics and Communication Engineering*, *5*, 1576–1578.

Pille, K. Maniyar, R., Bade, N. & Bakliwal, J. M. (2016). Solid Waste Management System, *National Conference on Internet of Things: Towards a Smart Future & Recent Trends in Electronics & Communication (IOTTSF-2016),* in Association with Novateur Publication.

Pushpa, M. K., Gupta, A., Shaikh, S. M., Jha, S., & Suchitra, V. (2015, May). Microcontroller Based Automatic Waste Segregator. *International Journal of Innovative Research in Electrical, Electronics, Instrumentation and Control Engineering*, *3*(5).

Silva, B. N., Khan, M., & Han, T. (2018) Towards substainable smart cities: A review of trends, architectures, components and open challenges in smart cities. *Substainable smart cities and society, 38,* 697-713

Singhal, S., & Pandey, S. (2001). Solid Waste Management India, Status and Future Direction. *TERI Information Monitoring on Environment Science*, *6*(1), 1–4.

Solanas, A., Patsakis, C., Conti, M., Vlachos, I. S., Ramos, V., Falcone, F., Postolache, O. … Martınez-Ballesté, A. (2014). Smart health: A context-aware health paradigm within smart cities. *IEEE communication society,* 74-81.

Talari, S., Shafie-khah, M., Siano, P., Loia, V., Tommasetti, A., & Catalão, J. P. (2017). A review of smart cities based on the internet of things concept. *Energies*, *10*(4), 421. doi:10.3390/en10040421

Chapter 2
Assessing the Impact of RFID Technology Solutions in Supply Chain Management

Kamalendu Pal
City, University of London, UK

ABSTRACT

The use of Radio Frequency Identification (RFID) in Supply Chain Management (SCM) is one of the promising innovations in recent decades. This chapter first presents an introduction to the concepts and principles of RFID. It then discusses advantages and disadvantages of this technology in a supply chain setting. Application areas of RFID in the context of supply chains are reviewed to demonstrate best practices and related important implementation issues. Different industries (e.g. automotive, transport, retail) are used to emphasizing the benefits of RFID technology. The chapter also highlights operational and strategic implications of adopting RFID-based technological solutions and summarizes available evidence. Finally, a theoretical framework that links RFID key benefits and information attributes used in decision making is proposed. This chapter also provides comprehensive guidance for those considering the implementation of RFID in their supply chains.

INTRODUCTION

All businesses today understand the value and importance of building effective supply chains, as part of sustainable growth and profitability (Pal, 2019). A supply chain consists of a network of business organizations, facilities, people and activities for delivering products or services to a final customer. The key to make a successful supply chain relies on an extended collaboration and integration of business partners involved in the supply chain network. Such collaboration relies in supply chain business processes effective automation. The first significant attempt to automate supply chain was in manufacturing, with the automation of the car assembly-line in the early 1900s. Since then, automation extended to many industries and business processes that includes procurement, manufacturing, customer and supplier relationship management.

DOI: 10.4018/978-1-5225-9246-4.ch002

Due to globalization, retail supply chain business activities are carried out in different locations around the world. Goods manufactured at a factory are transported to intermediate storage facilities (warehouses, distribution centers) for packing and shipping to retailers or customers. Activities therefore are performed in specific sequences, with the completion of each activity, usually determining the start of another. The signaling of different activities is termed business events. An activity will generate an event upon start, completion, but also at intermediate points with milestones of the activity are reached. Such events and milestones include the completion of product manufacturing, the loading of products on transportation units, the unloading and storage of products in the warehouse / distribution center, the subsequent reloading onto different transport units for shipment to the customer. Figure 1 shows a simple diagrammatic representation of a supply chain, which highlights some of the primary business activities.

Despite several socioeconomic and technological advances, the main goal of supply chain networks has remained the same, i.e., the efficient delivery of products and / or services from the producer to the ultimate customer. Efficiency can be achieved through elimination of unnecessary activities, the simplification of activities, and more importantly through automation of activities. Significant efforts have therefore been invested for the automation of activities on the factory and warehouse day-to-day business practice. Recent emphasis has also been placed on transportation activities that move the goods across the different supply chain locations.

Many researchers have proposed the existence of a positive correlation between supply chain business performances and corporate information systems integration and coordination of business partners activities (Pal & Karakostas, 2014) (Pal, 2018a) (Pal, 2018b). This is the reason why supply chain businesses are more and more attentive to the opportunity offered by both internal activity coordination and intra-company activity coordination through information sharing. Hence, information has been seen increasingly as a strategic asset for supply chain management. As a result, many supply chain businesses are investing in new technology in order to boost the information exchange and are mandating the adoption of interoperable supply chain solutions (Smith, 2005) (Karkkainen, 2003).

RFID is a recent information technology that promises to enable supply process integration through information sharing. It is an identification technology, that utilizes radio communication and consists of two main components: (1) transponders (commonly known as tags) carrying data and attached on the physical objects to be identified; (2) interrogators (also known as readers) able to receive the transmitted data. The benefits of RFID technology compared to older product identification technologies such as barcodes include: (i) no line of sight path between reader and tag is required; (ii) longer distances between transmitter and reader (up to several meters) are feasible; (iii) nearly simultaneous detection of multiple RFID tags is possible; and (iv) ability to store more data in the tag (up to several kilobytes) compared to barcodes.

Figure 1. Diagrammatic representation of a generic supply chain

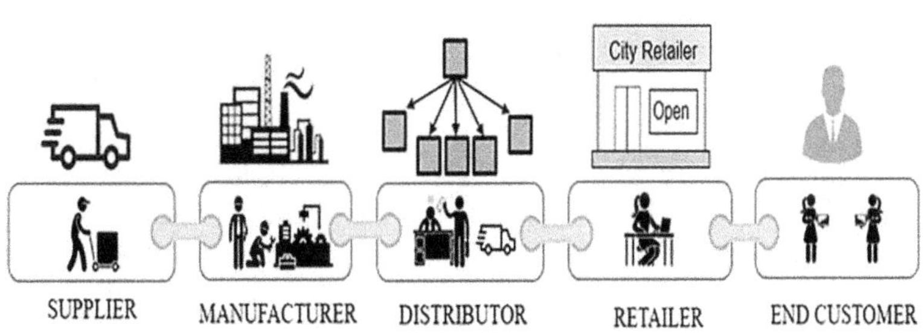

RFID tags, as shown in Figure 2, are attached to objects (e.g. machines, and merchandise items) for identification purpose. Each tag typically consists of an antenna constructed from a small coil of wires; a microchip used to store information digitally about the companion object; and encapsulating material to enclose the chip and the coil. Like there are distinct types of barcode, RFID tags are available with different memory sizes and encoding options. These tags can also incorporate sensors to record temperature, vibration, shock, or humidity, for example, providing the ability to track and report on an object's environmental characteristics dynamically.

Using an antenna, the information that is stored on the tag can be read or written from the tag. In RFID-based infrastructure, target tracking, object recognition, access control, and the quick and reliable identification of RFID tags to different locations, by using radio waves, greatly improves the business automation process along supply chain. Furthermore, a completely new product identification scheme, the Electronic Product Code (EPC), allows for a massively increased address space and have ushered in enormous opportunities to provide synchronized decision-making in supply chain operations management.

A RFID reader is the device generally communicates with the RFID tag(s). Reader, either as stationary or handheld devices, consists of a transmitter, receiver, antenna, microprocessor, controller, memory and power source. It discharges radio frequency (RF) signals to, and receives radio waves from, the tag via an antenna or antennas. The reader converts the received radio waves into digital information that is usually passed to a backend system, as shown in Figure 3.

The utilization of RFID as a high-level IT product identification in the supply chain operations has attracted a lot of attention of both researchers and logistics professionals. The RFID-based business solutions are well associated with the powerful and dynamic data capturing ability (Singh, 2003). The real-time availability of physical product identity allows other information, related to the product, to be drawn on in order to assess both the current state of the product and future actions required. In the context of supply chain operations, widespread introduction of such systems represents a major suitability to examine and enhance tracking and tracing systems, process control and inventory management. In this way, RFID-based business solutions can be an enriched business information source, which can usher to different level of organizational advantages. Thus RFID-based supply chain solutions should be viewed as an information facilitator that can directly enhanced the operational decision-making. However, there is very-little, or no investigation focus on research to evaluate what RFID-based technology solutions and their business implications in operational decision making.

Figure 2. A basic RFID tag and its interrogator

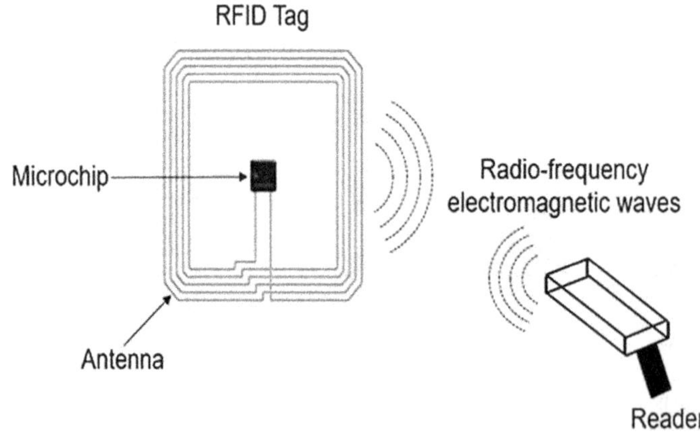

Besides its ever-broader diffusion across many supply chain industries, RFID has also become an attractive research topic, not only in electrical engineering and computer science, but also in business management. Several success stories have recently been emerged in the fields of information systems and operations management that not only support designers of RFID-based systems and processes but also explain how RFID can generate business value in organizations (Pal, 2019) (Roussos et al, 2002) (Jones et al, 2004a) (Choi et al., 2018). However, most of this work tends to consider RFID as a "next generation data capturing innovative technology" differencing only slightly from its predecessor (e.g. barcode reading systems) in its enhanced precision and the timeliness of collected data. This narrow view runs the risk of ignoring the various fundamental levels of data quality associated with the RFID-based technologies. To initiate the exploration of the main theme of this chapter, the next section presents background information on a simple RFID-based system.

RFID technology solutions have paved the great potential to improve supply chain performance, and the technology is still far from reaching mass adoption. Indeed, the RFID-based technology adoption rate has been depicting a grim picture except from the RFID mandates issued by major retail enterprises. Therefore, it is worth to review the RFID uses business landscape and shed some light regarding the causes of the barriers of RFID-based technology widespread diffusion in global supply chain operation.

Academics and practitioners research works highlighted that a lack of standards, hardware and software reliability, versatility related issues, and misaligned information technology diffusion within strategic objectives are creating barrier for supply chain industry-wide RFID technology adoption. Some of the issues of lack of standardization of this technology has been recently addressed by industry standards enterprises and groups such as global Electronic Product Code enterprise i.e. EPC Global (EPC, 2019). Unfortunately, RFID-based information system deployment advantages remain unclear. However, few success stories have been reported in the business world; and RFID technology solutions deployment managers are finding it very difficult to justify the potential advantages and how to achieve them in their own business.

Researchers provided policy guidance and literatures have shown that commercial world should access RFID-based technology infrastructure deployment with respect to their company strategic goal before adopting it. However, many business enterprises have been heavily influenced by Wal-Mart's highly publicized RFID-based technology success stories – branded by industry commentators as a '*bandwagon-effect*'. Business practitioners suggested that the initial success stories may have influenced companies to adopt RFID-based technology solution for their supply chain without judicious justification, but following others, possibly due to organizational blind-technology-savvy culture.

This chapter presents a framework for understanding the business value of RFID in supply chain management. The chapter consists of seven sections. Section 2 provides the background knowledge of RFID-based business solutions. Section 3 describes main business applications and use cases of RFID applications. Section 4 provides the theoretical foundation of technology adoption decisions as applied to RFID-based application systems. Section 5 explains RFID-derived benefits and information quality attributes. Section 6 provides some concluding remarks. Section 7 describes briefly the future research directions.

Background Information

Due to the rapid development in the field of wireless communication, RFID technology has been used widely in the field of logistics and supply chain management. This technology acts as enabler to seamless connectivity as the flow of information, materials, and final products along the supply chain. Seamless flows are enabled by two factors: (i) information systems connectivity, and (ii) physical connectivity. Information systems connectivity refers to the electronic linkage of partners up and down the supply chain. Different technologies can be used to collect and / or exchange information among supply chain partners. Among them are RFID, electronic data interchange (EDI), and various data mining and Big Data technologies.

A wide range of product information can be transmitted using RFID technology, including the identification of the product, location details, price, and dates of manufacturing, transportation, and purchase. EDI is a computer-to-computer data transmission system that collects and share information over the Internet. It is used, for example, to place purchase orders, generate billing and payment orders, transmit sales and inventory data, and give advanced shipping notice. Data mining refers to the process of extracting previous unknown and actionable information from massive amount of collected data (Big Data) and using the information generated out of algorithmic computation for decision making. Apply data mining technology to realize hidden knowledge, relationships, and trends from the Big Data accumulated in the supply chain can help companies to make real-time decision and improve supply chain management.

RFID Technology and Its Data Processing

There are three main types of RFID tag in use today:

Active RFID: The tags of active RFID systems, tags have their own transmitter and power source and broadcast their own signal to transmit the information stored on their microchips. Active RFID systems typically offer a range of up to 100 meters.

Battery-Assisted Passive RFID: This type of RFID tags uses an integrated power source (usually a battery) to power on the chip so that all the energy captured from the reader can be bounced back.

Passive RFID: In passive RFID systems, the tag is energized by the signal transmitted from the reader to power on and reflect a signal back. As they do not require a power source or transmitter, and only require a tag chip and antenna, passive RFID are cheaper, smaller, and easier to manufacture than active tags.

RFID System Solution Main Components

A simple RFID-based system solution consists of six main components as shown in Figure 3. These components are:

1. **Tag**: A tag is the data (or information) carrier part of RFID-based business solution. Generally, it contains a unique identification number, and specific *electronic product code* (EPC) programmed into the tag.
2. **Reader and Antenna**: A reader captures the data provided by the tag when tags come in the range of sensing area covered by the specific reader using its antenna (i.e. an electronic *signal receiving special circuit*).
3. **Middleware**: The middleware can be software as well as hardware dedicated to process data captured by the *tag reader*, then dispatch this information to backend servers. It helps data gathering and management for any RFID deployment in a supply chain. It consists of a set of services that allow multiple processes running on one or more RFID system to interact. RFID middleware as-

sists with the filtering, aggregation, and routing of RFID data. It has built-in business rules that monitor the data stream and direct data to appropriate enterprise systems. It is used to manage data flow between the RFID networks and the IT systems within an organization (Sarac et al., 2010) (Chuang & Shaw, 2005) (Asif & Mandviwalla, 2005).

4. **Backend Servers**: The backend servers hold the data collected from RFID-based application systems for processing purpose.
5. **Network Infrastructure**: This part of RFID-based solution infrastructure is basically providing data communication provision. It plays a very important role for RFID-based system solutions security and privacy related issues.
6. **Database for Tags Information**: This is the part of the enterprise information system.

RFID-based supply chain application systems can be divided into two major groups, as mobile and immobile applications. Immobile RFID systems include RFID readers, antennas (usually two or four antennas for each reader), hosts servers, middleware and external units such as light and sensors. These systems are also called RFID gates. In these systems, readers serve as gates, receive information from tagged objects and send the information to servers or controllers. Mobile systems use wireless communication to gather data and monitor objects. They are like fixed systems due to RFID system structure. They provide advantages such as data gathering and managing, reading/writing and communication of data. RFID readers receive information and transfer this information to controllers, servers, and database systems in supply chains.

Function of RFID Database Systems in Supply Chain Management

RFID can provide higher levels of automation in supply chain management. It also helps to reduce errors that manual scanning of barcodes can create, as an RFID reader can identify and record information about an entire shipment in one pass, without the need to scan each item manually. In recent decades,

Figure 3. A basic RFID-based technology solution architecture

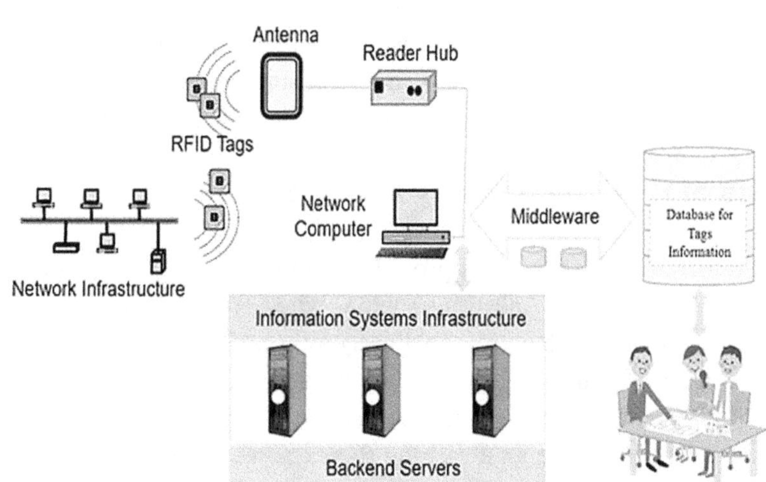

industry is progressively rallying around few worldwide standards for RFID technologies. In this effort, a leading role played by EPC Global consortium.

The EPC Global Network can virtually connect physical object and data via the Internet. Data about every product – its history or other product related information can be made available through a standardized infrastructure anywhere and anytime. Also important are RFID web service providing integration among heterogeneous systems (internal and external) and business-to-business communications, as Simple Object Access Protocol (SOAP) is fully interoperable between systems. With RFID web services, a customer or even a supplier can synchronize data between enterprise resources management (ERP) applications and other corporate legacy system and back-end systems for the RFID infrastructure.

RFID data can be classified into two categories: the evet data and the master data. The event data keeps real-time (or dynamic) information, which is about RFID tagged objects such as containers, pallets, material handling equipment, cases, automobiles and so on. The master data provides conditional information and verification about the event data.

Event data is related to a definite time, and it provides the communication about RFID tagged objects during supply chain processes. It is created whenever some sort of transaction occurs. It is captured in distributed data repositories, and only the relevant event data must be sent to the monitoring for further processing (Miles et al., 2008) (Angeles, 2005) (Brazeal, 2009). The processing and matching of automatically generated monitoring instructions with the event data gathered from distributed data repositories must be performed by an appropriate event processing engine (Veronneau & Roy, 2009) (Kwake et al., 2011). Event data creates information which is about investing the existence of items somewhere at some time. Event data is generally used for tracking and tracing applications to monitor items associated with transportation processes and transported goods (Ferrer et al., 2010). The combination of new technologies provides the potential to use RFID based on event data for the automatic and near real time monitoring of processes in supply chain networks to detect anomalies according to specific objectives. By using RFID widely, applications may need more information and more sensor observation (Werner & Schill, 2009).

Master data, also called reference data, describes an item and its general properties. It includes useful data about customers, products, employees, materials, suppliers, manufacturers and so on in a supply chain. It contains information such as source verification, product definition referenced by EPC (Electronic Product Code), manufacturer information, details about the object which event data is collected from, and storage information. It can define transactional processes and operations (Glover & Bhatt, 2006) (Kwak et al., 2011) (Tajima, 2007). Master data is a key information for quality-assurance, persisting demands, business operations applications. It provides processes for collecting, aggregating, consolidating, matching, and distributing such data throughout a company to ensure control and consistency in the ongoing maintenance and application of this information (Chuang & Shaw, 2005) (Asif & Mandvwalla, 2005).

Potential Benefits of RFID Technology in Supply Chains

Adopting RFID-based technological solution provides a number of advantages to business operations (Pal, 2018). It is tricky to answer basic questions such as whether a RFID adoption strategy will deliver sustainable stakeholder value in the changing business environment, and how to demonstrate the value a business creates after it adopts this ubiquitous computing for day-to-day operations.

The value of RFID-based supply chain operation solution in strengthen decision making has been endorsed by academics and practitioners publishing their research works. In this chapter, modelling of logistic processes and evaluation of their performances are done by using published business cases from academic literatures.

RFID technologies offer several benefits to supply chain operations through their advanced properties such as unique identification of products, easiness of communication and real-time information provision (Saygin et al., 2007). Thus, this technology can improve the traceability of products and the visibility throughout the entire supply chain, and can make reliable and speedy tracking, shipping, checkout and counting processes, which leads to improve inventory flows and information that is more accurate. Furthermore, in identifying RFID benefits, it is possible to filter-out appropriate information quality attributes that can be subsequently related to industry specific supply chain attributes that invariably enhance judicious operational decision-making.

Previous research works (Pal, 2019) (Angels, 2005) suggested that RFID technology has been important driver of business processes associated with supply chain operational areas that included retail distribution (e.g. good received, checking, put away, replenishment, shipping, and order taking) and transportation (product and asset tracking). However, these two-supply chain operational area do not include business processes associated with RFID technology benefits that may be applicable to store-based activities – or as additionally identified from the literature review, where the supply chain appears to be extending into post-store sphere and the realm described by Michael Porter (Porter, 1985) as "*added value*" for the customer. Drawing from the work of Angeles (Angeles, 2005) and adding additional supply chain operational areas (e.g. addressing retail and post store activities), it is possible to classify potential RFID technology related benefits and associated decision-making information attributes, as shown in Table 2.

RFID Applications Across the Supply Chain Business

Analysts are seen the early RFID-based technology adoption in supply chain management receptive to macro or large-scale focus applications. For example, the distribution and transportation areas have a predominated application of RFID tags at the pallet, role-cage and / or container level where it is appropriate to capture pertinent data variables that may be associated with location, shipment type or positioning of a large set of product items (Schindler, 2003) (Thompson, 2003) (Angeles, 2005). In addition, the information attributes identified with these applications are multi-faceted containing both time (O'Brien, 2001) (Cappiello et al., 2004) and content (O'Brien, 2001) (Pipino et al, 2002) based attributes.

Within the retail store application of the supply chain management, companies are found to benefit from RFID implementation across different business functions, which were related with in-store customer marketing (Juels et al, 2003). In-store customer marketing utilizes the tracking properties of RFID tags allowing customer specific product profiles, with the subsequent promotion of secondary product. Further in-store benefits of RFID technology allow management to gauge a real-time view of stock levels facilitating critical product ordering and replenishment.

The post-retail store business can be considered as extending the retail supply chain to the customer home. Within the customer home the technology appears to have the potential to add values with tagged products being able to interact with compatible RFID home-based appliances (Smaros & Holmstrom, 2000) (Juels et al, 2003). In this way, the aspect of home-based RFID utility could be considered as a stimulus for appliance suppliers to accommodate RFID technology in future products. An unexpected consequence of RFID technology adoption in the post-retail store business is one of addressing societal environmental values – where embedded tag enhances or even promote consumer product recycling (Juels et al, 2003). Like the retail store activity, the post-retail activity appears to have RFID-derived information attributes that have a content dimension. As the retail business extends to the post-retail area there tends to be more robust in RFID technology applications at the item level, resulting in item-specific benefits and information value. One of the techniques to justify the value of technology adoption is to analysis previously reported business case studies.

RFID Application Cases

Six business cases are used from different industries to analysis the business value. These cases are categorized according to their: industry type, business application area, and main advantages gained by RFID based technology adoption. The business cases are shown in Table 1.

A Framework for RFID Business Impact

The RFID-based technological solutions benefits can span the entire supply chain, associated with improved distribution and inventory control within warehouse as well as benefits associated with the retail store environment. Some benefits have a functional application in the business domain, while other benefits related directly to customer interaction. Furthermore, in identifying RFID benefits it has been possible to also filter-out appropriate information quality attributes that can be subsequently related to supply chain attributes that invariably enhance the decision-making process. Supply chain business-areas with relevant RFID benefits and information attributes are shown in Figure 4, which can be considered as an RFID uses information value chain - a variation of the Porter's original business analysis framework (Porter, 1985).

Table 1. Successful RFID-based business cases

Industry Type	Business Application Area	Main Advantages Gained
Retail	A research group (Delen et al., 2007) reported on actual RFID data gathered from Wal-Mart (a fast mover, implementing the technology across 1400 stores) to study the "*mean time between movement*". Placement of RFID tags at the pallet and case level, with items tracked from the time they enter the distribution centers until the pallets / cases were crushed at retailer outlets.	RFID technology can ensure the freshness of perishable products, identify redundant processes, assess supply chain partners' performance, reduce inventory, and improve business processes.
Transport	Researchers (Ngai et al., 2007) highlighted the experimental results on RFID application for managing a container depot.	Efficiency, and customer service.
Automotive	Gaukler and Hausman (Gaukler & Hausman, 2008) investigated how RFID applications can support a mass customization of assembly line.	Continuous improvement and quality assurance.
Third party logistics provider (3PL)	Kim and other researchers (Kim et al., 2008) reported about automate many labour-intensive barcode scanning processes.	Reduced its original seven business processes to four, thereby improving its order processing time. In addition, customers obtained more accurate data and reporting on their shipments.
Automotive supply chain	Mourtzis and fellow researchers (Mourtzis et al., 2008) described how RFID provided visibility to an upstream supply chain automobile industry. RFID readers and attaching RFID tags to major parts to gather data from upstream supply chain partners. A web service to monitor and analyze the parts available in the upstream supply chain.	Allowed dealers to check whether a customer's order was available and to determine a more accurate delivery date. Allowed OEM (Original Equipment Manufacturer) plants to develop more accurate production plans, leading to a reduction in costs and delivery time.

Figure 4. Proposed business impact framework

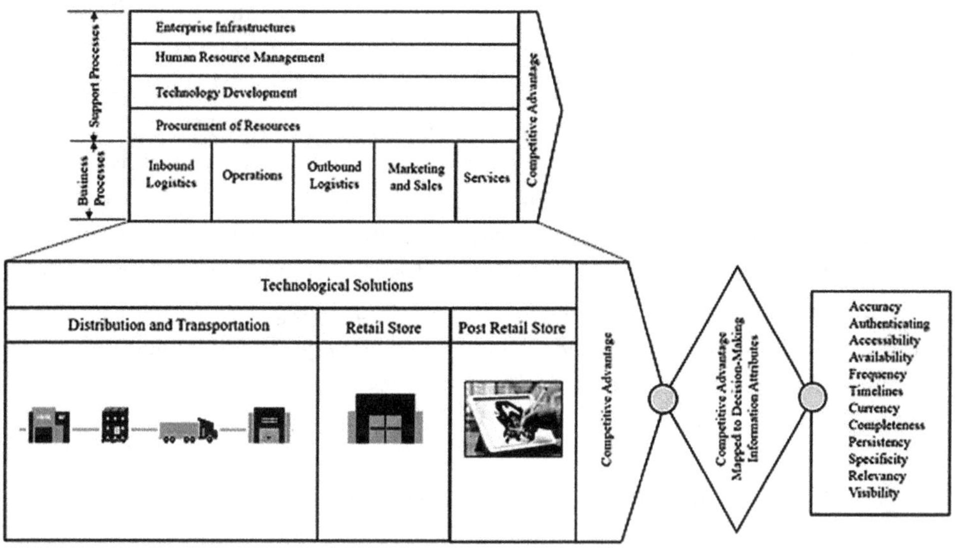

RFID-derived information identified early in the retailer supply chain tends to provide descriptive attributes that reflect an application of the technology to what is, in practice, a collectively enormous number of items. As one traverses to the retail sector there appears to be an accrual of benefits and associated information value that relies on using RFID at the carton and / or individual product level. Within the retail business, RFID technology tends to be a variable substitute for item-level barcodes allowing reported benefits such as enhanced retail inventory control as well as improved product identification and tracking ability (Roussos et al., 2002) (Juels et al., 2003) (Karkkainen, 2003).

Moreover, the RFID benefits and information value derived within the retail business are delineated from the previous business by being applicable at the case and item level compared to the physically large pallet and container. As the retail business extends to the post retail business there tends to be a more concerted applicability of RFID technology at the item level – resulting in commensurate item-specific benefits and information value.

It should be noted that as one traverses the benefits-information value chain (as shown in Figure 4) from left to right, the RFID-derived decision-making information associated with benefits becomes more item or product-level specific. Hence, identified information attributes (e.g. accuracy, completeness) that are similar in context, but affiliated with benefits in different business areas of the chain, will have a different point of reference for organizational decision makers.

RFID Technology and Associated Information Attributes

Generally, information can be considered as an organizational resource for decision-making; and this resource is heavily dependent on raw data collected from daily business operations. Individual information consists of different attributes and characteristics that reflect business processes and values (Miller, 1996). Information may have descriptive characteristics that relate, for example, to accuracy, completeness, relevancy, and persistency – attributes that have an information content dimension (O'Brien, 2001) (Pipino et al, 2000). Information may also be directly related to a time dimension (O'Brien, 2001) (Cappiello et al., 2004) that includes descriptive characteristics such as timelines, frequency and currency. Arguably, from organizational perspective it is the important decision-making attributes of RFID-based business solutions that allows supply chain operational managers to address *tactical* and *strategic* issues. Academics and practitioners tend to refer to distinct categories of RFID-based business solutions advantages; however, they appear to overlook the important intermediate process of using RFID-based information that is a key factor in deriving these advantages. For example, it is RFID-based information system that help operational managers to have a visibility of their business's supply chain (Jones et al, 2004a) (Jones et al, 2004b); or information relate to inventory stock-levels that permits a very exact gauge and understanding of a business's RFID-based information monitoring system (Angeles, 2005); or unique RFID-based business information that is crucial in being able to identify and authenticate products (Weiss, 2003) (Moon & Ngai, 2008). Hence, in assessing the advantageous impacts of RFID-based supply chain solutions, there is an associated process of also identifying the information value that the technology provides to a business.

Time and Content Value of RFID-Derived Information

This chapter proposes that visibility related to a time-based information attribute allowing organizational managers to be powerfully informed about the status of their supply chain. Moreover, RFID-derived benefits appear to be also associated with information that has a content dimension (O'Brien, 2001) (Pipino et al., 2002). Content-based information attributes including accuracy, completeness, relevancy and persistency appear to be important in supporting and generating RFID-derived benefits. Hence, the adoption of RFID technology can be expected to produce both time and content related organizational information, and information that will improve decision-making to realize business benefits.

Information quality is affiliated with different dimensions and has been related to the user perception that are embodied in business requirements and values (Miller, 1996). The capture of data facilitated by RFID addresses a time dimension (O'Brien, 2001) (Cappiello et al, 2004) that includes attributes such as timelines, frequency and currency. Hence, organizational personnel can consider decision-making processes at an earlier point in time than if RFID technology was not in place.

The ability to engage the decision-making process earlier because of RFID implementation is an example of information added value – the RFID-derived information having a higher quality value than information that may have been traditionally captured before RFID implementation (Turban et al., 2002). Academics and practitioners (Smaros & Holmstrom, 2000) (Singh, 2003) (Jones et al., 2004a) (Jones et al, 2004b) (Angeles, 2005) allude to RFID technology providing supply chain visibility, inferring that managers can see the status of the supply chain at any given point in time.

In this way, organizational managers to be informed about the status of their supply chain. Moreover, RFID-derived benefits appear to be also associated with information that has a content dimension (O'Brien, 2001) (Pipino et al., 2002). Content-based information attributes including accuracy, completeness, relevancy and persistency appear to be important in supporting and generating RFID-derived benefits. Hence, the adoption of RFID technology can be expected to produce both time and content related organizational information, and information – information that will improve decision-making to realize multi-sector benefits. The detailed classification of identified benefits and information attributes is shown in Table 2. The table also shows the number of instances that a benefit was considered in the literature.

Table 2. Information quality dimensions and references

Information Quality Reference / Dimension	(Singh, 2003) (A)	(Angeles, 2005)	(Smaros & Holmtrom, 2000) (B)	(Karkkainen, 2003) (C)	(Kinsella, 2003)	(Jones et al, 2004a, b) (D)	(Weiss, 2003)	(Thompson, 2003) (E)	(Roussos et al, 2002) (F)	(Juels et al, 2003) (G)	(Schindler, 2003) (H)	Total
Accuracy	1	1	1	1	1	1	1				1	8
Actionable											1	1
Authenticating							1			1		2
Accessibility								1		1	1	3
Availability										1		1
Frequency	1											1
Timeliness	1	1	1	1	1	1						6
Customised									1			2
Currency	1			1		1						3
Completeness		1		1	1				1	1		5
Persistency								1				1
Specificity								1				1
Relevancy									1	1		2
Visibility						1						1

RFID Adoption Decisions

Information Technology (IT) adoption decisions require the evaluation of a multitude of factors within a given set of constraints. It practically results in a complex decision-making situation for the enterprise decision-makers. Previous academic research has identified several technological, organizational and environmental factors (commonly classified under the TOE framework) as significant in determining IT innovation adoptions. These factors are important in determining the adoption of IT innovations, however, they provide little understanding regarding the actual cognitive processes involved in the adoption decision-making. In recent work researchers have been focusing on understanding the actual decision-making process and started using various psychological constructs and cognitive theories in innovation and strategic decision-making research to explain the cognitive process involved in organizational innovation adoption. Adoption of IT innovations constitutes a complex information processing scenario that involves making sense of an information technology that the enterprise is unfamiliar with and is typically characterized by uncertainty and ambiguity over the outcomes of the innovation process. Although IT innovations are believed to be able to confer strategic and competitive benefits to the adopting enterprise, they are often complex technologies that call for significant investment of organizational resources. Thus, managers are faced with the task of analyzing the ramifications of innovation of their enterprise. Under such circumstances, deciding on whether an innovation is a good thing for the enterprise, whether the timing of the innovation is appropriate, and how the adoption is best carried out requires organizational decision makers to attend to the innovation with reasoning grounded in their own facts and specifics (Fichman, 2004).

Mindful RFID Adoption Decisions

Mindfulness in the context of organizational adoption of IT innovations corresponds to an engagement with a given innovation based on facts and details which are unique to enterprise itself (Swanson & Ramiller, 2004). It has been suggested that mindfulness can reduce the possibility of failure when innovating with IT because mindfulness will result in a decision which is based on richer and contextually relevant interpretation of a given situation (Swanson & Ramiller, 2004; Fichman, 2004). Therefore, decision-maker mindfulness is a desirable property in the process of adoption of IT innovations in enterprises. Further, IT innovation adoptions are often prone to bandwagon behavior among enterprises (Swanson & Ramiller, 2004). It has also been shown that enterprises feel mimetic, normative and coercive institutional pressures when deciding on innovations to adopt. While adoption decisions resulting from normative and coercive forces can be explained as being a strategic choice or requirement, it is likely that mindfulness in organizational decision-makers will help in overcoming the influence of mimetic institutional forces and the resulting bandwagon behaviour.

A common way to assess the effectiveness of both supply chain strategies and organizations is to verify whether an organization utilizes supply chain practices that match its supply chain strategy. Organizations can expect significantly better results when they match; e.g., a common best supply chain practice for a lean supply chain is to reduce inventory, whereas a common best supply chain practice for an agile supply chain is to satisfy customer requirements.

CONCLUSION

This chapter presents the issues and potential industrial applications for a range of RFID-based information system solutions in supply chain management. It is worth to note that the use of RFID-based supply chain solution has grown across a variety of core industries, such as logistics, manufacturing, and retail business. Although each application has its own specific requirements, and information technology (IT) based solutions are considered as the main mechanisms to streamline the complex business processes along supply chain. This chapter highlights that RFID technologies can provide several advantages in supply chain management through better traceability and improved visibility of products and processes all along the chains. Increase of efficiency and speed of business processes, improvement on information accuracy, reduction of inventory losses is some of these advantages.

Although each application has its own specific requirements, and IT based solutions are considered as the main mechanisms to streamline the complex business processes along global supply chain. Real-world business cases are used to examine the commercial benefits of RFID-based solutions in supply chain management. The benefits are reflected in business functions such as improved inventory management, enhanced customer-engaged marketing facilities or cost saving exercises associated with improved retail supply chain operations. The significance of value-added information derived from the RFID-based technologies receive are appeared to be both an intuitive and tactical mechanism for its effectiveness in business decision-making.

This chapter identified advantages and information attributes related RFID-based solutions use within the supply chain that can lead to improved business responsiveness through improved decision-making capabilities. Particularly, the examples of RFID-derived information benefits taken from the academic literature provide improved lessons for enterprises that are currently experimenting or expecting to use RFID-based business solution.

Future Research Directions

This chapter proposed that identified RFID-based business solutions advantages are related to the sharing of pertinent information value and occurs across the different business areas of supply chain management. Although, the analytical and simulation models proposed in the literature are limited to one product, one retailer, or one manufacturer. In further investigation, it would be interesting to conduct research work for multiple items and multiple business partners of the supply chain business.

Further research needs to recognize the technology's importance as a significant producer of new or enhanced data; data that can be used as a basis for enhancing corporate decision-making. In addition, RFID-based information solution research needs to assess not just the overall retail supply chain – but also concentrate on each individual segment of the supply chain, to gain insights into the diverse nature of RFID-derived information within each business operations. Because each business operation has its own unique information requirements, there are diverse types of information attributes involved, which will provide a different point of reference for the RFID adoption process.

REFERENCES

Angeles, R. (2005). RFID technology: Supply-chain applications and implementation issues. *Information Systems Management*, *22*(1), 51–65. doi:10.1201/1078/44912.22.1.20051201/85739.7

Asif, Z., & Mandviwalla, M. (2005). Integrating the supply chain with RFID: A technological and business analysis. *Communications of the Association for Information Systems*, *15*(24), 393–427.

Brazeal, M. (2009). *RFID: Improving the Customer Experience*. New York, NY: Paramount Marketing Publishing.

Cappiello, C., Francalanci, C., & Pernici, B. (2004). Time-related factors of data quality in multichannel information systems, *20*(3), 71-91.

Choi, T. M., Yeunge, W. K., Cheng, T. C. E., & Yue, X. (2018). Optimal Scheduling, Coordination, and the Value of RFID technology in Garment Manufacturing Supply Chains. *IEEE Transactions on Engineering Management*, *65*(1), 1. doi:10.1109/TEM.2017.2739799

Chuang, M. L., & Shaw, W. H. (2005). How RFID will Impact Supply Chain Networks, *IEEE International Engineering Management Conference*, Newfoundland, Canada, 231-235.

Delen, D., Hardgrave, B. C., & Sharda, R. (2007). RFID for better supply management through enhanced information visibility. *Production and Operations Management*, *16*(5), 613–624. doi:10.1111/j.1937-5956.2007.tb00284.x

EPC. (2019). Available at https://www.gs1.org/epcglobal

Ferrer, G., Dew, N., & Apte, U. (2010). When is RFID right for your service? *International Journal of Production Economics*, *124*(2), 414–425. doi:10.1016/j.ijpe.2009.12.004

Fichman, R. G. (2004). Going Beyond the Dominating for Information Technology Innovation Research: Emerging Concepts and Methods. *Journal of the Association for Information Systems*, *5*(8), 314–355. doi:10.17705/1jais.00054

Gaukler, G. M., & Hausman, W. H. (2008). RFID in a mixed-model automotive assembly operation: Process and quality cost saving. *IIE Transactions*, *40*(11), 1083–1096. doi:10.1080/07408170802167654

Glover, B., & Bhatt, H. (2006). RFID Essentials. Sebastopol, CA: O'Reilly Publishing.

Jones, P., Clarke-Hill, C., Shears, P., Comfort, D., & Hillier, D. (2004a). Radio frequency identification in the UK: Opportunities and challenges. *International Journal of Retail & Distribution Management*, *32*(3), 164–171. doi:10.1108/09590550410524957

Jones, P., Clarke-Hill, C., Shears, P., Hillier, D., & Comfort, D. (2004b). Radio frequency identification and privacy and public policy issues. *Management Research News*, *27*(8/9), 46–56. doi:10.1108/01409170410784563

Juels, A., Rivest, R. L., & Szydlo, M. (2003, October). The blocker tag: selective blocking of RFID tags for consumer privacy. *Proceedings of the 8th ACM Conference on Computer and Communications Security,* ACM Press, Washington, DC, 103-110. 10.1145/948109.948126

Karkkainen, M. (2003). Increasing efficiency in the supply chain for short shelf life good using RFID tagging. *International Journal of Retail & Distribution Management*, *31*(10), 529–536. doi:10.1108/09590550310497058

Kim, C., Yang, K. H., & Kim, J. (2008). A strategy for third-party logistics systems: A case analysis using the blue ocean strategy. *Omega*, *36*(4), 522–534. doi:10.1016/j.omega.2006.11.011

Kwak, C., Cho, Y., Ko, J. M., & Kim, C. O. (2011). Adaptive Product Tracking in RFID-Enabled Large-Scale Supply Chain. *Expert Systems with Applications*, *38*(3), 1583–1590. doi:10.1016/j.eswa.2010.07.077

Miles, S. B., Sarma, S. E., & Williams, J. R. (2008). *RFID Technology and Applications*. New York, NY: Cambridge University Press. doi:10.1017/CBO9780511541155

Miller, H. (1996). The multiple dimensions of information quality. *Information Systems Management*, *13*(2), 79–83. doi:10.1080/10580539608906992

Mourtzis, D., Papakostas, N., Makris, S., Xanthakis, V., Xanthakis, V., & Chrysolouris, G. (2008). Supply chain modelling and control for producing highly customized products. *CIRP Annals – Manufacturing Technology, 57*(1), 570-586.

Ngai, E. W. T., Cheung, T. C. E., Au, S., & Lai, K. (2007). Mobile commerce integrated with RFID technology in a container depot. *Decision Support Systems*, *43*(1), 62–76. doi:10.1016/j.dss.2005.05.006

O'Brien, J. A. (2001). *Management Information Systems: Management Information Technology in the Internetworked Enterprise* (5th ed.). Boston, MA: McGraw-Hill.

Pal, K. (2018a). Building High Quality Big Data-Based Applications in Supply Chains. In A. Kumar, & S. Saurav (Eds.), Supply Chain Management Strategies and Risk Assessment in Retail Environments (pp. 1–24). Hershey, PA: IGI Global. doi:10.4018/978-1-5225-3056-5.ch001

Pal, K. (2018b). Ontology-Based Web Service Architecture for Retail Supply Chain Management. In *the Proceedings of 9th International Conference on Ambient Systems, Networks and Technologies,* Porto, Portugal. *Procedia Computer Science*, *130*, 985–990. doi:10.1016/j.procs.2018.04.101

Pal, K. (2019). Radio Frequency Identification Systems Security Challenges in Supply Chain Management. In J. Rodrigues, A. Gawanmeh, K. Saleem, & S. Parvin (Eds.), Smart Devices, Applications, and Protocols for the IoT. Hershey, PA: IGI Global. doi:10.4018/978-1-5225-7811-6.ch010

Pal, K., & Karakostas, B. (2014). A multi agent-based service framework for supply chain management, Procedia 53-60. *Computer Science*, *32*, 53–60.

Pipino, L., Lee, Y. W., & Wang, R. Y. (2002). Data quality assessment. *Communications of the ACM*, *45*(4), 211–218. doi:10.1145/505248.506010

Porter, M. E. (1985). Technology and Competitive Advantage. *The Journal of Business Strategy*, *5*(3), 60–78. doi:10.1108/eb039075

Roussos, G., Koukara, L., Kourouthanasis, P., Tuominen, J., Seppala, O., & Frissaer, J. (2002). A case study in pervasive retail. Proceedings of the 2nd International Workshop on Mobile Commerce, Atlanta, GA. New York, NY: ACM Press, 90-94. 10.1145/570705.570722

Saygin, C., Sarangapani, J., & Grasman, S. E. (2007). A Systems Approach to Viable RFID Implementation in the Supply Chain. *Springer series in advanced manufacturing,* 1.

Schindler, E. (2003). Location, location, location: what effect will tracking technologies like RFID and GPS have on connected businesses? *networker, 7*(2), 11-14.

Singh, N. (2003). Emerging technologies to support supply chain management. *Communications of the ACM*, *46*(9), 243–247. doi:10.1145/903893.903943

Smaros, J., & Holmstrom, J. (2000). Viewpoint: Reaching the consumer through e-grocery VMT. *International Journal of Retail & Distribution Management*, *28*(2), 55–61. doi:10.1108/09590550010315098

Smith, A. D. (2005). Exploring radio frequency identification technology and its impact on business systems. *Information Management & Computer Security*, *13*(1), 16–28. doi:10.1108/09685220510582647

Swanson, B. E., & Ramiller, N. C. (2004). Innovating mindfully with information technology. *Management Information Systems Quarterly*, *28*(4), 553–583. doi:10.2307/25148655

Tajima, M. (2007). Strategic value of RFID in supply chain management. *Journal of Purchasing and Supply Management*, *13*(4), 261–273. doi:10.1016/j.pursup.2007.11.001

Thompson, N. B. (2003). The big showdown? Will RFID technology eventually phase out bar code? Paper. *Film and Foil Converter*, *77*(8), 38–40.

Turban, E., Mclean, E., Wetherbe, J., Bolloju, N., & Davison, R. (2002). *Information Technology Management: Transforming Business in the Digital Economy* (3rd ed.). New York, NY: Wiley.

Veronneau, S., & Roy, J. (2009). RFID benefits, costs, and possibilities: The economical analysis of RFID deployment in a cruise corporation global service supply chain. *International Journal of Production Economics*, *122*(2), 692–702. doi:10.1016/j.ijpe.2009.06.038

Weiss, A. (2003). Me and my shadow: RFID tags polarize the debate over privacy vs. efficiency, *networker, 7*(3), 24-30.

Werner, K., & Schill, A. (2009). Automatic Monitoring of Logistics Processes Using Distributed RFID based Event Data. *International Workshop on RFID Technology,* Milan, Italy, 101-108.

KEY TERMS AND DEFINITIONS

Active Tag: A tag with its own battery that can initiate communications.

Auto-ID: Automatic Identification (Auto-ID) systems automatically identify physical objects through optical, electromagnetic, or chemical means.

EPC: Electronic Product Code. A low-cost RFID tag designed for consumer products as a replacement for the UPC (Universal Product Code).

Linear Barcode: A one-dimensional, optical bar code used for auto-ID.

Passive Tag: A tag with no on-board power source that harvests its energy from a reader-provided RF signal.

Reader: An RFID transceiver, providing real and possible write access to RFID tags.

RF: Radio Frequency.

RFID: Radio Frequency Identification. Describes a broad spectrum of devices and technologies and is used to refer both to individual tags and overall systems.

Supply Chain Management: A supply chain consists of a network of *key business processes* and facilities, involving end users and suppliers that provide products, services and information. In this chain management, improving the efficiency of the overall chain is an influential factor; and it needs at least four important strategic issues to be considered: supply chain network design, capacity planning, risk assessment and management, and performances monitoring and measurement. Moreover, the details break down of these issues need to consider in the level of individual business processes and sub-processes; and the combined performance of this chain. The coordination of these huge business processes and their performance improvement are the main objectives of a supply chain management system.

Tag: An RFID transponder, typically consisting of an RF coupling element and a microchip that carries identifying data. Tag functionality may range from simple identification to being able to form ad hoc networks.

Chapter 3
Smarter Phone

Rushit Dave
North Carolina A&T State University, USA

Brinta Chowdhury
North Carolina A&T State University, USA

Evelyn R. Sowells-Boone
North Carolina A&T State University, USA

ABSTRACT

Over the last two decades, the evolution of mobile technologies has led to an unprecedented adoption of cellphones in mainstream society. Consumers have moved from having the ability to make telephone calls from anywhere at any time using the traditional cellphones to having an all-access pass to cyberspace using today's smartphones. This portable device has become a reliable support system for the user because of the accessibility and flexibility it offers to maintain the users' daily routine. Authors in this chapter have chosen software optimization techniques to increase battery efficiency because these techniques are more robust. This chapter introduces a novel idea of an automated system for smartphones that prioritize application access based on the owner's usage patterns and daily routine to conserve battery life. This system will serve two purposes: save battery power and improve the smartphone's artificial intelligence.

INTRODUCTION

Power conservation has become an increasingly important issue among modern digital system designers. As the digital technology evolution takes us into the 21st century coupled with ground breaking system performance, the power consumed by these systems are at record highs. Smartphones and other portable devices have been most affected by this growing trend since users depend on the mobility that battery powered electronic devices offer. Currently, smartphone users are faced with limited battery lifetime which is an inverse process. The more we use the smartphones, the less battery life our smartphones have.

If you own a smartphone, then you probably cannot imagine a day without it. In fact, you are probably more dependent on this portable device than you realize. Today's smartphones offer features that allow its user to accomplish a task more efficiently by providing endless access to the World Wide Web. The

DOI: 10.4018/978-1-5225-9246-4.ch003

unlimited applications and built-in features are seamlessly embedded into our daily routine and tasks that once took hours to complete can now be completed with the touch of a button. Have you actually thought how many ways your smartphone is assisting your daily life?

The first handheld cellular mobile phone was developed by Motorola in 1973 and weighted 2kg (about 4.5 lbs.) ('Mobile phone', 2018). In later years, mobile phones began to add more features, which in turn, increased its value to users. Nokia, Motorola, and Blackberry introduced phones which allowed users to read and response email from anywhere. This feature was extremely popular with business professionals. This was considered the birth of the smartphone genre. Although, the first commercially available smartphone was the IBM Simon personal communicator in 1994; the Nokia 900 communicator introduced in 1996 was more popular. It had the same features as Simon but showcased a graphical web browser (Pothitos,2016). It was widely considered the first smartphone. Even so, the size of the keyboard and inputs limited its popularity. In 2006, the Blackberry was introduced which featured a full keyboard. Its popularity grew but mostly with business professionals. It became more common when Apple and Google launched their smartphone with iOS and Android in 2007 and 2008, respectively. These phones were easy to operate because they featured touch screens which displayed applications. Simply touching an application icon on the screen made the smartphone more user-friendly which led to widespread usage. In the last decade, smartphones have reached unprecedented popularity because they place the world in the palm of your hand instantly.

The advancements of smartphone functionalities have allowed it to become an indispensable part of our daily life. In fact, smartphones are replacing personal computers. For 2007-2017, the number of smartphones sold worldwide was 1.54 billion and it is projected that by 2020 this number will be 1.7 billion (Number of smartphones sold, (2018)). By 2021, 40% of the world population is predicted to own a smartphone ('Smartphones industry: Statistics & Facts'. (2018).). If you closely examine the features incorporated in the smartphones, you will notice that they use RAMs comparable to computers, the ROMs is in the scale of Gigabytes and they use a variety of sensors. Additionally, by the end of the first quarter of 2018, smartphone users have the options of choice among 3.8 million and 2 million applications in the marketplace like Google Play and Apple Appstore respectively ('Number of apps available',2018). The downside to this growing trend of increased usage is the smartphone's limitation of battery power. Currently, smartphone users are faced with limited battery lifetime or limited battery interaction which is an inverse process. The more we use the smartphones, the less battery life our smartphones have.

This growing concern of conserving battery life is the focal point of this proposed chapter. There have been several hardware-based and software-based optimization techniques introduced over the years to combat this issue. Hardware-level modifications have been highly successful. However, research has proven that techniques that are effective for one smartphone architecture may be ineffective for different architectures (Zaman & Almusalli, 2017). Software-based optimization techniques are more robust which removes platform restrictions. For these reasons, we have chosen a software level optimization technique to extend battery life.

This project will advance knowledge in artificial intelligence of portable device like smartphones, wearable devices, laptops etc. Not only that, this project's outcome will contribute to the power efficient device design which is unique for every user. The focus of this research is saving power consumption in portable device through an automated system which will be different from user to user. This research will benefit research areas of artificial intelligence and power optimization for smart devices at the same time. Other research benefitting from this sort of work include data analytics and automation.

Our project's outcome will be a prototype that showcases the merits of a self-dependent power efficient portable device with strong potential for national and even global impact. Efficient power consumption of smart devices directly confronts the challenge to extend the battery life and reduce or eliminate the accumulation of electronic trash and hazardous incidents due to overheating of the battery. Along with this, the intelligence of the device reduces manual efforts of the user which saves time and increases work efficiency. In addition to that, this prototype can be converted to a generalized system and used in smart grid components where user-based power efficient systems can be developed. Eventually it will help to save energy globally.

Background

Researcher and design engineers have developed various methods for optimizing power usage in mobile phones. In comparison to older model mobile phones, smartphones are capable of performing complex task which requires power hungry sensors like the camera and GPS. As a result, this new generation of smartphones require more efficiency in battery power conservation. These power optimization can be in hardware components, software components or both. It has been noted that 47% of battery draining issues are caused by smartphone applications (Nikzad, Chipara, & Griswold, 2014). In (Guo, Wang, & Chen, 2017), a large scale empirical study was implemented with 80,000 smartphones. They found that about 75% of smartphones can last more than 12 hours with an average batter life of about 22.6 hours. This means that smartphone users need to charge their phones at least once or twice per day which is more than most of the standard mobile phones.

Decreasing the vitality utilization (power usage) that is consumed by Android advanced cells is a vital task for application engineers. While this issue can be address from different levels of abstraction, it is imperative to decrease the vitality utilization of individual applications to fundamentally operate with more efficient behavior patterns. The most straightforward strategy for decreasing the vitality utilization of smartphone applications is to reduce and solve bugs. Vitality profiling techniques can identify application tasks that are consuming an absorbent amount of vitality and implement strategies to diminish their general utilization. Existing techniques gauge vitality utilization of the whole advanced mobile phone by utilizing information retrieved from the OS (e.g. CPU time, measure of access to record frameworks, activity, and so on.). Be that as it may, these techniques cannot distinguish explicit task where excessive vitality is consumed or represent all conceivable utilization examples and conditions. As a result, application designers can only plan constrained profiling experiments.

Over the past century, PC manufactures, programming architects and programming designers were chiefly searching for simple electronic gadgets to program. Over the past 10 years, this trend has definitely shifted as a result of the advancements in cell phone technologies. Currently, execution time is not the main concern for cell phone designers. Truth be told, the principle processing bottlenecks is control utilization. In addition, cell phone manufactures and their clients are just as concerned about the execution of their gadget as they are with battery utilization/lifetime. This developing mindfulness on vitality proficiency is in turn changing the manner in which software engineers design their product. Shockingly, creating vitality for power efficient architecture programming is as yet another troublesome undertaking. While the programming language network has progressed and broadly utilized programming apparatuses, as for instance, debuggers and blame confinement instruments, memory profiler devices, testing devices, benchmark and runtime observing systems, compiler improvements, and so on; there are no comparable devices/structures to profile/upgrade control utilization.

With all of the cell phone technology advancement, vitality remains the basic asset bottleneck for most cell phone usage. Although there has been moderate advancement in battery innovations and size requirements for hand-held electronic gadgets, battery power still limits cutoff points for powerful cell phone utilization between charges. In addition, the present day smart phones are normally outfitted with quad-center or even octa-center processors. Incredible multicore processors put further strain on the limited battery vitality of a smart phone. Therefore, multicore processors are not vitality relative: the main running CPU causes higher power cost than each extra processors does. This can be credited to two reasons. To start with, modern day processors are great at power gating. At the point when the framework is totally inert, most parts of the CPU can be shutdown bringing about least vitality utilization. Second, the sharing of equipment assets on a multicore implies that the principal running center must enact the greater part of shared assets while extra centers can use the effectively initiated assets at much lower cost. This vitality disproportionality proposes that a multicore processor is more vitality proficient when a greater amount of its centers is used simultaneously. Interestingly enough, typical cell phone applications are occasion driven, UI-driven system and serve just a solitary client. They do not have adequate parallelism to use different CPU centers at the same time. Ongoing examinations on Android applications demonstrate an absence of string level parallelism crosswise over applications and an over-provisioning of center assets crosswise over gadgets. This infers cell phone multicore processors frequently work at low center use bringing about poor vitality effectiveness. Meanwhile, when one CPU center is being used, figuring assets at different centers are accessible at a profound vitality markdown.

Additionally, a typical understanding of each nation's electronic eco-system aids in vitality increases, carbon decrease and ecological insurance. To act on these vitality approaches, the most imperative requests for new electrical gadgets are decreasing power utilization, high lumen effectiveness, decreasing life expectancy and lower contamination. Force control, a power electronic gadget known as dimmer, is utilized in different applications incorporating mechanical and furthermore in private lighting control. Dimmer gives a control to alter the fake lighting level in addition to saving vitality. In various process control applications, inserted remote framework has been broadly utilized for observing and control. The improvement of brilliant power sparing framework dependent on android has been introduced in this chapter.

Presently cell phones, but mostly smartphones are assuming an indispensable positons in our daily lives. Since smartphones have many of the functionality of personal computers or laptop computers, they are financially savvy, promising and solid gadget that helps in performing daily errands utilizing the android application programming. The principle objective of applications is to simplify everyday task with the simple touch of a button. The user is no longer concerned with how to complete a task instead they only think about what they want to do. An easy to use application is modified in the Massachusetts Institute of Technology (MIT) App Inventor 2 programming. The Bluetooth MAC address of client PDA is made to be matched with the android application programming which makes an individual region arrange. In view of the direction send from client PDA to the Bluetooth interface Microcontroller (Atmega 328P) at the less than desirable end, the sixteen-diverse force level is gotten and after that these sixteen dimensions are utilized to control the power dimension of the knob to spare power. Additionally, a present sensor is associated with Microcontroller (Atmega 328P) to screen the distinctive current dimensions.

While developing eDoctor in (Ma et al., 2013) authors studied 213 real world battery drain issues for smartphones and they concluded that about 47.9% issues are for applications. Other issues that can lead to abnormal battery drain are system bugs (22.1%), configuration change (11.8%) and environmental conditions (8.2%). After a brief period of idle time device, components move to suspend state unless they get any pending requests to act. These requests keep the system awake even if the user is not interacting and so one of the big reasons behind unnecessary battery drain. To address this problem, an Android OS mechanism named TAMER is designed in(Martins, Cappos, & Fonseca, 2015). TAMER interposes the background activities which are the cause of wakeups-alarms, wakelocks, broadcast receivers and service invocations. It actually controls the frequency at which background tasks are handled based on some defined flexible policies and eventually lowers the energy consumption. Their design has three different agents: Observer- which detects event frequency, Arbiter- which compares the policy defined threshold and notifies the next agent Actuator when it crosses the limit. Actuator reduces the frequency by force. As described in (Li, Lyu, Gui, & Halfond, 2016) Hypertext Transfer Protocol (HTTP) request consume almost 80% of the network related energy usage and on average 40% of the non-idle state energy of an app is consumed by network communication. These authors proposed a method which can optimize HTTP request. This method can be described in three phases: Detection is done by static analysis to identify HTTP request which are eligible for bundling; then they identify the relationships between the requests by doing string analysis and configure the bundling rules and then using a proxy server HTTP requests are optimized in the third phase Runtime optimization. Their method is proven to save 15% power in the application level.

When applications are designed, programming language, runtime, complier and implementation choices have significant impact on energy consumption (Chen & Zong, 2016). In that paper, the authors showed that native languages like C, C++ has better energy efficiency than Java and O3 optimization can improve the performance of native languages more. Java codes can also be improved using ART runtime environment. They also showed that parallel implementation has better performance in terms of energy consumption over serial implementation and it is 2.6 times energy efficient. In (Nikzad et al., 2014), the authors developed an Annotation language for developing energy efficient mobile applications. They developed power management policies which will defer the execution of power-hungry code segment until the device enters into cost minimization state. This desired state can be described using an abstract model based on timed automata. Though their approach could save 63.7% energy saving over an application, they did not define any range for operation time delay which has a trade-off with power saving and eventually related to users' experience. They addressed this problem in their later work (Nikzad, Radi, Chipara, & Griswold, 2015) where the authors presented a new approach "Tempus" in which developers assign a delay budget and power management policies will not delay an object more than that. They used static analysis to determine the impact of power related delays in applications. In most cases applications do not have access to the battery level or status so that they can modify their operations automatically based on the information. In (Datta, Bonnet, & Nikaein, 2014) authors developed a framework for Android application development in which applications will be aware of battery and context based on which applications will be modified at the real time run. They designed an analyzer engine which will take the information and decide which profile is too active. All three self-adaptive profiles (Light, medium, Strong) comes with four levels: Hardware and software resource adaptation, user features adaption, additional optimization. The main purpose is to check the battery level and if it is critically low then to maximize the batter life until next charging. By developing applications in this framework, they reduced 40% power consumption. Optimizing memory data access had been taken into consideration by the authors of (Almusalli, Zaman, & Rasool, 2017). They provided Data Layout Transformation tool to convert the code of an applications from the Array of structures source code in the form of two Structures of Arrays. Their transformed code takes less time to access memory which eventually results in low power consumptions.

Although high resolution display in smartphones has become one of the trendiest features, this does not improve user experience all the time. There is always a trade-off between display density and power consumption (He, Liu, & Zhou, 2015). They proposed Dynamic Resolution Scaling (DRS) which adjusts the user-interfaced resolution on the screen based on the viewing distance while keeping user experience unaffected. With this approach they reduced energy per frame on average 30.1% and maximum 60.5% for 15 apps by decreasing resolution to half of before. Ultrasonic sensors used in their hardware-based prototype may have an issue while integrating in smartphones but can be overcome with micrometer level ultrasonic devices. But their prototype used only one kind of processor which may vary from phone to phone. Another limitation they mentioned is their approach can work on only one app in the foreground but cannot act on the multiple apps running in the background.

In the later sections, we will describe our methodology, results we achieved so far, discussion where we will analyze our findings, our limitations of the research and future research. We hope rest of the chapter will be an enjoyable journey for the readers.

Technical Description

We have developed a power saving application, Elina that utilizes unique user features to optimize smartphone battery vitality and improve the user's experience.

Features of the Application

We used two different methods to classify work that can save energy (power) and can keep good battery life of an android smart phone as shown in Figure 1.

Figure 1. Elina's application LOGO

- **Location-based:** Once a user opens the application, user will be able to find two different optimization block options- location-based and time-based events. In location-based event option see Figure 2, as the user travels to new and different locations, the new location is added to the application see Figure 2. The application will be able to save the location in the database once added. The user can also enter the location manually and will be able to search for that location on the screen of the android smart phone using google maps which is already connected to the location-based event see Figure 3 and Figure 4. Once the user is done with searching of the desired location, the user will see three different services which are WIFI, Mobile Data and Bluetooth see Figure 5. User will be able to make preferences for these services whether the user wants to have those services in active mode or not. Additionally, the application does show the mode of the phone as Ring, Silent, DND see figure 5. User will also be able to select any mode for that location and it will be saved. There is also an option to edit all preferences for the services and editing option for the mode of the phone which is saved for the location. Therefore, users do not need to manually activate or de-activate those services for the saved locations as it will retrieve preferences given to those services for saved locations each time the application is launched.
- **Time-based:** This option performs more like a scheduler, as a user will be able to add a new task and that task will be saved once added see Figure 6. The user will be able to assign a name to that task see Figure 7, choose dates using calendar see Figure 8 and can choose the start and end time for that task see Figure 9, as well. Once the user has completed assigning names, date and time for the task, the user will be able to choose those same services which was available in location-based option (WIFI, Mobile Data, and Bluetooth) and user will be able to make preferences for that task whether user wants those services to be activated or de-activated. Additionally, in this event, it too has the phone modes such as Ring, Silent and DND which a user will be able to select for that task and after selecting all preferences of services and mode of the phone, user will be able to save the task. Once saved, then for whatever date and time, user has saved that task, all services will start working or will get de-activated based on their preferences saved for that task and the phone will turn on the mode automatically which was saved in the task for that date and time. So, this does save power for the android smart phone as a user do not need to go and make the services activate or de-activated provided by the application with preferences.

Figure 2.

Figure 3.

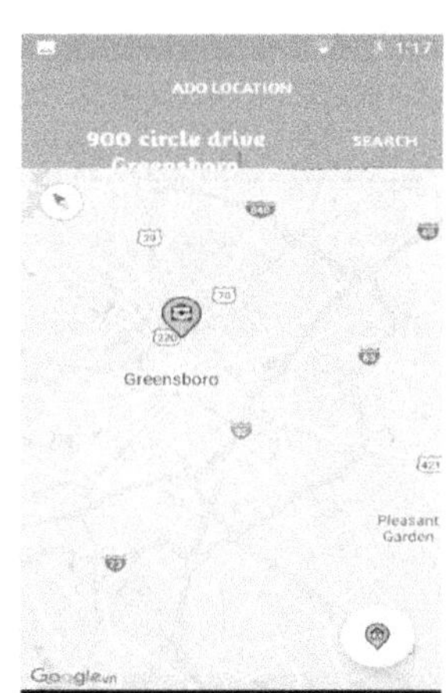

Figure 4.

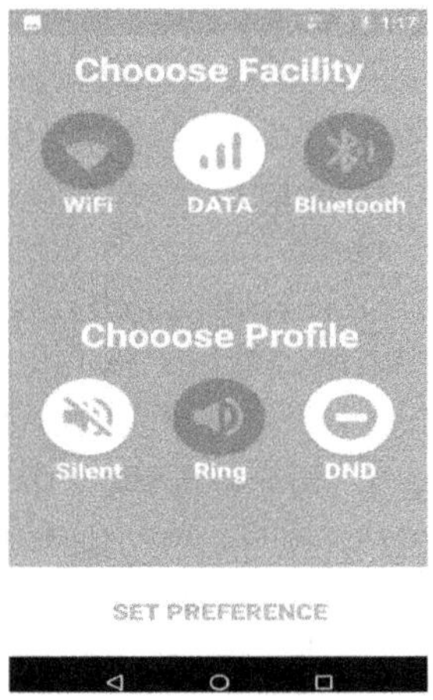

Figure 5.

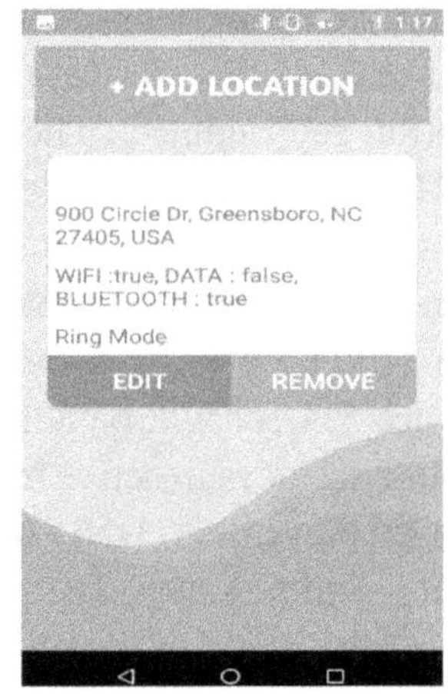

Figure 6.

Figure 7.

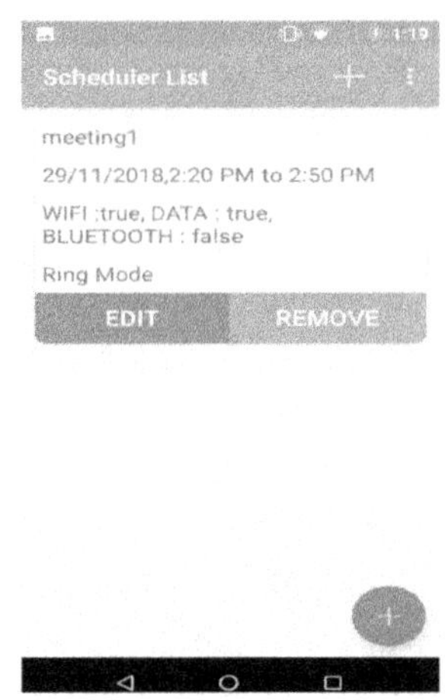

Figure 8.

Figure 9.

The combination of location-based and time-based automated power saving optimizations for smartphone applications are novel and will be fully examined in this chapter.

PROCESS

Phase One

Realistically, there are several applications which have power saving options or mainly developed for saving energy of android smart phones. In the discussion section of this chapter you will find that we have analyzed and compared several power saving applications with our application. When any developer starts developing an application, the first and most important process is to research similar applications and distinguish what is novel about the idea proposed in the framework or architecture. Has this been done before? If so, how is this proposed project better? We then analyze the differences between all applications with similar objectives and formulate realistic ideas which can improve the routine battery life of any android smartphone user. Once you have gathered all requirements and specifications information for your application, the next step is to decide the platform to develop it which can include analyzed features of the application with different scenarios.

Phase Two

We assembled all the requirements with their sub levels in order to start working on the development part. The important component of this step is developing the collected features to produce desired outputs. We chose android studio to work on as we have developed an android application which can work with any android phones. Android studio version 3.2 includes the major features and the major bug were fixed. At present, over 76.6% of the smartphone's, including HTC, LG and Samsung Models utilize Android as their working framework (OS), and expecting that Android will be in intelligent watches, PCs, vehicle soon. Android controlled gadgets including tablets have turned into the preeminent need of all the well-informed individuals over the world and the prime reason is that it gives an open source stage for the advancement of incredible applications in addition to permitting application engineers to instantly distribute them (Rajput, Mehul, 2015). In the near future, more smartphone design engineers should increase their knowledgeable about Android application due to its extraordinary development and promising future.

Phase Three

The last phase is completing the development of the application and testing to get desired outcomes. We have used different web services to complete main functionalities. The main goal of the application is savingpower. We have used priority checker to work for the schedulers, even when it is the same for the location-based feature in order to prioritize location. In Figure 10, Each event has one major and common function which helps to save the energy; this function includes services like WIFI, Internet Data and Bluetooth, where a user will be able to assign a preferences the application. Once the development component was completed, we conducted dummy testing using this application (refer to results section). For our testing and analysis, we used TracFone android smart phones. The model number is

Figure 10. General process Diagram of the system

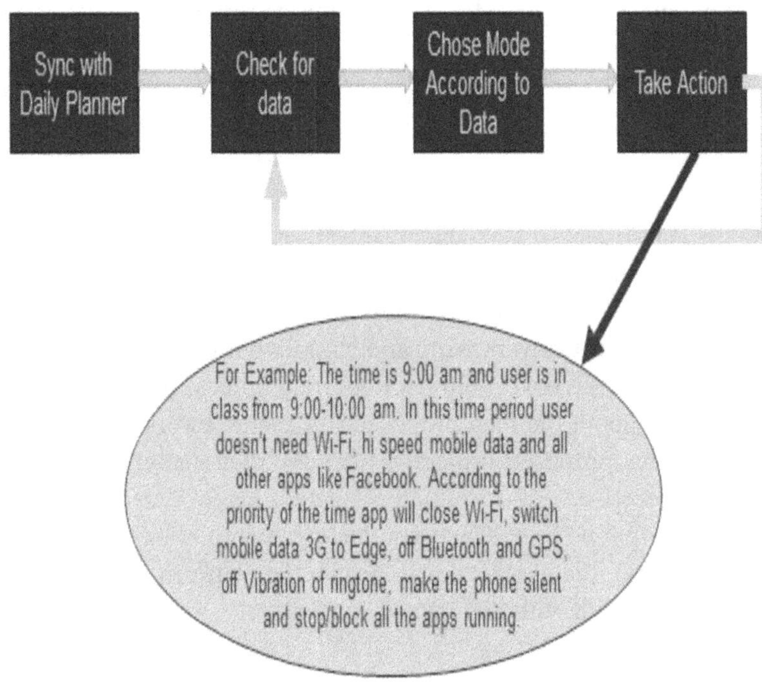

STALA502DCP. Some features which it has are as follows: Design- 9.62*4.51*0.61 in and weight 5.29 oz, Memory- 2GB/ 16 ROM and SD support up to 32GB, it has NT6739WM quad-core A53 1.1 GHz with Android 8.1 Oreo, Display- size-5.43 inch, Connectivity- GSM/HSPA?LTE cat4, UMTS B2/4/5, LTE B2/4/5/12(MFBI)/13/66/71/GSM and Battery- standby(3G): 547 hours.

The demand for power saving applications is increasing daily. Presently, more ground-breaking smartphone applications with power hungry advances like GPS, 3G and Wi-Fi are being developed and used globally. All these applications require more power for different applications. YouTube, a video streaming applications is one of the more popular applications. Most smartphone user utilizes these sorts of applications. If you realize how your smartphone's battery is used, you can implement a power saving technique to extend battery life. It has been noted that 47% of battery draining issues are caused by smartphone applications. There are also a significant number of applications that are embedded in Android smartphones which further increase battery power drainage. Many of these embedded application are installed because of two reasons: Standard functionality such as messaging, email, calendar and maps or as it very well may be found in consent of the operation system. Most of the end clients experience issues figuring out which applications are more vitality productive application. Architects have a motivation to create vitality proficient smartphone programming. Their primary obstacle is the trouble of deciding the effect of programming plan choices on framework vitality utilization, however that boundary can be survived. This section depicts the battery watcher methods. This model gives an exact ongoing application control utilization on every application. The most famous smartphones power consumption components are WIFI, Internet Data and Bluetooth.

The limit of smartphone battery is constrained due to the size and weight of the battery. Therefore, the vitality effectiveness of the smartphone is basic for their utilization. Currently, smartphones have numerous capacities in addition to making telephone calls. They act as cameras, video recorders, and GPS's just to name a few. This builds the prerequisites for battery limit and the requirement for vitality proficiency. For understanding the vitality productivity of the smartphone, it is vital to analyze which parts of the framework utilize the vitality and how much vitality is consumed in various conditions. Customary, smartphone that does not utilize any intelligent applications can operate for a few days without charging. In any case, when utilized effectively using applications, a smartphone can barely operate for little over a day without charging the battery. Smartphones utilize more vitality than customary cell phones regardless of whether intelligent applications are seldom utilized. Even though cell phone battery limits have improved, the life span of the battery is shorter in contrast to a standard cell phone.

Novelty

After completing a literature review on technologies developed to save battery life, we found that our application has location-based feature is unique, in which a user is able to store location with preferences of services like WIFI, Internet Data and Bluetooth. When using this feature, a user will be able to save power in saved locations as all saved preferences will be enforced and working automatically eliminating the need to make changes manually. The combination of schedulers which works based on saved time-based tasks and location-based features is something new that no other application has implemented until now. These features save power on android phones which shows significant results for power saving applications .

Table 1. Comparative Analysis of Two Phones (With And Without Our Application) In Different Scenarios

Event	Factor	WIFI-Off Bluetooth-Off	WIFI-Off Bluetooth-On	WIFI-On Bluetooth-Off
Location Based	Location- 900 circle Drive, Greensboro, NC, USA, 27405	Phone -1 - 77% Phone-2 - 75% **Saved Energy-2%**	Phone-1- 76% Phone2- 75% **Saved Energy-1%**	Phone-1- 76% Phone-2- 75% **Saved Energy-1%**
Time Based	Start time-8 am End time-11 am	Phone -1 - 78% Phone-2 - 76% **Saved Energy-2%**	Phone -1 - 77% Phone-2 - 76% **Saved Energy-1%**	Phone -1 - 77% Phone-2 - 76% **Saved Energy-1%**
Time Based	Start time-5 pm End time-7 pm	Phone -1 - 79% Phone-2 - 77% **Saved Energy-2%**	Phone -1 - 78% Phone-2 - 77% **Saved Energy-1%**	Phone -1 - 78% Phone-2 - 77% **Saved Energy-1%**
Time Based	Start time-8 am End time-12 pm	Phone -1 - 78% Phone-2 - 75% **Saved Energy-3%**	Phone -1 - 77% Phone-2 - 75% **Saved Energy-2%**	Phone -1 - 77% Phone-2 - 75% **Saved Energy-2%**
Time Based	Start time-10 am End time-10:30 am	Phone -1 - 80% Phone-2 - 80% **Saved Energy-0%**	Phone -1 - 80% Phone-2 - 80% **Saved Energy-0%**	Phone -1 - 80% Phone-2 - 80% **Saved Energy-0%**

RESULTS AND ANALYSIS

Modern application designers must consider multiple functional components of the smartphone. Mainly performance and its effect on vitality proficiency and power usage. This developing mindfulness on vitality proficiency is likewise changing the manner in which software engineers build up their product. Our application design team has utilized a time-based power savings technique and couple it with an automated location-based optimization to create a novel approach to increasing power saving. Our results are more than promising saving up to 3% battery life or 43 minutes of battery life in a 24-hour period.

In Table 1, The first event is based on the saved location in the application. Phone-1 has the application installed and phone-2 does not have the application installed in it. We have three different scenarios listed in result Table1 for that location. When the WIFI and Bluetooth are off and saved in the application, phone-1 saved 2% of battery life comparing to phone-2. In the second scenario, WIFI is off and Bluetooth is on. The application shows phone-1 has 1% more battery life than phone-2. In the third scenario WIFI is on and Bluetooth is off, and it is saved as preferences in the application; which shows phone-1 has 1% more battery life than phone-2.

Second event in the Table1is based on the saved time tasks in the application. Phone-1 has application installed and phone-2 does not have the application installed in it. We have taken time duration of three hours which is 8 am to 11 am. We have three different scenarios listed in result Table1, when we had WIFI and Bluetooth off saved in the application, phone-1 saved 2% of battery life comparing to phone-2. In the second scenario, WIFI is off and Bluetooth is on and it is saved as preferences in the application which shows phone-1 has 1% more battery life than phone-2. In the third scenario WIFI is on and Bluetooth is off, and it is saved as preferences in the application which shows phone-1 has 1% more battery life than phone-2.

In the third event which is again based on the saved time tasks in the application, we considered time duration of two hours which was from 5 pm to 7 pm and we got the same results as we got for the last event. Fourth event is like the second event, but we tried to take different time than we took in second event as we considered time duration of four hours which was 8 am to 12 pm. We have three different scenarios listed in result Table1, when we had WIFI and Bluetooth off saved in the application, phone-1 saved 3% of battery life comparing to phone-2. In the second scenario WIFI is off and Bluetooth is on and it's saved as preferences in the application which shows phone-1 has 2% more battery life than phone-2. In the third scenario WIFI is on and Bluetooth is off, and it's saved as preferences in the application which shows phone-1 has 2% more battery life than phone-2.

In the last fifth event which is based on the time tasks saved in the application, we considered lesser time than other events as we took half an hour of time duration which was from 10 am to 10:30 am. We found that for lesser amount of time we were not able to save battery life as after comparing in each scenario we were able to notice that difference between both phone-1 and phone-2 was 0%. This can be a major limitation of the application that for lesser amount of time, using this application there are lesser chances of saving energy of an android smart phone.

DISCUSSION

Research in the area battery power saving is not keeping pace with the ever growing demand for high performance technologies. It is genuinely is hard to discover an application that really reduces your battery power consumption since most battery saver measures are manual, including reducing the brightness on your smartphone screen, reducing the recurrence that applications synchronize information also known as background refresh, and other reliable techniques. As a rule, you see greater upgrades when chip producers like Qualcomm, screen makers like Samsung, and battery makers enhance battery proficiency of the equipment. Nevertheless, there are a couple of applications that can assist with battery savings. Let's compare the best battery saver applications for Android.

Amplify (Amplify, 2018) is a root application with some OK includes. Its primary usefulness is distinguishing and ceasing wake locks and is the point at which an application keeps your telephone from going into rest mode. This application encourages put a stop to the majority of that. You can control applications, cautions, and different administrations that could be depleting your battery. It looks great and it's anything but difficult to utilize. You can get it for nothing. There is additionally an ace form that includes more highlights. It's unquestionably extraordinary compared to other battery sparing applications.

Greenify is a standout amongst the most well-known battery sparing applications. It distinguishes applications that awaken your telephone more as often as possible (Kingsley-Hughes, Adrian, 2016). It can likewise help shield them from doing that so frequently. The application additionally has present day highlights for Android Nougat and past with Aggressive Doze and Doze modes. This application is helpful for both root and non-root gadgets. Be that as it may, you'll get greater usefulness and power with root.

GSam Battery Monitor is a standout amongst the most well-known battery sparing applications. It distinguishes applications that awaken your telephone more as often as possible. (SAM battery Monitor, 2018) It can likewise help shield them from doing that so frequently. The application additionally has present day highlights for Android Nougat and past with Aggressive Doze and Doze modes. This application is helpful for both root and non-root gadgets. Be that as it may, you'll get greater usefulness and power with root.

Servicely is one of the better root-only battery saver apps. It works by halting administrations that keep running out of sight. (Hindy, Joe, 2018) It keeps rebel applications from going bananas and shields them from matching up constantly. This is extraordinary for applications you like having, however you would prefer not to match up constantly. You may get a deferral with stuff like notices, however, so do utilize this apparatus precisely. This application functions admirably with wake bolt locators as an intense one-two punch.

Wake lock Detector (Hindy, Joe, 2018) is one of the best battery saver apps. As the name infers, this application identifies wake locks. It can identify both halfway and full wake locks. You can likewise get a rundown of all the applications that are causing it. From that point, you can make moves to uninstall the applications.

Elaina, the application which we developed also has major features of time and location with different services which can have preferences based on saved tasks and locations. Once the application is installed and it is easy to save battery power. Resetting and de-activating service that are no longer needed at stored times or locations is simple as well.

FUTURE RESEARCH

So far, we have discussed our idea of making a smarter phone which not only work as our personal assistant but also work on increasing battery efficiency. We have discussed primarily a proof of concept with result discussed above. But this is just the beginning of the idea. To make this successful we have fixed our goal in two categories: long term and short term.

Short term goal: Our short-term goal is to make an application which will convert our smart phone into a self-dependent device. With this application a phone will be able to think on its own and make decision for the benefit of the user and eventually for saving battery life. Our target is not only phone. We want to make our tablets, laptops, wearable gadgets smarter than before with this application. Every portable device will act like a personal assistant to its user.

Long term goal: Our goal is not limited to design an application only. In future we want to design a system which will be embedded to device systems. Users won't have to install any applications to their device. Device itself can understand user's requirement and hence will work accordingly.

CONCLUSION

Our main goal is to optimize smartphone power usage which doesn't need any manual change of settings. Phone will be the head of any decision-making regarding battery life efficiency. Our work is still in progress. But the results we have shown above can make change in everyone's think. And it is obvious from the result that our phone can be made self-dependent device. Present world is working how to make life easier with smart cars, smart homes, smart grids and so on. Now the time has come where we can think of our life with smarter device. Our research is first step towards this. On the other hand, with the growing technology battery efficiency is one of the biggest concerns. Our research addresses two problems together: one is saving power, and another is giving intelligence to the portable devices. At the end, we think readers have gained a full insight of how we can make a smarter world in future.

This project advances knowledge in power efficient electronic system design and implementation. The project's outcome contributes to the state of the art in digital system design through understanding how efficient power system design strategies improve portable device performance and reliability, thereby leading to a national model tailored toward power aware electronic system designs. The focus of this research is the development of power-aware algorithms and techniques for electronic system optimization. Other research areas benefiting from this sort of work include high performance computing and computational science.

ACKNOWLEDGMENT

We are thankful to Computer System and Technology department of North Carolina A&T State University to support us conducting our research.

REFERENCES

Almusalli, F. A., Zaman, N., & Rasool, R. (2017). *Energy efficient middleware: Design and development for mobile applications.* Paper presented at the 2017 19th International Conference on Advanced Communication Technology (ICACT). 10.23919/ICACT.2017.7890149

Amplify. (2018). *Amplifyandroid.* Available at: http://amplifyandroid.com/

Chen, X., & Zong, Z. (2016). *Android app energy efficiency: The impact of language, runtime, compiler, and implementation.* Paper presented at the 2016 IEEE International Conferences on Big Data and Cloud Computing, Social Computing and Networking, Sustainable Computing and Communications. 10.1109/BDCloud-SocialCom-SustainCom.2016.77

Datta, S. K., Bonnet, C., & Nikaein, N. (2014). *Self-adaptive battery and context aware mobile application development.* Paper presented at the IWCMC. 10.1109/IWCMC.2014.6906452

Guo, Y., Wang, C., & Chen, X. (2017). Understanding Application-Battery Interactions on Smartphones: A Large-Scale Empirical Study. *IEEE Access: Practical Innovations, Open Solutions*, *5*, 13387–13400. doi:10.1109/ACCESS.2017.2728620

He, S., Liu, Y., & Zhou, H. (2015). *Optimizing smartphone power consumption through dynamic resolution scaling.* Paper presented at the Proceedings of the 21st Annual International Conference on Mobile Computing and Networking. 10.1145/2789168.2790117

Hindy, J. (2018). *5 best battery saver apps for Android and other ways too!* Retrieved from https://www.androidauthority.com/best-battery-saver-android-apps-266980/

Kingsley-Hughes, A. (2016). *The best battery saver app for Android: Greenify.* Available at https://www.zdnet.com/article/the-best-battery-saver-app-for-android-greenify/

Li, D., Lyu, Y., Gui, J., & Halfond, W. G. (2016). Automated energy optimization of http requests for mobile applications. In *Proceedings of the 38th international conference on software engineering.* 10.1145/2884781.2884867

Ma, X., Huang, P., Jin, X., Wang, P., Park, S., Shen, D., . . . Voelker, G. M. (2013). *Edoctor: Automatically diagnosing abnormal battery drain issues on smartphones.* Paper presented at the NSDI.

Martins, M., Cappos, J., & Fonseca, R. (2015). Selectively Taming Background Android Apps to Improve Battery Lifetime. In *USENIX Annual Technical Conference*, (pp. 563-575).

Nikzad, N., Chipara, O., & Griswold, W. G. (2014). *APE: An annotation language and middleware for energy-efficient mobile application development.* In Proceedings of the 36th International Conference on Software Engineering, Hyderabad, India. (pp. 515-526). ACM. 10.1145/2568225.2568288

Nikzad, N., Radi, M., Chipara, O., & Griswold, W. G. (2015, November). Managing the energy-delay tradeoff in mobile applications with tempus. In *Proceedings of the 16th Annual Middleware Conference (pp. 259-270). ACM.* 10.1145/2814576.2814803

Pothitos, A. (2016). Mobile Industry Review. Retrieved from http://www.mobileindustryreview.com/2016/10/the-history-of-the-smartphone.html

Rajput, M. (2015). *Why Android Studio Is Better For Android Developers Instead Of Eclipse.* Retrieved from https://dzone.com/articles/why-android-studio-better

SAM battery Monitor. (2018). *Google Play store.* Available at https://play.google.com/store/apps/details?id=com.gsamlabs.bbm&hl=en_US

Statista. (2018). *Number of apps available in leading app stores as of 3rd quarter 2018.* Available at https://www.statista.com/statistics/276623/number-of-apps-available-in-leading-app-stores/

Statista. (2018). *Number of smartphones sold to end users worldwide from 2007 to 2017 (in million units).* Available at https://www.statista.com/statistics/263437/global-smartphone-sales-to-end-users-since-2007/

Statista. (2018). *Smartphones industry: Statistics & Facts.* Available at https://www.statista.com/topics/840/smartphones/

Wikipedia. (2018). Mobile phone. Available at https://en.wikipedia.org/wiki/Mobile_phone

Zaman, N., & Almusalli, F. A. (2017, April). Smartphones power consumption & energy saving techniques. In *2017 International Conference on Innovations in Electrical Engineering and Computational Technologies (ICIEECT) (pp. 1-7). IEEE.*

Chapter 4
Evaluation of RFID Tag Anti-Collision Algorithms in Supply Chain Automation

Kamalendu Pal
City, University of London, UK

ABSTRACT

Radio Frequency Identification (RFID) is a technology that uses radio frequency signals to identify tagged objects. RFID is an important technology used by the Internet of Things (IoT) applications. This technology enables communication between the main devices used in RFID system, the reader, and the tags. The tags share a common communication channel. Therefore, if more than one tag tries to send information at the same time, the reader will be incapable of differentiating these signals in the case of radio signals interference. This phenomenon is known as tag collision problem. The problem of tag collision is one of the major disadvantages for fast tagged-object identification in supply chain management. This chapter describes four different types of binary search algorithms for avoidance of tag collision, and then presents a performance measurement mechanism for RFID application system. Finally, simulation-based experimental results on the performance of these algorithms are presented.

INTRODUCTION

Production in modern economies is organized around supply chains, which involve business processes ranging from product design to customer delivery. In a typical supply chain, raw materials are purchased from suppliers and products are manufactured at one or more production plants (Pal, 2017) (Pal, 2018). Then, they are transported to intermediate storage facilities (e.g. warehouse, distribution centers) for packing and shipping to retailers or customers. In this way, a supply chain consists of business entities in the chain and these are the suppliers, manufacturers, distributors, retailers, and end-customers. The ultimate performance of any supply chain is governed by the business practices and corporate behaviors of the involved stakeholders, such as suppliers, manufacturers, transport service providers, technological infrastructure suppliers, and enabled by public policies and business environment.

DOI: 10.4018/978-1-5225-9246-4.ch004

Moreover, all supply chains share the following characteristics: (i) the supply chain comprises all business activities in order to supply a product or service to its end-customers; (ii) any number of supply chain partner organizations can be linked in the supply chain; (iii) a customer can be a supplier to another customer within the supply chain, which means that the total network of activities can consists of a number of supplier / customer relationships; (iv) the path from supplier to customer, depending on the products and markets, can include a number of intermediaries (distributors) such as wholesalers, warehouses, and retailers. Product or service flows from supplier to customer are called downstream flows while demand information from customer to supplier is called upstream flows.

In this way, a supply chain creates a complex business network. Given process decentralization, the efficient performance of a supply chain requires a high degree of visibility – defined as the capability of sharing on time and accurate data throughout the entire supply chain, and coordination among supply chain partners. In today's global business environment, companies recognize the strategic important of well-managed supply chains. For example, companies such as Dell, Nokia, Intel, Toyota, Wal-Mart, Zara, and Li & Fun have based their corporate strategy around achieving supply chain superiority over competitors (Copacino & Anderson, 2003). These global companies have gained competitive advantages by effectively managing the complex web of supply chain process interactions that extend across continents and across enterprises in product procurement, manufacturing and distribution. The requirements of modern supply chains are:

- **Connectivity**: With emphasis on the ability to make and maintain connection between business partners. It characterizes the ability to exchange information within supply chain partners in a way which provides inter-organizational collaboration.
- **Integration**: The ability to connect and coordinate business processes in a seamless way. It improves supply chain business processes performance by establishing collaborative connections among supply chain partners. For example, seamless integration increases information transparency among partners and allows pooling of inventories and sharing resources.
- **Visibility**: Visibility refers to the capability to access or view data or information related to logistics and the supply chains. For example, visibility is the ability of knowing where raw materials inventory for manufacturing, semi-finished products in the production line, and finished products are, at any time. But it is also actionable information that can help support customers at different interface points along the supply chain and improve business processes performance.
- **Responsiveness**: It is supply chain ability to react quickly to customer needs or specifications by delivering a product of the right quality, at the right time, at the right place at the lowest possible cost.
- **Lean and Agility**: As an aspect of lean production, lean supply chain depicts the state of business in which there is a dynamic competition and collaboration of equals in the supply chain, aimed at adding value at minimum total cost, while maximizing end customer service and product quality. Agility in contrast, refers to the ability to reconfigure supply chains with minimum effort.

The increasing interdependence of supply chain processes and the multiplicity of actors involved in them suggest that the full benefit of information and communication technology (ICT) adoption, for example in terms of end-to-end visibility, flexibility and global optimization, can only be captured if all supply chain stakeholders are aligned and coordinated in their efforts towards digitally transforming the system.

In the context of the digital transformation, supply chains are characterized by a high degree of cyber-physical interconnection, enabled by RFID tags (or sensors) that collect business functional data (known as Big Data) for large-scale, real-time decisions to optimize supply chain performance. The large-scale deployment of RFID tags (or sensors) and Big Data Analytics (BDA) enables preventive maintenance, avoiding disruptions from unexpected failures. Moreover, the use of sensors in conjunction with artificial intelligence provides automated inventory management, thus diminishing human error, input shortages and the high cost of unnecessary inventory carrying. Similarly, the use of sensors network, Big Data and artificial Intelligence in transport operations and infrastructure management allows for real-time route and asset optimization, hence improving reliability and efficiency in logistic operations.

Many researchers have proposed the existence of a positive correlation between supply chain business performances and corporate information systems integration and coordination of business partners activities (Pal & Karkostas, 2014) (Pal, 2018a) (Pal, 2018b). Information has been considered increasingly as a strategic asset for supply chains, in recent decades. As a result, many supply chain businesses are investing in new technology in order to harness the quicker information sharing provision in supply chain operations (Smith, 2005) (Karkkainen, 2003). RFID technology is a promising supply chain information integration mechanism. It is a mainly a wireless technology that enable automatic identification and data capturing technology that gives the opportunity to monitor objects by using a tag that carries information.

In RFID systems, there are different software and hardware requirements for data gathering and management. One of the most important components of RFID systems is tag. A tag can be identified as a microchip that has an electronic circuit and antenna on it. For the purpose of tracking the movement of goods, tags can be placed anywhere, such as containers, pallets, materials handling equipment, cases or even on individual products. Tags can be classified as passive (no battery), active (with battery) or semi-passive according to their power supply. While active tags use an energy source that is integrated to a tag physically, passive tags obtain this energy from the readers in the communication field. The other component of RFID is reader which connects the tags to external world. Although readers can be classified as portable and mobile, all of them consist of same components. In every reader, there are some parts that read tags, gather data and handle communication. While the reader antenna receives / sends the radio waves, it builds the signal and decrypts the signal which is sent from tags. RFID tags are available with different memory sizes and encoding options. These tags can also incorporate sensors to record temperature, vibration, shock, or humidity, for example, providing the ability to track and report on an object's environmental characteristics dynamically.

RFID readers and tags are being increasingly used in various application areas in supply chain operations. The RFID-based business solutions are well associated with the powerful and dynamic data capturing ability (Singh, 2003). The real-time availability of physical product identity allows information related to the current state of the product and future actions required (Smaros & Holmstrom, 2000; Choi et al., 2018; Singh, 2003; Sabbaghi & Ganesh, 2008; Kärkkäinen, 2003); and this helps in accurate asset tracking and monitoring (Bansal, 2003; Johnson, 2002; Kinsella, 2003). In the context of supply chain operations, widespread introduction of such systems represents a major suitability to examine and enhance tracking and tracing systems, process control and inventory management. In this way, RFID-based business solutions can be an enriched business information source, which can bring to different level of organizational advantages. Thus RFID-based supply chain solutions should be viewed as an information facilitator that can directly enhanced the operational decision-making.

The motivation behind RFID-based supply chain applications is to eliminate the barriers by enabling the synchronization and sharing of valuable information among trading partners. Therefore, it has become an important objective in the development of RFID technology-based applications that how to ensure the safe, reliable and simultaneous recognition of multiple targets along supply chain operations. In RFID application systems, tag is mainly used to store the information encoded by the marked object and security encryption, the reader used to read, change and verify the label information. However, there are several problems in the application of RFID system. In the process of RFID identification of multiple tags simultaneously, the tag collision is a key issue affecting the efficiency of RFID-based information systems.

Several algorithms have been proposed to deal with tags' collisions in RFID systems. There are not any perfect anti-collision algorithms in RFID environments in use today; therefore, all improvements on existing algorithms are appreciated. This chapter focuses on tree-based algorithms, which work by polling the collided tags with queries in order to distinguish them from one another. This can be done by various methods and is further presented in literature survey section. The main objective of this chapter is to compare the performance of the simple binary search algorithm, dynamic binary search algorithm, and backtrack binary search algorithm in RFID tag collision purpose.

The remainder of this paper is organized as follows. Section 2 introduces a simple RFID-based business application architecture. Section 3 provides a brief survey of related research works. Section 4 explains the operation of simple binary tree-based anti-collision algorithm and the use of the Manchester coding for collision detection purpose. It also includes examples of binary search algorithm, dynamic binary search algorithm, backtracking binary algorithm, and improved binary search algorithm. Section 5 presents the simulation based experimental result. Finally, section 6 provides concluding remarks.

A SIMPLE RFID-BASED APPLICATION SYSTEM

A RFID system consists of three main components: (i) radio frequency (RF) tags (or transponders); (ii) reader (or transceiver); and (iii) middleware, which can make up the communication sequence with the reads and to transfer data between reads and application systems. A tag has a unique identification (ID) number that can be identified by networked electromagnetic readers and is attached to the object that

Figure 1. Representation of the constituent's parts of a RFID application system

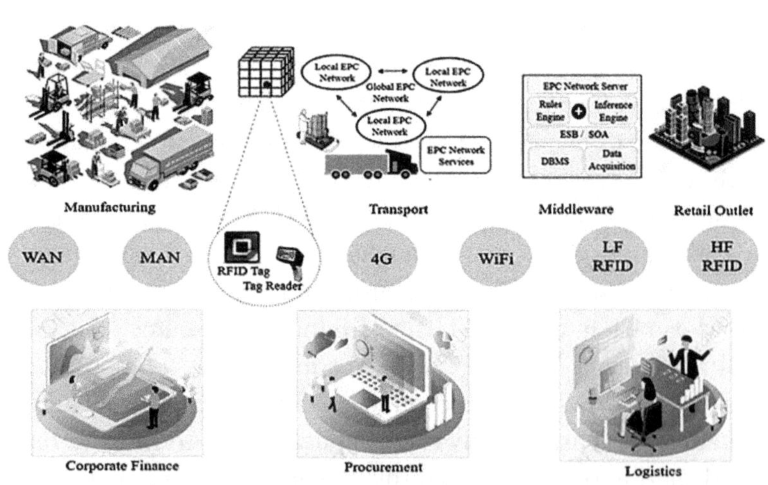

needs to be identified or to be counted. A reader consists mainly of a radio frequency (RF) module to make active the tags, and a control unit to makeup the communication sequence with the tags and to transfer data between tags and the RFID middleware. A RFID middleware is an application system or database, depending on the supply chain automation.

In industrial supply chain and its business process (e.g. manufacturing, transport, retail) automation often uses different types of data communication mechanisms (e.g. wide area network – WAN, metropolitan area network – MAN, fourth generation (4G), fifth generation (5G), different classes of wireless protocols – WiFi, low frequency RFID communication, high frequency RFID communication). The supply chain business processes (e.g. Corporate Finance, Procurement, and Logistic Activities) are getting their data from RFID-based systems data sharing network. A simple diagrammatic representation of the components of supply chain automation is shown in Figure 1.

In recent years, the Electronic Product Code (EPC) Network is becoming a global RFID data sharing infrastructure based on standards that are built around the EPC, an unambiguous numbering scheme for the designation of physical goods. The EPC network has become the subject of enormous interest, not only in research but also in several industries in general. The rapid and escalating diffusion of the EPC was previously driven by the Auto-ID Center, a project to develop RFID standards founded in 1999 at the Massachusetts Institute of Technology (MIT) with cooperation from numerous industrial partners. The Auto-ID Center created the EPC to ensure RFID interoperability in supply chain wide applications.

Use of RFID Data in Supply Chain Management

RFID system captured data can be classified into two categories: the event data and the master data. The event data keeps real-time (or dynamic) information that is about RFID tagged objects such as containers, pallets, materials handling equipment, cases, and so on. The master data provides conditional information and verification about the event data.

Event data is related to a definite time, and it provides the communication about RFID tagged objects during supply chain processes. It is created whenever some sort of transaction occurs. It is captured in distributed data repositories, and only the relevant event data must be sent to the monitoring for further processing. The processing and matching of automatically generated monitoring instructions with the event data gathered from distributed data repositories must performed by an appropriate event processing engine. Event data creates information which is about investigating the existence of items somewhere at some time. It stores the identity, location and time information. Event data is currently used for tracking and tracing software applications to monitor items associated with transportation processes and transported goods. The combination of new technologies provides the potential to use RFID based on event data for the automatic and near real time monitoring of processes in supply chain networks to detect anomalies according to specific business process objectives.

Master data, also called reference data, describes an item and its general properties. It includes useful data about customers, products, employees, materials, suppliers, manufacturers and so on in a supply chain. It contains information such as source verification, product definition referenced by EPC, manufacturer information, details about the object which event data is caught from, and storage information. Master data is a key information for quality assurance, and business operational applications. It provides processes for collecting, aggregating, consolidating, matching, and distributing such data throughout a business to ensure control and consistency in the ongoing maintenance and application of this information.

The tag collision problem can impose a major problem in RFID-based information systems data gathering processes, and it affects the performance supply chain operation management. Many approaches are proposed to reduce the impact of collisions and minimize business data collusion.

Problem Identification

In the RFID system, the reader and the tag communicate with each other through radio frequency signals. The channel is a signal transmission medium, and its role is to transmit the signal carrying information from the sender to the receiver. When multiple readers or multiple tags transmit radio frequency signals at the same time, the signals will collide with each other in the wireless channel, causing collision problems. The reader cannot receive the correct signal information, or the signal fails to be sent. The collision types of RFID systems include tag collision and reader collision, in which reader collision includes reader-tag collision and reader-reader collision.

Tag-Tag Collision

Tag-Tag collision refers to the collision problem between multiple tags, usually occurs when multiple tags send data signals to the reader at the same time, as shown in Figure 2.

Pluralities of tags within a reader identification range respond to the reader's command after receiving the reader's command. When two or more tags send data signals to the reader at the same time, the data signals interfere with each other in the wireless channel. At this time, the reader cannot correctly receive the data signals of these tags, causing communication failure.

Figure 2. Tag-tag collision diagram

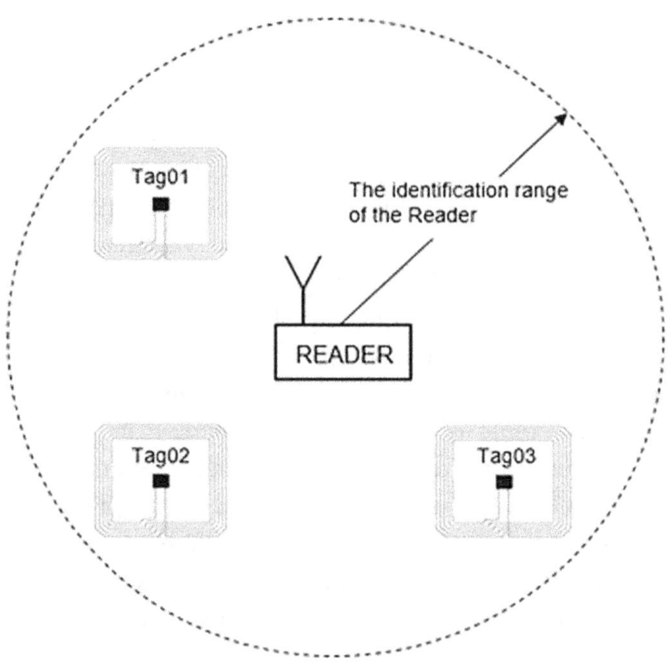

Figure 3. Tag-Reader Collision diagram

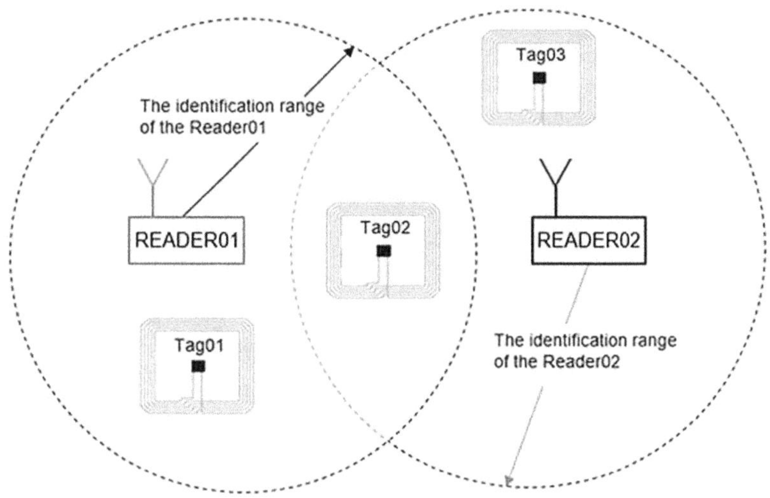

Tag-Reader Collision

Tag-reader collision occurs in a multiple readers identification scenario, as shown in Figure 3.

Tag-reader collision means that when a tag is in the identification range of two or more readers at the same time, multiple readers will send a command to a tag and the tag fails to process the signal after it receives signals from multiple readers simultaneously. Tag02 in Figure 3 is in the identification range of READER01 and READER02 at the same time and cannot respond to either of them when both signals are received at the same time. As a result, REARDER01 and READER02 cannot read the Tag02 information.

Reader-Reader Collision

Reader-reader collision refers to the fact that when one tag is within the recognition range of two or more readers simultaneously, one or more readers cannot receive the tag signal due to the frequency interference among multiple readers.

In industrial production, commodity storage, and other RFID applications, it is necessary to fix multiple readers in one space to form a small identification area network to ensure that objects in this space can be identified. At this time, the recognition areas of multiple readers overlap each other to form interference, as shown in Figure 4. Due to the restriction of the radio frequency identification method, the readers that are not covered by the identification area also interfere with signals due to the electromagnetic waves, and the more readers are arranged, the greater the probability of collisions between readers. This chapter is mainly focused on tag-to-tag collision problem and related research issues.

Figure 4. Reader-to-reader collision diagram

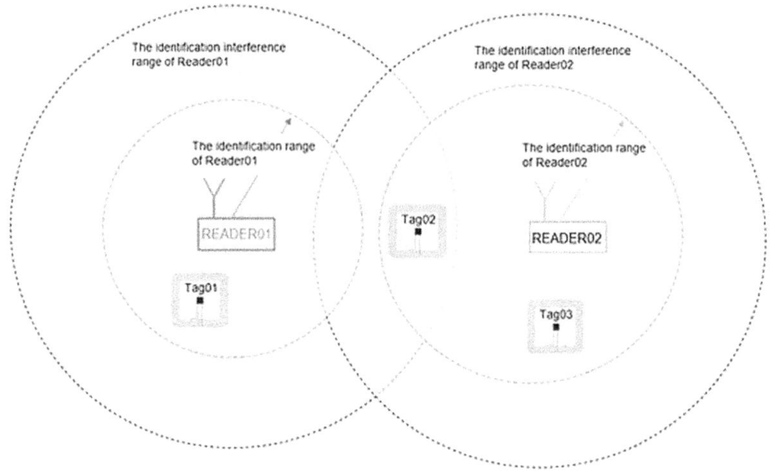

RELATED RESEARCH WORKS

The problem of tag collision creates a barrier to fast tag identification process. Many approaches were recently proposed to reduce the impact of tag collisions, minimize interference, and maximize the read range. There are two major approaches to handle tag collision problem: *probabilistic* and *deterministic* collision solution. The first ones are ALOHA-based algorithm (Abramson, 1970) (Bonuccelli, et al., 2006) (Schoute, 1993) and the other is tree-based algorithm (Finkenzeller, 2003). There are also hybrid approaches, where haphazardness is considered in tree-based schemes (Hush & Wood, 1998) (Micic et al., 2005) (Ryu et al., 2007).

The ALOHA-based algorithms, such as ALOHA, Slotted ALOHA, Framed Slotted ALOHA, and Dynamic Framed Slotted ALOHA, reduce the probability of the occurrence of tag collisions since tags try to answer at distinct times. Tree-based tag anti-collision algorithms can have a longer identification delay than Slotted ALOHA-based ones, but they are able to prevent the tag starvation. Among tree-based algorithms, there are binary search algorithms and query tree protocols.

These anti-collision algorithms can be suitable to minimize the amount of the response time to the tag identification process. The central theme of tree-based algorithms is to exhaust all possible tag identification process with the help of a tree-based search facility to detect tag-collision. There are different types of binary search algorithms for anti-tag collision problem: simple binary search algorithm (Finkenzeller, 2003), dynamic binary search algorithm (Shih et al., 2006) (Law & Siu, 2000), and backtracking binary algorithm (Du et al., 2006).

TREE-BASED ALGORITHMS

The tree-based anti-collision algorithms transform the RFID tag label number to a binary string consisting of '0' and '1'. When a collision occurs, it is divided into "0" and "1" branches according to the label number. In each branch, the algorithm repeats the query process until it can correctly identify a label.

Figure 5. (a) Manchester coding 5 (b) The theory of binary tree algorithm

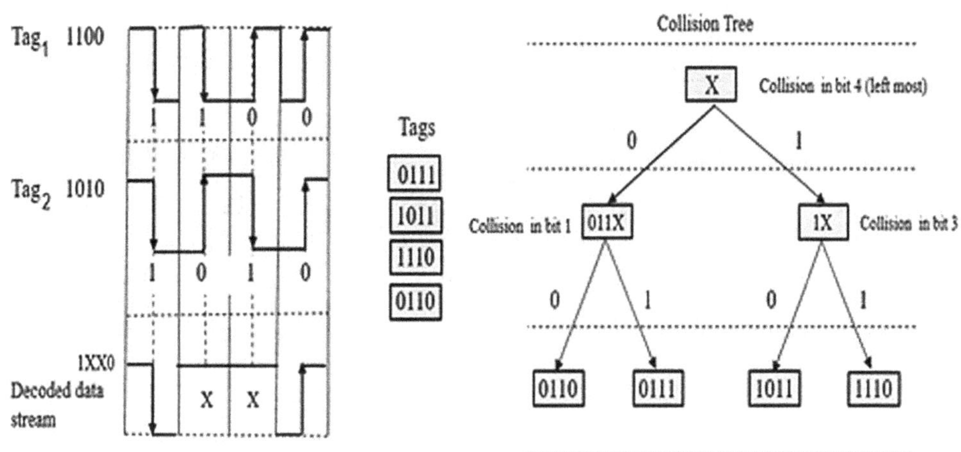

The Manchester code is used in the tree-based algorithms to recognize the bits where there is a collision. In Manchester code, the ascending edge is coded as logical "0", and the descending edge is coded as logical "1". Therefore, when the reader receives an identification code and the states of the code do not change in some bits, the reader can know where collision occur. Let us assume that the binary electronic product code (EPC) or identification number (ID) is set to the length of the four, with the format $D_3D_2D_1D_0$. As shown in Figure. 5(a), when the Tag_1, with ID = 1100, and Tag_2, with ID = 1010, return their digit bit at the same time, the rising edge and falling edge will counteract. This results that the reader will encounter collisions at the second and third bits. The main idea behind tree-based algorithms consist in browsing through all possible nodes in the tree, and tags will be divided into two subsets when collision bit is detected, as shown in Figure. 5(b).

Binary Search Algorithm

The idea behind the binary search algorithm is to check the number of collision bits read, and if there are two or more collision bits the reader cannot uniquely identify all the tags without querying the tags. The highest collision bit is set to 0 while all other collision bits are set to 1. For example, if 0101 and 0011 is read with collision as 0XX1 the reader sets its query number to 0011 where the first X is 0 and the second is 1. The query is sent to the tags and all tags with an identification number less than or equal to the query reply with their identification number back to the reader. If no collision bits are returned, one tag is identified, and it is set to sleep. This process is repeated from the start until there are no responses from the initial query. All tags respond in the first query and several collisions are detected. Figure 6 shows the binary search process identifying four tags with four bits identification number.

Dynamic Binary Search Algorithm

An alternative to the regular binary search algorithm is the dynamic binary search algorithm, which helps with reducing the amount of data sent between the reader and tags. The regular binary search algorithm sends and receives data packets containing the same number of bits as the whole identification numbers.

Figure 6. Execution process of the basic binary search algorithm with four tags

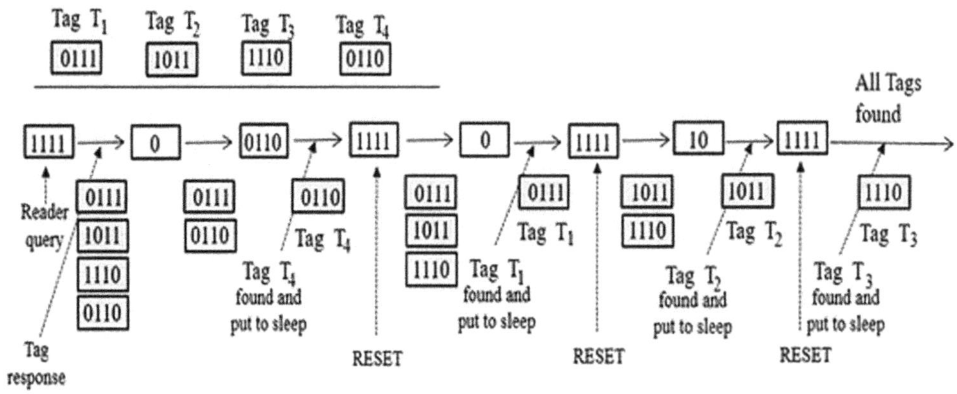

Figure 7. An example of the dynamic basic binary search algorithm

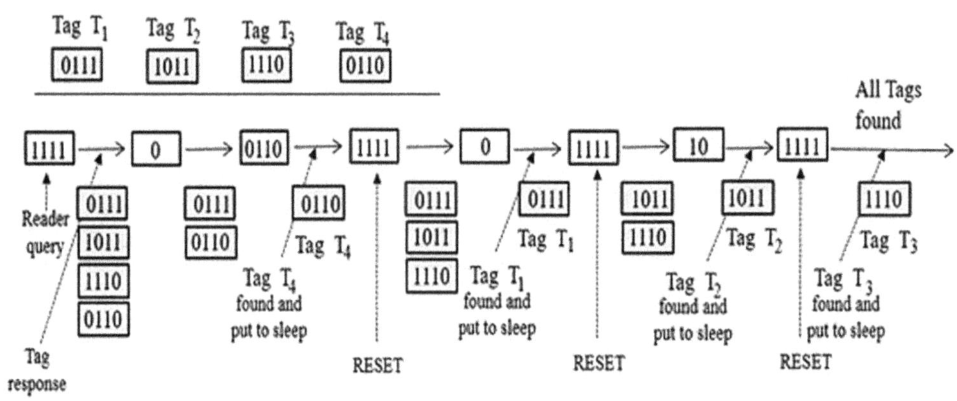

For example, if the identification number is 4 bits the regular binary search algorithm sends 4 bits as the query. The dynamic binary search algorithm recognizes that this not a necessity. Its main idea is the same as the regular binary search algorithm but instead of the reader sending the whole query, it only sends the known bits from a query and receives the bits it does not know from the tags. The total amount of data sent is lowered using this version of the binary search algorithm. Figure 6 shows the same scenario as the regular binary search but using the dynamic binary search algorithm. After the initial read the reader queries a 0 and receives 111 and 110, this means that tags with identification number 0111 and 0110 responded. The following query 0110 is sent because the first three bits 011 are shared between the two responding tags and therefore the first collision bit is the fourth. Figure 7 shows the dynamic binary search process identifying four tags with four bits identification number.

Backtracking Binary Search Algorithm

This algorithm is an improvement on binary tree searching algorithm. When there is no collision, the reader can acquire next request signal from superior layer. In other words, the backtracking binary algorithm is based on binary search algorithm with a different identification process. When the reader

Figure 8. An example of the backtracking binary search algorithm

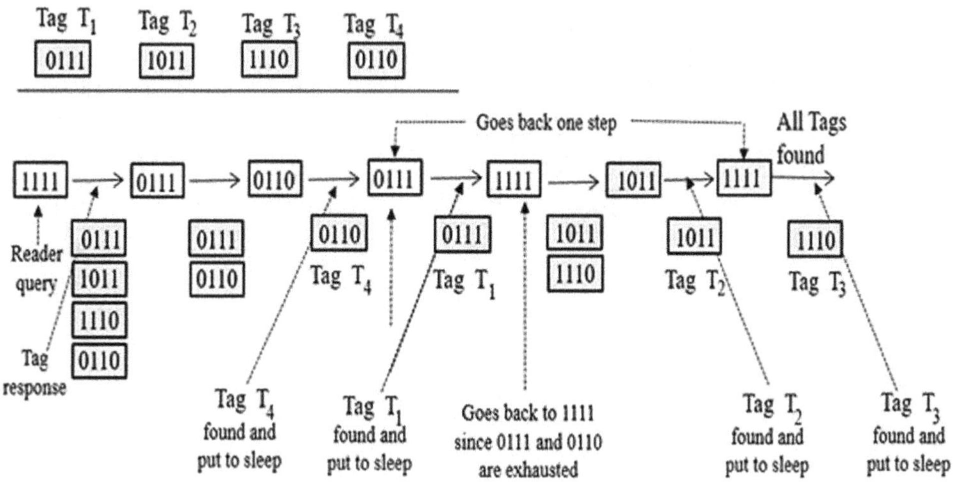

uniquely identified a tag using the binary search algorithm, it repeats the query process from the very start. The backtracking binary algorithm utilizes the fact that when a tag is identified the algorithm already has query results from previous queries. In order to capitalize on these results, the backtracking binary algorithm restarts the query process by moving back just one-step to the previous query result. This results in fewer queries than in a regular binary search algorithm. Figure 8 shows the backtracking binary algorithm-based tag identifying processes for four tags (i.e. 0111, 1011, 1110, 0110) with four bits identification process. All tags respond in the first query and several collisions are detected.

Improved Binary Search Algorithm

Improved binary search algorithm is presented by Djeddou and his colleagues (Djeddou et al., 2013), and is based on the regular binary search algorithm. It is named *improved* because benchmarks showed an overall increase in performance compared to binary search and dynamic binary search. The idea in this algorithm is to use present values on the first two collision bits and using a loop over these values, which will be 00, 01, 10 and 11. If there are fewer than two collision bit one can do a static check and identify the tags. If the number of collision bit is two or more then present loop will be run on the first two of these collision bits. This process is represented in each branch until one or fewer collision bits are read. Figure 9 shows the flow of the algorithm when four tags are being identified by a reader.

After the initial query where every tag sends a response three collision bits are detected. The two highest collision bits are subjects to the preset loop values and the first value is 0 and 0 in the two places. The third collision bit is set to 1. The first query of the preset loop gets no response from any tag. The next value in the present loop is 0 and 1. This query gets a collision bit in its responses, which is fewer than 2, and a static check with a regular binary is used. The static check sets the new collision bit first to 0 and then to 1 in two queries and two tags are uniquely identified. The branch with 0 and 1 as its preset loop value is done and the original third collision bit is set to 1 and the loop values continues with 1 and 0. The last two preset values each uniquely identifies a tag and all tags have been found.

Figure 9. An example of the efficient binary search algorithm

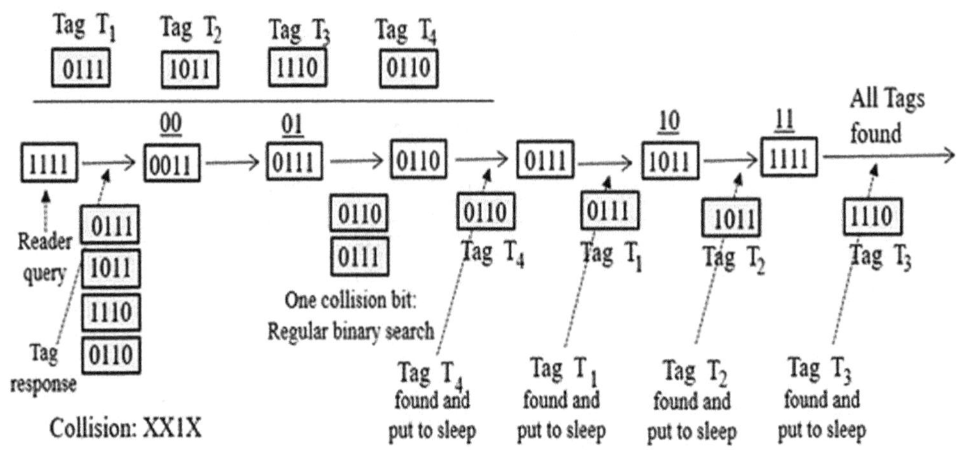

SIMULATION RESULTS

When measuring performance of anti-collision algorithms two metrics are usually used, the number of requests sent by the reader and the number of bits sent between the tags and the reader. In this research a software-based experiment is used as a tool to simulate the communication between the reader and tags. Real physical readers and tags are not needed when measuring performance with request and bit throughput. It does not matter on what system the experiment is run, since physical time is not measured.

Basing on the theory of the binary search-based algorithms, it is known that the searching time depends on two factors. The first one is the number of the tags within the interrogation zone of a reader and the other one is 'Ubiquitous Identification'. In this chapter, simulation experiments are compared the performance of the binary search algorithm (BSA) with backtracking binary search algorithm (BBSA), dynamic binary search algorithm (DBSA), and improved binary search algorithm (IBSA) in terms of the number of reader's request and the total number of queries per tag during the identification process. A sample of data collected in the experimental simulation (for BSA, DBSA, BBSA) is shown in Table 1.

The above tabular data represents the total length of the transmitted binary data of the three discussed algorithms against the number of identified tags. This also shows the performance of these algorithms; and the performance is calculated by the equation (1), where Nbits represents the total number of bits.

$$Percentage\,of\,channel\,saving = \left\{1 - \frac{Nbits\left(BBA\right)}{Nbits\left(DBSA\right)}\right\} \times 100\% \tag{1}$$

Table 1. The results of simulation experiment

Number of Tags	20	60	100	140	180
Total bits of BSA	7158	2330	5778	8600	144880
Total bits of DBSA	5952	21578	53118	85134	130840
Total bits of BBSA	3987	17512	37340	59864	62240
Performance (%)	33.02	18.85	29.71	29.69	52.44

Figure 10. Total number of queries used

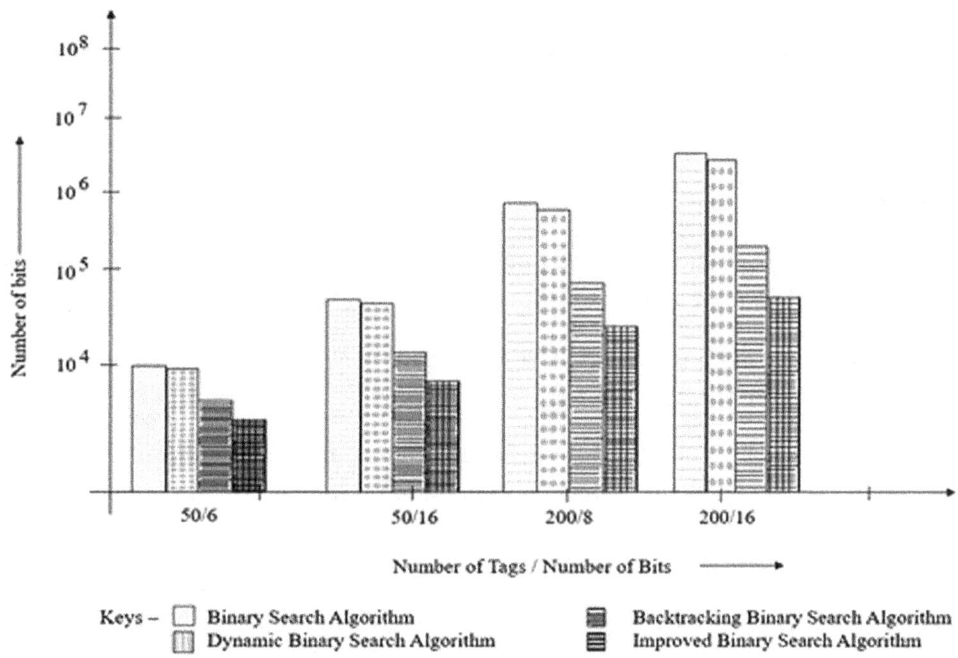

The above result shows that approximately 29.69 percent and 52.44 percent channels are saved by the algorithm (BBSA) than the DBSA when the number of tags is 140 and 180 respectively. The BBA algorithm can achieve better performance in terms of the number of transmitted binary data to identify tags. The experimental data for all four algorithms using number of queries is shown in Figure 10.

Figure 10 shows BSA and DBSA are transmitting considerably more bits than the rest of the algorithms. The difference between the backtrack, dynamic and improved algorithms are significant but improved algorithm shows the best overall performance.

CONCLUSION

The use of RFID technology has increased across a variety of business processes (e.g. manufacturing, warehousing, transportation, retail outlet) along supply chain networks. Although each of these business processes has its own specific requirements, weakness due to tag collision problems will be always an essential feature when deploying RFID-based solutions. RFID-based business solution is also depending on wireless telecommunication technology. While the non-contact and non-line-of-sight characteristics of RFID systems enhance the convenience and efficiency of their applications, these characteristics also increase the corporate information systems vulnerability, and tag collision problem is one of the important issues.

This chapter introduces the basic concept of tag collision problem in RFID-based supply chain network. It also presents a brief review of research work in tree-based anti-collision problem; and it explains different variation of binary search algorithmic solutions for tag collision. A simulation experiment compares the performance of different binary search algorithms. The result shows that the improved binary search algorithm proposed performs much better than the regular binary search algorithms. It scales well with increased number of tags in a collision and is a feasible algorithm to use in large scale RFID environments.

REFERENCES

Abramson, N. (1970). The ALOHA system-Another alternative for computer communications, In *Proceedings of Fall Joint Computer Conference*, (pp. 281-285). ACM.

Bansal, R. (2003). Coming Soon to a Wal-Mart Near You. *IEEE Antennas & Propagation Magazine*, *45*(6), 105–106. doi:10.1109/MAP.2003.1282186

Biao, L. Ai qun, H., & Zhong-Yuan, Q. (2006, June). Trends and Brief Comments on Anti-collision Techniques in Radio Frequency Identification System. In 6th International Conference on ITS Telecommunications Proceedings (pp. 241-245). IEEE.

Bonuccelli, M. A., Lonetti, F., & Martelli, F. (2006, June). Tree slotted ALOHA: a new protocol for tag identification in RFID networks. In *Proceedings of IEEE International Symposium on a World of Wireless, Mobile and Multimedia Networks* (pp. 603-608). IEEE Computer Society.

Cha, J. R., & Kim, J. H. (2005, July). Novel anti-collision algorithms for fast object identification in RFID system. In *ICPADS (2)* (pp. 63-67).

Choi, T., Wing-Kwan, Y., Cheng, T. C. E., & Yue, X. (2018). Optimal Scheduling, Coordination, and the Value of RFID Technology in Garment Manufacturing Supply Chains. *IEEE Transactions on Engineering Management*, *65*(1), 72–84. doi:10.1109/TEM.2017.2739799

Copacino, W., & Anderson, D. (2003). Connecting with the Bottom Line: A Global Study of Supply Chain Leadership and its Contribution to the High Performance Business, *Accenture,* p.1.

Djeddou, M., Khelladi, R., & Benssalah, M. (2013). Improved RFID anti-collision algorithm. *International Journal of Electronics and Communications*, *67*(3), 256–262. doi:10.1016/j.aeue.2012.08.009

Du, H. T., Xu, K. L., & Wang, W. L. (2006). An anti-collision algorithm based on binary-tree searching of backtracking. *Journal of Yunnan University*, *28*, 133–136.

Finkenzeller, K. (2003). *RFID Handbook: Fundamentals and Applications in Contactless Smart Cards and Identification* (2nd ed.). New York, NY: John Wiley. doi:10.1002/0470868023

Hush, D. R., & Wood, C. (1998, August). Analysis of tree algorithms for RFID arbitration. In *Proceeding of IEEE International Symposium on Information Theory,* 107. IEEE.

Jiang, L. F., Lu, G. Z., & Xin, Y. W. (2007). Research on anti-collision algorithm in radio frequency identification system. *Computer Engineering and Applications*, *15*, 29–32.

Johnson, D. (2002). RFID tags improve tracking, quality on Ford line in Mexico. *Control Engineering*, *49*(11), 16.

Kärkkäinen, M. (2003). Increasing efficiency in the supply chain for short shelf life goods using RFID tagging. *International Journal of Retail & Distribution Management*, *31*(10), 529–536. doi:10.1108/09590550310497058

Kinsella, B. (2003). The Wal-Mart factors. *Industrial Engineering (American Institute of Industrial Engineers)*, *32–36*(November).

Law, C., Lee, K., & Siu, K. Y. (2000). Efficient memoryless protocol for tag identification (extended abstract). In *Proceeding of 4th International workshop on Discrete Algorithms and methods for mobile computing and communications (DIALM 2000)*, New York, NY, 75-84. 10.1145/345848.345865

Liu, L., & Lai, S. (2006). ALOHA-Based Anti-Collision Algorithms Used in RFID System. In International Conference on Wireless Communications, Networking and Mobile Computing (pp.1-4), Wuhan, China. IEEE.

Micic, A., Nayac, A., Simplot-Ryl, D., & Stojmenovic, I. (2005). A hybrid randomized protocol for RFID tag identification. In *Proceeding of 1st IEEE International Workshop on Next Generation Wireless Networks* (WoNGeN 2005).

Pal, K. (2017). Supply Chain Coordination Based on Web Services. In H. K. Chan, N. Subraanian, & M. D. Abdulrahman (Eds.), *Supply Chain Management in the Big Data Era* (pp. 137–171). Hershey, PA: IGI Global. doi:10.4018/978-1-5225-0956-1.ch009

Pal, K. (2018). A Big Data Framework for Decision Making in Supply Chain. In M. V. Kumar, G. D. Putnik, K. Jayakrishna, V. M. Pillai, & L. Varela (Eds.), Emerging Applications in Supply Chains for Sustainable Business Development, Chapter 1, 1-22. Hershey, PA: IGI Global.

Pal, K. (2019). Quality Assurance Issues for Big Data Applications in Supply Chain Management. In P. K. Gupta, T. Oren, & M. Singh (Eds.), Predictive Intelligence Using Big Data and the Internet of Things, Chapter 3, 51-76. Hershey, PA: IGI Global.

Prater, E., Frazier, G. V., & Reyes, P. M. (2005). Further impact of RFID one-supply chain in grocery retailing, Supply Chain Management. *International Journal (Toronto, Ont.)*, *10*(2), 134–142.

Ryu, J., Lee, H., Seok, Y., Kwon, T., & Choi, Y. (2007, June). A Hybrid Query Tree Protocol for Tag Collision Arbitration in RFID systems. In *Proceeding of IEEE International Conference on Communications*, 5981-5986. IEEE.

Sabbaghi, A., & Ganesh, V. (2008). Effectiveness and Efficiency of RFID Technology in Supply Chain Management: Strategic Values and Challenges. *Journal of Theoretical and Applied Electronic Commerce Research*, *3*(2), 71–81. doi:10.4067/S0718-18762008000100007

Schoute, F. C. (1983, April). Dynamic frame length Aloha. *IEEE Transactions on Communications*, *COM-31*(4), 565–568. doi:10.1109/TCOM.1983.1095854

Shih, D., Sun, P. L., Yen, D. C., & Huang, S. M. (2006). Taxonomy and survey of RFID anti-collision protocols. *Computer Communications*, *29*(11), 2150–2166. doi:10.1016/j.comcom.2005.12.011

Shih, D. H., Sun, P. L., Yen, D. C., & Huang, S. M. (2006). Taxonomy and survey of RFID anti-collision protocols. *Computer Communications*, *29*(11), 2150–2166. doi:10.1016/j.comcom.2005.12.011

Singh, N. (2003). Emerging technologies to support supply chain management. *Communications of the ACM*, *46*(9), 243–247. doi:10.1145/903893.903943

Smaros, J., & Holmstrom, J. (2000). Viewpoint: Reaching the consumer through e-grocery VMT. *International Journal of Retail & Distribution Management*, *28*(2), 55–61. doi:10.1108/09590550010315098

Want, R. (2004). RFID: A key to automating everything. *Scientific American*, *290*(January), 56–68. doi:10.1038cientificamerican0104-56 PMID:14682039

KEY TERMS AND DEFINITIONS

Active Tag: A tag with its own battery that can initiate communications.

ALOHA: A type of RFID anti-collision algorithm which works by telling each tag in the collision to idle for a randomly selected amount of time.

BSA: Binary Search Algorithm, a type of RFID anti-collision algorithm. Distinguishes tags by sending queries to the set of tags.

Collision: When multiple tags are being read by a reader during the same time frame and part of their identification number collides.

Collision Bits: The colliding bit positions of the response the reader receives when collisions occur.

DBSA: Dynamic Binary Search Algorithm, a type of RFID anti-collision algorithm like the Binary Search Algorithm but send and receives only fragments of queries.

EPC: Electronic Product Code. A low-cost RFID tag designed for consumer products as a replacement for the UPC (Universal Product Code).

Inventory Control: Inventory control can be defined as the process of managing enough materials which are held by the company in the supply chain for the purpose of customers' demand satisfaction.

Manchester Coding: A technique used in readers to identify whether there exist any collision bits.

Passive Tag: A tag with no on-board power source that harvests its energy from a reader-provided RF signal.

Reader: Reader reads and communicates with tags in its operating range by transmitting radio waves.

RFID: Radio Frequency Identification. Describes a broad spectrum of devices and technologies and is used to refer both to individual tags and overall systems.

Supply Chain: A supply chain is the series of integrated corporates that coordinate the physical execution and share information in order to assure a smooth flow of products and/or services, cash and information through the entire chain. The functions of supply chain are - the supply of products to the manufacturer, work in process, inventory and finished product distribution to the final customer.

Supply Chain Management: A supply chain management (SCM) is the management and planning of all the activities that are involved in sourcing, procuring, conversion and all the activities of logistics management. It also encompasses channel partner's collaboration and co-ordination. And the essence of SCM is integrating demand and supply management, in and across companies.

Tag: An RFID transponder, typically consisting of an RF coupling element and a microchip that carries identifying data. Tag functionality may range from simple identification to being able to form ad hoc networks.

Warehouse: A warehouse can also be called storage area and it is a commercial building where raw materials or goods are stored by suppliers, exporters, manufacturers, or wholesalers, they are constructed and equipped with tools according to special standards depending on the purpose of their use.

Chapter 5
Design of a Home Automation App to Assist Elderly and Limited Mobility People

Bianca Stephanie Guimarães Morais
Federal Rural University of the Semi-Arid, Brazil

Álvaro Sobrinho
Federal Rural University of Pernambuco, Brazil

Helder Oliveira
Federal Rural University of the Semi-Arid, Brazil

Lenardo Chaves e Silva
Federal Rural University of the Semi-Arid, Brazil

Robson Locatelli Macedo
Federal Rural University of the Semi-Arid, Brazil

ABSTRACT

This research focuses on a solution to assist elderly and limited-mobility people. It aims to improve the autonomy, and, consequently, the quality of life of this target audience by automating daily tasks conducted at home, such as turning on the lights and manipulating electronic devices. However, it is important to consider the costs and quality attributes (e.g., usability) related to the design of solutions to automate a specific environment, that may include hardware platforms and physical adaptations. In this context, the authors present in this chapter the software requirements discovery and elicitation of a home automation app considering the real needs of the elderly and limited-mobility people. Additionally, we conduct the requirements specification using the unified modeling language (UML) to improve completeness, along with graphical user interface (GUI) prototypes. Finally, we present a mobile app prototype using the Android and Arduino platforms to illustrate a use scenario of the solution.

DOI: 10.4018/978-1-5225-9246-4.ch005

INTRODUCTION

The system automation involves computerized or mechanical techniques to reduce (or remove) manual tasks conducted by human beings (Hock, 2005). It is an operational process established in an environment controlled and executed by means of a set of different devices. The automation is already a reality in many industries, including the field of manufacture, to decrease human interactions. Another example of a specific type of automation can be applied at home (known as home automation). The home automation represents the use of technologies in the domestic environment to provide more comfort, practicality, productivity, efficiency, and profitability (Manda, Kushal & Ramasubramanian, 2018). The goal is to integrate technologies with easy access to information, Internet, and security. Additionally, integrating it with the data network, voice, images, and multimedia. Therefore, the home automation is strongly related to the concept of the Internet of Things (IoT), where home devices are interconnected to achieve specific tasks (Ahlgren, Hidell & Ngai, 2016).

In the last years, the main goals of most of the world population are to make the everyday life simpler, easier, and safer. The home automation is an example of an area that can enable these goals by assisting residents to conduct tasks such as turn on the lights and manipulate electronic devices. Even these being simple tasks (for some people), the home automation may increase the autonomy and quality of life of residents, depending on the target audience. This is the case of elderly and limited mobility people. This population is growing, and consequently, the wish and need of people to be independent (impaired by age and physical limitations) (Nouy, Virone, Barralon, Ye, Rialle & Deongeot, 2003; Ramlee, Tang & Ismail, 2012). Considering a specific scenario, according to the Brazilian Institute of Geography and Statistics, 46 million of Brazilians have some kind of disability (24% of the total population) (IBGE, 2010). Therefore, it is relevant to design new solutions to enable them to be more active in the society, improving quality of life and autonomy (what may also have psychological implications).

However, it is important to consider the costs related to the design of solutions to automate a specific environment, that may include hardware platforms and physical adaptations. In this context, mobile devices and electronic prototyping platforms may be alternatives to enable low-cost automated home, considering small and medium-sized environments. Some mobile applications (apps) have been developed to assist users in their everyday life tasks at home (e.g., eWeLink[1], IFTTT[2], Smart Life[3], OpenHab[4], and Android Accessibility Suite[5] apps). Nevertheless, the apps available on the market have some limitations regarding software quality attributes such as usability, and usually, do not focus on the real needs of a specific target audience (in the case of this work, elderly and limited mobility people). Currently, mobile devices are widespread in the world population. For example, in Brazil, up to the present day, there are 280 million mobile devices with Internet connectivity (Sobrinho, Silva, Perkusich, Pinheiro & Cunha, 2018). This justifies the development of new technologies focusing on mobile platforms. Low-cost solutions have the potential to decrease barriers between financially disadvantaged people, automation technologies, and IoT's.

On the other hand, another issue that is faced by apps' designers to create solutions to assist elderly and limited mobility people is the usability (Zapata, Fernández-Alemán, Idri & Toval, 2015). The use of new technologies is not a simple task for this target audience. The "ease of use" characteristic of apps is a factor that must be exploited and studied. Usually, they have to deal with the size of the screen, graphical user interface (GUI), virtual keyboard, drag and drop features, and so on. This indicates that an app, disregarding the type of disability of the user, may provide (among other features) adaptable layouts without loss of information, customization resources, sound and vibratory alerts, and images.

Therefore, the main issue addressed in this chapter is related to the design of a low-cost home automation environment to assist elderly and limited mobility people in a way that is mobile and ease of use, considering the actual needs of this target audience. In this work, the term "ease of use" stands for the ability of target users to manipulate the home automation solution without losses with respect to time and satisfaction. When this type of quality characteristic is not considered during the design and development of a solution, it may result in adverse situations that discourage the target audience to keep using it.

The main objective in this chapter is present the design of a mobile app to assist elderly and limited mobility people in their most common tasks (based on their real needs) at home by means of automation. In order to achieve this goal, we identify and discuss the actual needs of this target audience by carrying out the software requirements discovery and elicitation of the app using literature reviews, analyzes of related apps currently available in the market, and use cases. Afterward, we specify the elicited requirements using the unified modeling language (UML) 2.0 to document the findings and to decrease ambiguities on the solution (Rumbaugh, Jacobson & Booch, 2004). In this case, the specified models incorporate concepts of the architectural standard model-view-controller (MVC) and the data persistent standard data access object (DAO). Finally, we present a GUI prototype along with an Android and Arduino prototype to illustrate the desired and proposed solution, integrated using a Bluetooth communication. It makes this chapter useful for readers who wish to create new home automation solutions that take into account the main characteristics of an ease of use app. Therefore, the main contribution presented in this chapter is the design of a home automation mobile solution that can be used to improve the accessibility of elderly and limited mobility people.

We carried out the literature review aforementioned based on the following two main research questions (RQ): what are the main needs to be met with the app? (RQ01) and what functionalities should we implement in the app to be useful to target users' demands? (RQ02). These questions were used to identify the primary requirements for the app. Research articles related to the home automation to assist elderly and limited mobility people were selected and analyzed from the databases IEEE Xplore Digital Library, PubMed, Scielo, Web of Science, ACM Digital Library, and Scorpus. The Google Academic was also used during this research. The terminology Disability, Application, and Automation were addressed in the process of keyword analysis, taking into account their synonyms and subject scope.

Background

We introduce in this section concepts about the main subjects discussed in this chapter. More specifically, we present concepts related to the platforms Android and Arduino, and the communication platform.

Android

Android is a Linux-based operating system (OS) design to run in many devices (e.g., smartphones and tablets). Software designers focus on the development of mobile apps by means of a rich and flexible GUI and application programming interface (API). The Java programming language is used to develop Android apps. Android was developed by the open handset alliance (OAH), a group led by Google. Each platform has an identification code called API level. In order to develop an app with an API, using this code, it is possible to know which devices are able to run it (Murphy, 2009).

The integrated development environment (IDE) Android Studio is used to develop Android apps. Developers can use this basic program to create apps for a large range of Android devices, considering the number of available tools. This platform is integrated with the Android software development kit (SDK) that is useful to develop apps along with a simulator or a physical device to conduct tests. In this work, the Android app developed is integrated with the Arduino platform to present an example of usage.

Arduino

The Arduino is a small programmable computer that enables one to process input and outputs between devices/components connected to it. It is called a platform of embedded or physical computation, being a system that interacts with the environment by means of hardware and software (McRoberts, 2010). The main objective is to be a low-cost, useful, and practical device, being accessible to students and amateur designers. This platform was adopted by the concept of open hardware, meaning that anyone can assemble, modify, and personalize it, using the same basic software.

An example of an Arduino board is the mega model (ATmega2560) (Arduino, 2018). It is composed of a microcontroller, that has 54 input/output pins, 15 being used as pulse width modulation (PWM) output, 16 analog inputs, 4 UARTS (serial hardware ports), 16 MHz crystal oscillator, USB connection, power connector, in-circuit serial programming (ICSP) connector (communication protocol) and a reset button. In this work, the Arduino and Android platforms are integrated by means of a Bluetooth communication.

Communication Platform

Bluetooth is a wireless communication technology considered as a standard protocol that has low costs when considering the power supply, at the cost of having low range (Miler & Bisdikian, 2001). It is possible to transfer data between devices, being able to connect up to eight devices simultaneously without the need for close proximity and alignment. The communication is conducted by electromagnetic waves, whereas it can reach a range up to 100 meters, being a communication standard based on low cost transmitting microchips that use the radio.

The Bluetooth architecture is divided into piconet and scatternet. The piconet can be considered a network with up to eight devices connected to each other (as highlighted above), while the scatternet represents the association between two or more piconets (i.e., it enables one to make a more complex network of devices).

HOME AUTOMATION APP

We describe in this section the activities conducted to design the proposed app. More specifically, the requirements discovery and elicitation, UML modeling, GUI, Arduino, and Android prototyping.

Requirements Discovery and Elicitation

We discovered and elicited the software requirements by means of a literature review, analyzing existing apps in the market, and use cases. In the first case, we used the software named *Start* to search, manage and analyze the articles consulted using a set of keywords and search strings. As highlighted above, we used the databases IEEE Xplore Digital Library, PubMed, Scielo, Web of Science, ACM Digital Library, and Scorpus. The Google Academic was also used during this research. This search was important to

Table 1. Related Apps

Apps	Description
OpenHab	It is a Java-based open source home automation platform that integrates and combines a wide range of different smart home systems and technologies into one solution.
IFTTT	It is available for Android and iOS. The app supports more than 300 social networks, cloud storage, music, and more.
eWeLink	It is a free software that turns a home into a smart place, only with the use of a remote control.
Android Accessibility Suite	It is a collection of services for accessibility in which users manipulate devices without the use of the eyes.
Smart Life - Smart Living	It is an app used to transform a home into a smart location, remotely controlling home components from anywhere.

obtain an overview of the field of home automation. Once the articles were analyzed, it was also possible to select some related apps to collect the requirements (see Table 1). Finally, we used the literature review and the related apps to generate use cases of the app to refine the requirements discovery and elicitation process.

Therefore, we concluded that (in general) an app to assist elderly and limited mobility people should have the following main characteristics:

- The app should allow one to insert, update, and delete users and home components using a database.
- The user registration should contain username and password. The user may also choose to access the app by a fingerprint (in this case, it must be recorded in the database).
- The app should allow one to insert, update, and delete the home components by providing a Bluetooth paired device listing.
- The app should provide the Android talkback function, vibration alert, lightening alert, and task scheduling.
- The app should have visuals resources that are easy of use, simple GUI, and information that are easy to understand in order to improve usability.

The app was fully documented using the standard ISO/IEC/IEEE 29148:2011[6], based on the adaptations proposed by Sommerville (2011). The designed app is named *AutBility*, influenced by automation and accessibility. It includes the basic functionalities presented by related apps, complemented, for instance, by vibrating and luminous alerts, and the clear and intuitive GUI (where the user can modify some visual elements, considering his/her comfort). Considering a specific scenario, the comfort may be improved by changing colors of the GUI and by increasing the size of the font for better visibility. In the next section, the requirements of the app along with more detailed descriptions are described by means of UML diagrams and prototypes.

Design of the App Using UML

The UML is considered a "unifying" language of notations that uses different modeling techniques. It is a graphical language that models object-oriented systems, that is, defines graphical elements that are used to represent specific characteristics (perspectives). It can be applied to a large number of systems,

Figure 1. Overall UML use case diagram of the app

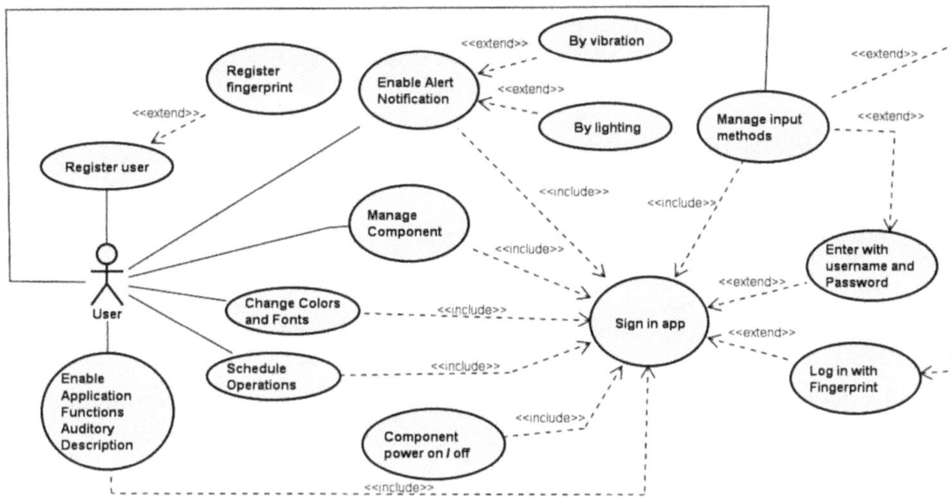

regardless of the programming language and development process, and it is adaptable to specific characteristics of it. The UML provides a set of diagrams to represent different perspectives of systems, such as classes, use cases, objects, sequence, communication, state, activity, and component. However, only a sample of the UML diagrams specified in this work is described due to space constraints.

Use Case Diagram

One of the techniques used to conducted the requirements discovery and elicitation was the specification of use cases. However, besides the textual description of use cases, UML diagrams were applied during the design phase to clarify the use scenarios of the app. The overall UML use case of the AutBility app is illustrated in Figure 1. It is possible to observe that the app is composed of 14 use cases that include, for example, *Manage Component*, and *By vibration*. This shows an external perspective of the app, considering interactions between the stakeholder *User* and the functionalities elicited.

Package Diagram

The packet diagram is often used to illustrate the system's logical architecture, which may include layers, subsystems, packets, and so on. It provides a way to group elements, such as classes, other packages, and use cases. One can show dependencies (coupling) between packets, displaying large-scale coupling of the system to present an architectural view to developers. In this case, the dependency lines (dashed arrows) point to the dependent package.

The four packages of the app are shown in Figure 2 (called *Model*, *View*, *Control* and *DatabaseDAO*). These packages follow the MVC development model and the data persistent standard DAO. They aim to separate business logic from the GUI, and database instructions, respectively. Thus, the programmers can easily make modifications to the required parts without affecting others. More specifically, the package *Model* groups classes related to the application domain. The *View* contains classes related to the GUI. The *Controller* defines the way the *View* behaves, according to the data entered by the user, updating

Figure 2. Overall UML package diagram of the app

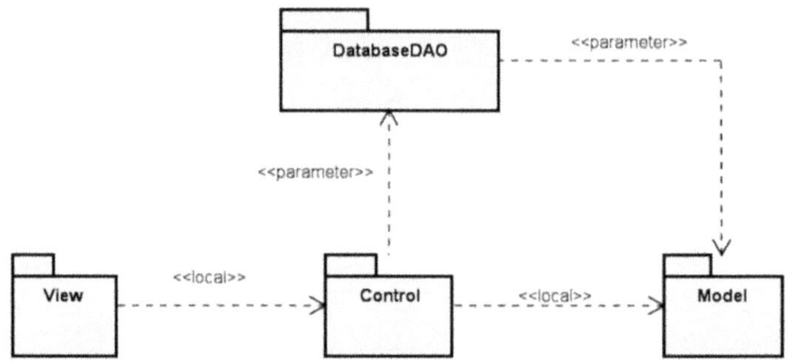

the *Model*. Finally, the *DatabaseDAO* package is composed of classes used as communication interfaces between the app and the structured query language (SQL) instructions.

Class Diagram

The class diagram is useful to show the overall structure of the system. The classes are represented by boxes, with the first compartment representing the name of the class, and the second representing the attributes (data that an object stores) of that class.

The overview of the class diagram and the relationships between classes and packages are shown in Figure 3. It is important to highlight that the relationships between classes of different packages occur by means of dependency arrows (as stated by the package diagram). The dependencies have descriptions of the local or parameter type (UML stereotypes). In the case of the local variables, some operations of the class, where the arrow starts, have an action that needs a local variable of the class that receives the arrow. On the other hand, parameter dependency indicates that some operation of the class, where the arrow starts, has a parameter of the class that receives the arrow.

Figure 3. Overall UML class diagram of the app

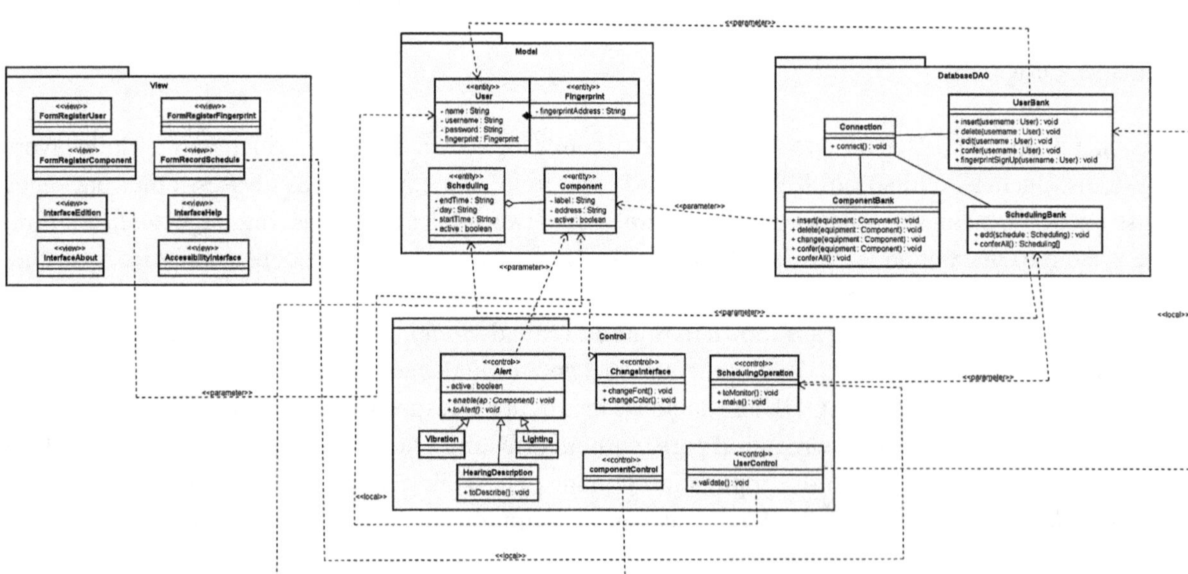

Object Diagram

In order to illustrate how the relationship between some of the defined classes occurs, the object diagram of the component scheduling operation is illustrated in Figure 4. The operation is started with the controller object of the class *UserControl*, which calls the view object of the class *FormRecordScheduling* to enable users of the app to state the home component he/she wants to schedule to execute. The data inputted by users are passed through parameters to a controller object of the class *ComponentControl* that manages the data persistence in the database. This is conducted using model objects of the classes *Scheduling* and *Component*, respectively.

Sequence Diagram

However, up to this section, we have not illustrated the interactions between objects by means of messages exchange in order to run a specific use case. Sequence diagrams that expose the messages exchanged between objects considering the order of execution using timelines can illustrate this type of relationship.

The sequence diagram of the *Component Scheduling* operation is presented in Figure 5. The actor in this scenario is the *User* and the classes that provide the needed behaviors are the *FormRecordScheduling*, *ComponentControl*, *ComponentBank*, *SchedulingOperation*, *Scheduling*, and *SchedulingBank*. The sequence starts with the actor *User* requesting the scheduling by the method *register()* of the class *FormRecordScheduling*. This action implies that an object of the class *ComponentControl* is requested

Figure 4. UML object diagram of the component scheduling operation

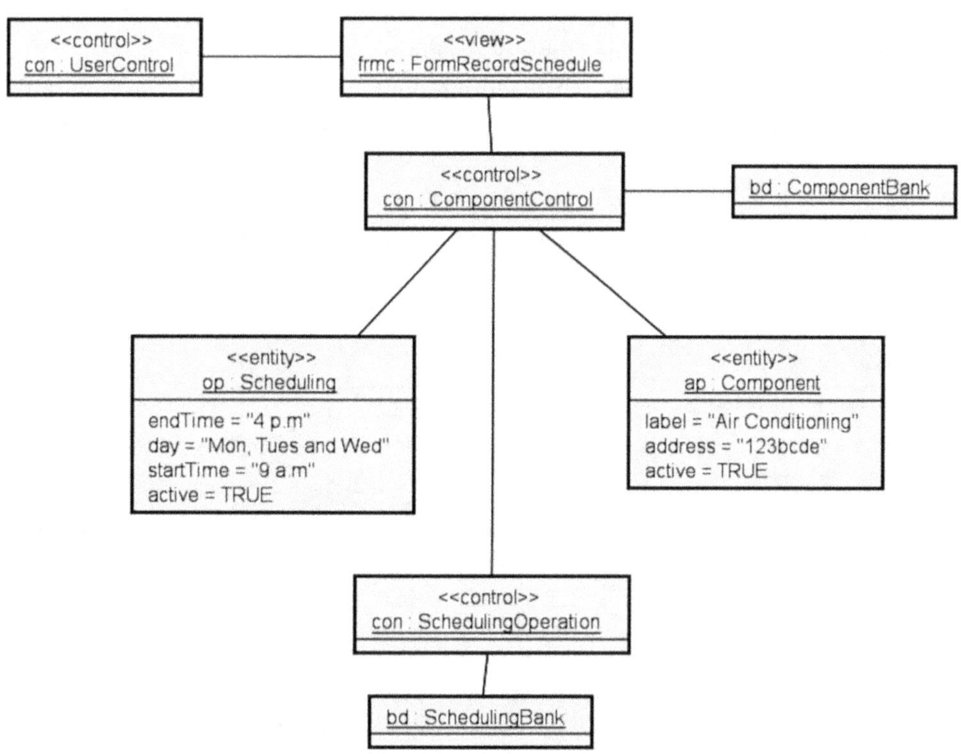

Figure 5. UML sequence diagram of the component scheduling operation

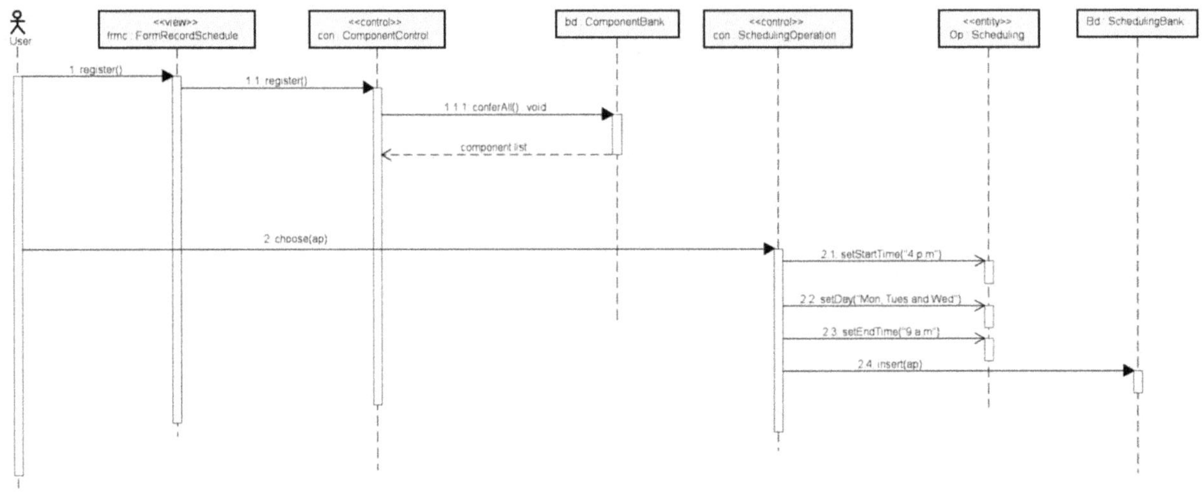

to return all the components already recorded in the database, by means of the communication with an object of the class *ComponentBank*. It enables the user to choose the component he/she wants to schedule to be executed later using objects of the classes *SchedulingOperation*, *Scheduling*, and *SchedulingBank*.

Activity Diagram

Once the order of messages exchange is stated, it is also necessary to show (step-by-step) how the functionalities of the app are executed. This perspective can be illustrated by means of an activity diagram. It describes the actions conducted and the flows of executions.

The activity diagram of the *Component Scheduling* operation is illustrated in Figure 6. The initial action is related to the choice of type of access, where the user will choose the input method (fingerprints

Figure 6. UML activity diagram of the component scheduling operation

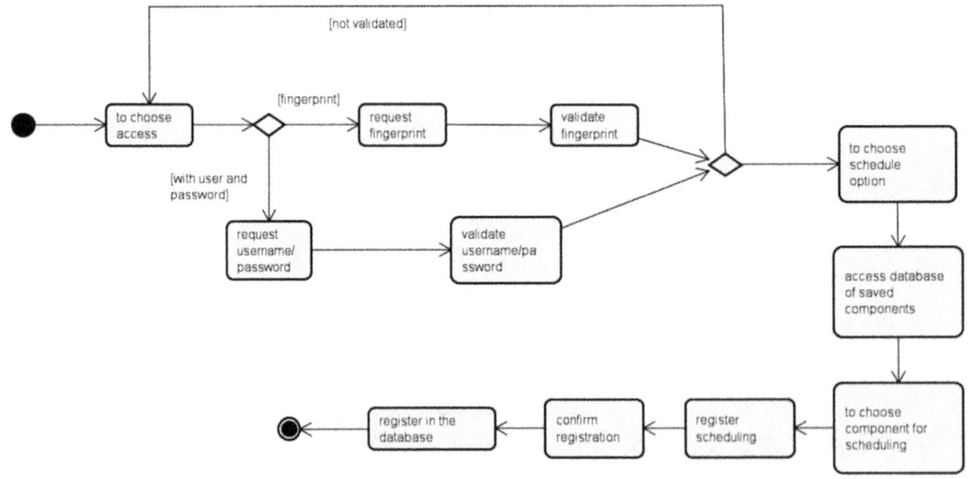

or virtual keyboard). If the choice is the fingerprint access, the fingerprint is requested and validated. If the choice the manual access, the user is asked to input the username and password, and the validation is carried out. A branching structure shows that, for any of the two choices, if the user is validated, the database of saved components is accessed, the schedule of the component is chosen, and the schedule is resisted.

State Diagram

The state diagram aims to describe the behavior, instead of a step-by-step execution sequence of the operation. This is important to enable analysis of the correct behaviors of the app. The representation of the app, component, or object is specified using states and transitions. A state is related to the current situation of the element under specification, whereas a transition relates to the event that triggers changes between states.

The component scheduling is also discussed in this section to illustrate the different modeling perspectives of the same operation. The state diagram of the component scheduling is shown in Figure 7. The initial state of the component is the inactive state, which indicates that the component will eventually be activated. The event *Schedule* represents that a user needs to schedule the component to be activated defining a date, time to start executing, and time to finish executing. When the app identifies that the date and de time to start has arrived, a state transition occurs between the states *Scheduling* and *Active*. Afterward, the user may turn off the component manually, or it is finished by the app considering the time specified. Thus, the component returns to the initial state.

Deployment Diagram

Once the structural, interactive, and behavioral perspectives are presented, it is useful to illustrate an overview of how the app should be deployed. The deployment diagram determines the way a specific system should be executed. It may define software and hardware components, along with the communication infrastructure.

Figure 7. UML state diagram of the component scheduling

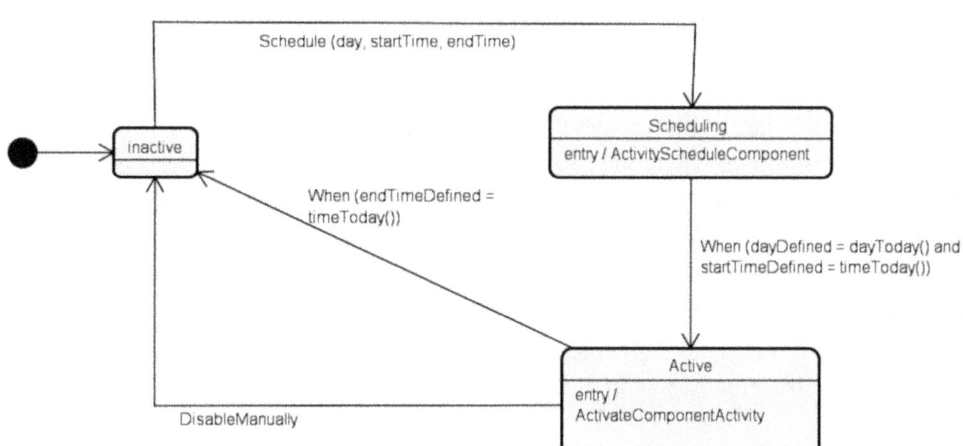

Figure 8. UML deployment diagram of the app

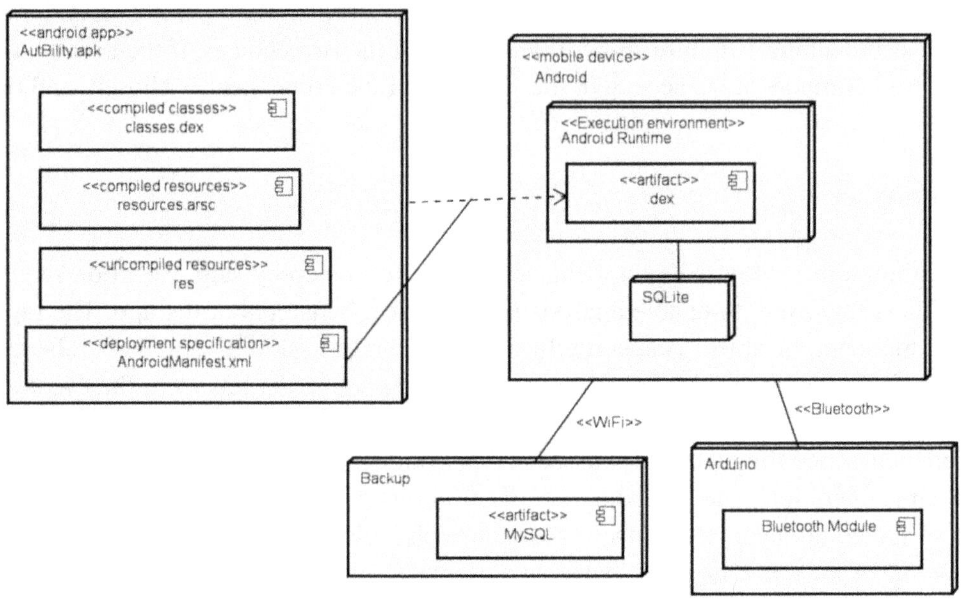

The deployment diagram of the AutBility app is shown in Figure 8. According to the requirements specification, the app is executed on the Android platform. The data persistence is conducted primarily using the SQLite database. Additionally, an external database is used to keep backups of the app. Finally, the Arduino platform is integrated with the Android device by means of a Bluetooth communication to control home components.

Prototyping of GUI

Considering the requirements discovery and elicitation, and the UML models, a prototype of the GUI of the app was designed to illustrate how one could develop the proposed solution. The sample of the GUI, illustrated in this section, was created using the Marvel tool[7].

The initial screen of the app is presented in Figure 9a. It is composed of options for manual access, fingerprint access, and registration of new users. In Figure 9b, the main menu is illustrated (considering that the access of the user was granted), with options presented in well-distributed and large boxes, providing comfort to touch events. It includes options to insert, edit, list, and delete a component, settings, help, and about function (data about the developer and the app's version).

An auxiliary menu is shown when one of the options is displayed, if the user wants to quickly access another functionality (see Figure 10a). In Figure 10b, the GUI related to the insertion of a new component is illustrated. It shows a list of devices connected to the Arduino and enabled via Bluetooth. If the user chooses one or more components, he/she should press the register button to add it to the list of saved devices. An option to receive a notification is also available, if a specific activity is going to happen. This is useful when the devices are triggered or the proximity sensor has some variation. In these cases, the user will receive a vibrating or bright alert depending on the choice of the settings.

Figure 9. a) Initial screen of the app. b) Main menu

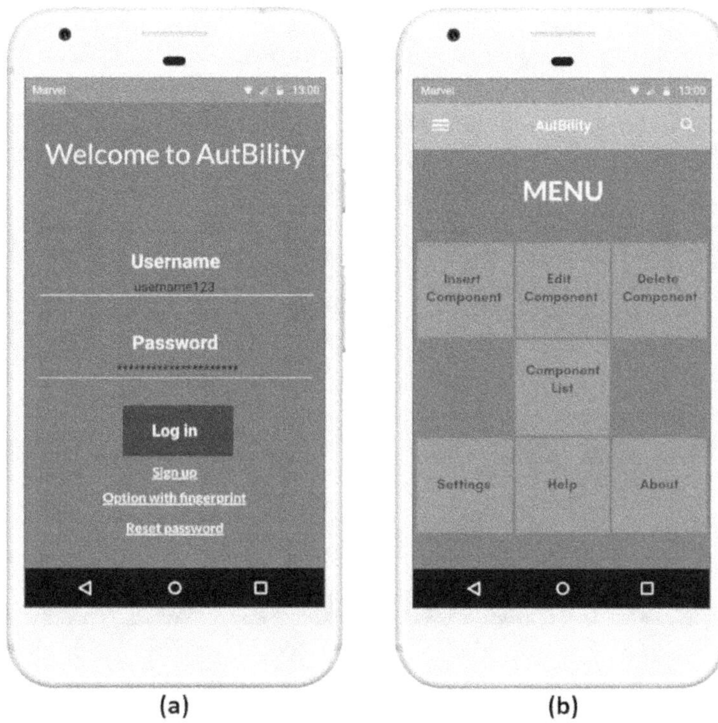

Figure 10. a) Navigation menu of the app. b) Insert component

Figure 11. a) Edit component b) Configuration menu

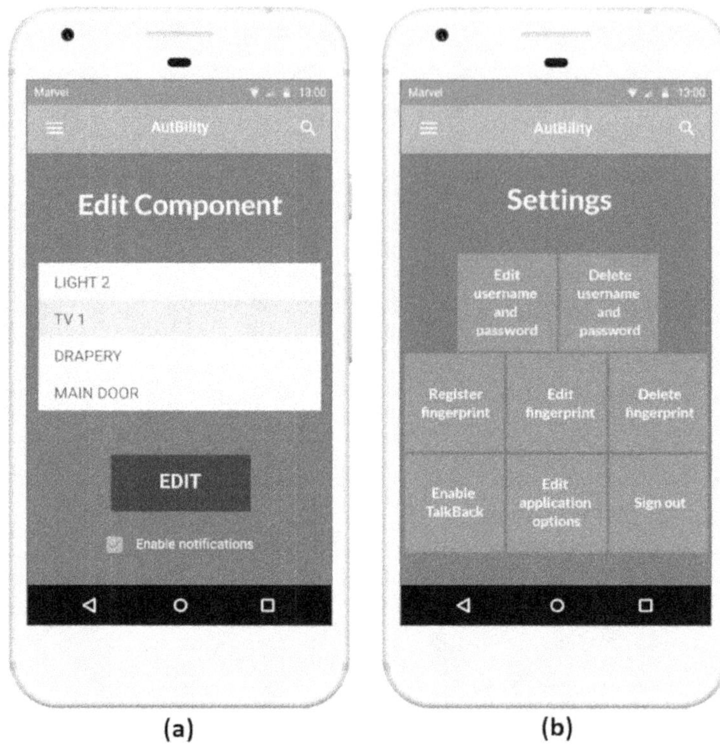

The GUI designed to update the saved devices, modifying some desired information, is presented in Figure 11a. If the modification has been completed, the user will again have the notification option, if there is an activity to be executed. In Figure 11b is shown the settings menu, where options are available to make registrations, deletions and updates the username, password, fingerprint, talkback, update features (change colors, font size, and alerts of the app), and to exit.

Android App and Arduino Integration

We used the Android Studio to develop an Android prototype of the AutBility app according to the requirements discovery and elicitation, UML design, and GUI prototypes. The AutBility prototype runs on Android devices from the Android platform 6, and it is designed according to the GUI prototypes previously illustrated using the Marvel tool. Additionally, as highlighted above, the Arduino Mega along with the Bluetooth module HC-05 were integrated with the app to enable the development of a use scenario to test it.

Therefore, a use scenario of the app related to the activation of a home component is presented in Figure 12. We choose a simple use scenario that allows the user of the app to turn on the lights of a house. It shows a smartphone running the AutBility app, the Arduino mega ATmega2560, the Bluetooth module HC-05, and a protoboard. The protoboard is connected to a small led. In this case, the app is showing that the led is currently disabled by means of the message “LED OFF”.

Figure 12. AutBility app prototype running a use scenario related to the activation of the lights of a house (disabled in this moment)

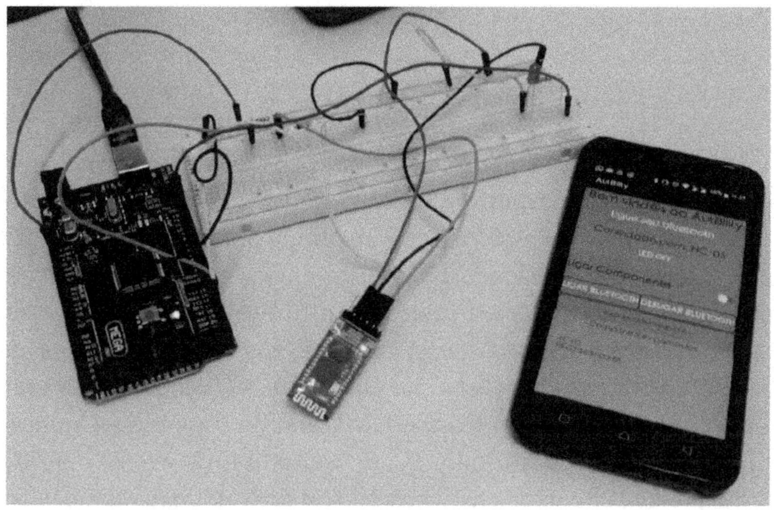

Figure 13. AutBility app prototype running a use scenario related to the activation of the lights of a house (enabled in this moment)

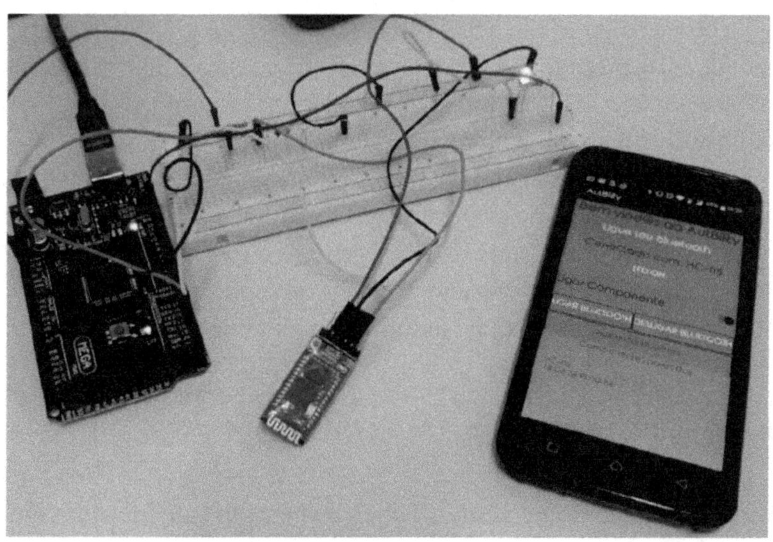

On the other hand, the same use scenario is shown in Figure 13, however, it considers a user event requesting the activation of the led connected to the protoboard. In this case, the app is showing that the led is currently enabled by means of the message "LED ON".

In order to enable the use of a Bluetooth device, it is required the initialization of the Bluetooth service as a function of the app. The button *Pair New Component* is used to send the components found in the environment to a list named *Paired Components*. The text *Bluetooth Status* is displayed when the smartphone device is paired with the Bluetooth previously connected to the Arduino Mega. Once the app is connected to the Arduino, the user can press the button *Turn on Component* to enable it. We used the Android version 8.1.0 to conduct the tests.

CONCLUSION

The task of designing and developing a home automation app is arduous and meticulous, especially when it is necessary to analyze all feasible features in order to meet the real needs of the target audience. This is the case of the AutBility app, used to assist elderly and limited mobility people to conduct home tasks, aiming to provide comfort and independence to the user (envisioning to improve the quality of life).

Therefore, we designed and developed a prototype of the AutBility app by means of well-known techniques in order to achieve the real needs of users, considering quality attributes (e.g., usability). More specifically, during the requirements discovery and elicitation, we used literature reviews, related apps, and use cases. On the other hand, we applied the UML and prototyping techniques to model the app before coding the final solution. These techniques are useful to improve the confidence in the solution and to decrease costs that eventually happens when an app does not meet the needs of users.

The literature review and the study of related apps was importance to conclude that there is a lack of apps aimed specifically to assist the elderly and limited mobility people (decrease daily obstacles, mainly at home). The UML diagrams were also useful to elaborate and understand the requirements, considering completeness. It is possible to observe that design and development of this type of app should be conducted guided by a broad study of the necessities of the target audience (mainly: functionalities and usability), given that this influences in the acceptance of the solution. Therefore, the main contribution of this work is the overview of a solution to assist the elderly and limited mobility people, that can be used as a roadmap for readers that aim to contribute in this field.

However, it is important to highlight some of the limitations of this work. The prototype of the app presented in this chapter only meet some of the elicited requirements. This limited the range of the tests conducted. Additionally, in the current state of this work, we only considered the integration of the app with devices that have embedded software and a Bluetooth communication interface. In this case, it limited the range of home devices that can be automated using the AutBility app.

FUTURE RESEARCH DIRECTIONS

In this context, we envision some future works. The first one is to fully develop the AutBility app considering all the discovered and elicited requirements presented in this chapter. In order to achieve this goal, we envision to develop the app for multiple mobile platforms using the react native framework[8] provided by Facebook. This has the potential to increase the range of users that can be benefited by the use of the app. Afterward, we also envision to improve the solution enabling the integration of home devices that are not originally composed of Bluetooth modules.

REFERENCES

Ahlgren, B., Hidell, M., & Ngai, E. C.-H. (2016). Internet of Things for Smart Cities: Interoperability and Open Data. *IEEE Internet Computing*, *20*(6), 52–56. doi:10.1109/MIC.2016.124

Arduino. (2018). Mega 2560 Reference Design.

Hock, G. (2005). Highly available computer-based automation systems. *Computing & Control Engineering Journal*, *16*(3), 17–20. doi:10.1049/cce:20050302

IBGE. (2010). Census 2010 of the Brazilian Institute of Geography and Statistics.

Manda, V. L. K., Kushal, V., & Ramasubramanian, N. (2018). An Elegant Home Automation System Using GSM and ARM-Based Architecture. *IEEE Potentials*, *37*(5), 43–48. doi:10.1109/MPOT.2016.2515644

McRoberts, M. (2010). Beginning Arduino. New York, NY: Apress. doi:10.1007/978-1-4302-3241-4

Miler, B. A., & Bisdikian, C. (2001). Bluetooth revealed: the insider's guide to an open specification for global wireless communication. Upper Saddle River, NJ: Prentice Hall.

Murphy, M. L. (2009). *The Busy Coder's Guide to Android Development.* USA: CommonsWare; Revised & enlarged edition.

Nouy, N., Virone, G., Barralon, P., Ye, J., Rialle, V., & Deongeot, J. (2003, June). New trends in health smart homes. In *Proceedings 5th International Workshop on Enterprise Networking and Computing in Healthcare Industry (pp. 118-127). IEEE.*

Ramlee, R. A., Tang, D. H. Z., & Ismail, M. M. (2012). Smart home system for Disabled People via Wireless Bluetooth. *International Conference on System Engineering and Technology*. 10.1109/ICSEngT.2012.6339347

Rumbaugh, J., Jacobson, I., & Booch, G. (2004). *The Unified Modeling Language Reference Manual* (2nd ed.). Pearson Higher Education.

Sobrinho, A., Silva, L. D., Perkusich, A., Pinheiro, M. E., & Cunha, P. (2018). Design and evaluation of a mobile application to assist the self-monitoring of the chronic kidney disease in developing countries. *BMC Medical Informatics and Decision Making*, *18*(7), PMID:29329530

Sommerville, I. (2015). Software Engineering. Pearson, 10th edition.

Zapata, B. C., Fernández-Alemán, J. L., Idri, A., & Toval, A. (2015). Empirical Studies on Usability of mHealth Apps: A Systematic Literature Review. *Journal of Medical Systems*, *39*(2), 1. doi:10.100710916-014-0182-2 PMID:25600193

ADDITIONAL READING SECTION

Asai, Y., Ueda, Y., Enomoto, R., Iwai, D., & Sato, K. (2016), ExtendedHand on Wheelchair. *Proceedings of the 29th Annual Symposium on User Interface Software and Technology.*

Banks, A., & Porcello, E. (2017). *Learning React: Functional Development with React and Redux*. O'Reilly.

Banzi, M. (2008). *Getting Started with Arduino*. Make.

Bloch, J. (2018). *Effective Java* (3rd ed.). Pearson.

Chiarelli, A. (2016). *Mastering JavaScript Object-Oriented Programming*. Packt Publishing.

Conde, M. Á., Garcéa-Penalvo, F. J., & Matellán-Olivera, V. (2014), Mobile apps repository for older people. *Proceedings of the Second International Conference on Technological Ecosystems for Enhancing Multiculturality*.

Eisenman, B. (2017). *Learning React Native: Building Native Mobile Apps with JavaScript*. O'Reilly.

Fowler, M. (2003). *UML Distilled: A Brief Guide to the Standard Object Modeling Language* (3rd ed.). Addison-Wesley Professional.

Freeman, E. & Robson, E. (2004), Head First Design Patterns: A Brain-Friendly Guide. O'Reillly.

Friedman, D. P., & Christiansen, D. T. (2018). *The Little Typer*. MIT Press.

Hashizume, S., Suzuki, I., Takazawa, K., Sasaki, R., & Ochiai, Y. (2018), Telewheelchair: the Remote Controllable Electric Wheelchair System combined Human and Machine Intelligence. *Proceedings of the 9th Augmented Human International Conference*. 10.1145/3174910.3174914

Mandel, T. (1997). *The Elements of User Interface Design*. John Wiley & Sons.

McLaughlin, B. D., Pollice, G., & West, D. (2011). *Head Fist Object-Oriented Analysis & Design*. O'Reilly.

Pressman, R., & Maxim, B. R. (2014). *Software Engineering: A Practitioner's Approach*. McGraw-Hill Education.

Schuddinck, S. (2018), Spring MVC made easy: A quick introduction to the Spring MVC framework for beginner (programming made easy).

Siebra, C., Gouveia, T. B., Macedo, J., Silva, F. Q. B., Santos, A. L. M., Correia, W., ... Florentin, F. (2017), Toward accessibility with usability: understanding the requirements of impaired uses in the mobile context. *Proceedings of the 11th International Conference on Ubiquitous Information Management and Communication*. 10.1145/3022227.3022233

Singh, H., Pallagani, V., Khandelwal, V., & Venkanna, U. (2018), IoT based smart home automation system using sensor node. *International Conference on Recent Advances in Information Technology*. 10.1109/RAIT.2018.8389037

Smyth, N. (2017). *Android Studio 3.0 Development Essentials – Android* (8th ed.). CreateSpace.

Zakas, N. (2014). *The Principles of Object-Oriented JavaScript*. No Starch Press.

KEY TERMS AND DEFINITIONS

Automation System: It is an operational process established in an environment controlled and executed by means of a set of different devices.

Data Access Object: It is a persistence standard used to design systems in a way that the developer can separate the database instructions from the application logic and graphical user interfaces.

Home Automation System: A specific type of automation can be applied at home.

Limited Mobility People: A set of people who have some disability that prevents them to perform daily tasks.

Model-View-Controller: It is an architectural standard used to design systems in a way that the developer can separate application logic from graphical user interfaces.

Unified Modeling Language: It is a modeling language used to design object-oriented systems that integrates different user perspectives.

Usability: A non-functional software requirements used to measure the ability of users to easily and fully manipulate a software.

ENDNOTES

1 https://play.google.com/store/apps/details?id=com.coolkit&hl=pt_BR
2 https://ifttt.com/
3 https://play.google.com/store/apps/details?id=com.tuya.smartlife&hl=pt_BR
4 https://www.openhab.org/
5 https://play.google.com/store/apps/details?id=com.google.android.marvin.talkback&hl=pt_BR
6 https://ieeexplore.ieee.org/document/6146379/
7 https://marvelapp.com
8 https://facebook.github.io/react-native/

Chapter 6
Smart Homes and Offices

Nikita Jain
Bharati Vidyapeeth's College of Engineering,India

Rachna Jain
Bharati Vidyapeeth's College of Engineering, India

Vaibhav Kumar
Bharati Vidyapeeth's College of Engineering, India

ABSTRACT

Smart Homes and Offices (SHO) are composed of interlinked components with constant data transfer and services targeted at increasing the lifestyle of the people. This chapter describes about the smart components and how SHO are direct implementation of Internet of Things (IOT). The major paradigm in this chapter is appliances supporting smart aspects of SHO, their applications and change in technology in context of smart Homes and Offices. Here we have also discussed the standardization and personalization of gadgets and how it has been increasing our standard of living. Finally, the chapter focuses on privacy preserving mechanisms, its essence over smart cities, strong architecture related to privacy, preserving mechanism, and various approaches available that can retaliate these issues in a smart city environment.

INTRODUCTION

At the turn of this century, the evolution of intuitive and sustainable environments has gained pace, and having smart homes and offices is the first step on this path. You can now own a fully connected house well stocked with sensors to control temperature, moisture, air quality, energy consumption, and even more, and you will be notified about the status of nearly all of the appliance running in your SHO from anywhere in the world with just a Smartphone or any mobile device. But all these are partially smart, they are still managed by us. These appliances don't truly understand and support human behaviour, they are just more comfortable to use. So, the state we are at today is just the beginning of a greater path towards the automation of the world we live in today, a world with great potential for growth, be it in our personal lives, or in our society. Building automation for homes and offices helps with supervising

DOI: 10.4018/978-1-5225-9246-4.ch006

and controlling human environments, aiming to help people live a safe, secure and comfortable life. We can live in a world where our impulses and desires are interpreted and understood by machines and carried out to the best degree of their ability. That is what would make the appliances truly intelligent.

Smart homes and office technology, generally alluded as home automation, provides buyers much needed safety, comfort, convenience and energy efficiency by allowing them to control intelligent appliances, often by an app on the phone or some other device connected to the same system.

The basic principle of all the automation remains the same as it has ever been for all the companies at the forefront of the field - providing services for consumers in smart and practical ways. Investing in where you live is always deemed as a shrewd investment, but the question of home automation is not all about only comfort, it's also about elevating humanity while exploring the future of the world. The increasing trend of us burgeoning comfort with technology gives researchers a medium to once again explore these facets of development with a much wider audience at the receiving end. So, in this chapter, we will attempt to understand the purpose of smart homes and offices through the perspective of people on both ends of the spectrum, the creators and the consumers.

Although the primary concerns here is about comfort, but not the only one. Smart homes can be beneficial for people with busy lives. Houses with a whole lot of appliances working together in a synchronous manner make the day to day chores easier for everyone but the parallel nature of these task saves time and can provide us with energy to focus and prioritize our work efficiently. Used correctly, these can provide us with peace of mind, as well as time to pursue more fulfilling endeavours. Similarly, when used in offices, automation can save us a lot of manpower and money, and be very beneficial for small businesses. Buyers nowadays are more open than ever to the concept of smart homes but doubts about technology controlling their lives can cloud their judgment and that is why they needed to be convinced of the benefits these homes can have in their lives. This is one of the other things to be discussed in this chapter, the technological and ethical challenges to be faced in the search for a truly smart home, and precautions and remedies to be taken to minimize these risks.

As there are various interlinked appliances, virtual assistants, and ample amount sensors that one installs in their modern SHO, but does these "smart" devices really are "smart" enough for today's market standards where things are changing at such a fast pace. It's perhaps best then to think of today's smart home and offices (SHO) as a remote area where things are in our control, but our SHO does not stand as a smart entity in itself. If you have to manually control certain things just because it requires assistance from outside world, we cannot say the SHO smart enough then it doesn't matter how much well connected your house is inside. So technologies that can make the whole SHO work as. Smart entity is much needed.

In Recent years have seen great progress in the field of human activity recognition which deals with observing and interpreting the actions of humans to help with the process of monitoring and fulfilling their needs.

The question of why we even need smart Homes and offices should be discussed along with question that are SHo only for luxury or they provide something solid, or they improve something or not. we will see in this chapter how SHO can be beneficial and are better version ordinary housing in every aspect. The **motive** of the chapter is to provide a holistic knowledge of smart homes and offices how IOT and different components of SHO helps an ordinary to transform into a SHO.

Some basic importance smart homes and offices are as follows:

1. Promotes technology-based design and development of things which were used by mechanical means for a long time.
2. Provides analytical framework for sectors like security, robotics etc over non analytical

Framework of ordinary homes.

3. Makes electric devices more efficient and incorporates devices with more control, data sharing and Artificial intelligence.
4. generates evidence-based recommendation for future design, development and research and provides policy on smart Home technologies.

We will talk about different appliances and their application along with the technology involved in this chapter. The role of IOT and AI in home automation is also discussed in detail. Various other aspects like cloud and personalization are also touched upon in this chapter.

History

The definition of smart changes with time and technology the smart of 10 years ago wouldn't be considered as smart today. Here we inspect briefly how the smart developed over ages. The very first move towards automation in homes started with the inception of idea of providing electricity to the masses and this very small change concludes all the development in the modern era. It later turned into something huge in early twentieth century with the invention new Home appliances that gave the buyers a cause to invest money in something like never before and industries rapidly invest in providing these services to the small group of consumers present at that time. Northern America saw the invention of washing machine and vacuum cleaners which were big things for home owners. Mid 20th century saw a real move in Smart Home industry when alert-based House appliances for elderly and children were sold with a great success, including Fall alarm for elderly and premise alarm, at the same time the closed circuits televisions (CCTV) were invented in Nazi Germany which are a major part of the security sector till today. The commercial sale of CCTV changed the whole arena of crime detection. Early 2000s marks for affordable Smart Homes and Offices. Current emphasis is on the interactivity between humans and appliances and the connectivity between different appliances. Air conditioning have evolved so much in recent times, automated air conditioning systems are now visible in every developing as well as developed nation. The technology of SHO are changing and can be categorized in different generations, the three generation of Home automation can be labelled as:

1. Wireless technology
2. Home automation
3. Robo buddy

As our definition of smart changes, smart appliances with new technologies keep coming in market that are easier to use and more efficient than ever.

Figure 1. Appliances in different part of smart homes
Source: https://www.moseleyelectronics.com/2017/04/27/10-of-the-most-exciting-smart-home-appliances-in-2017/smart-home-appliance-features/

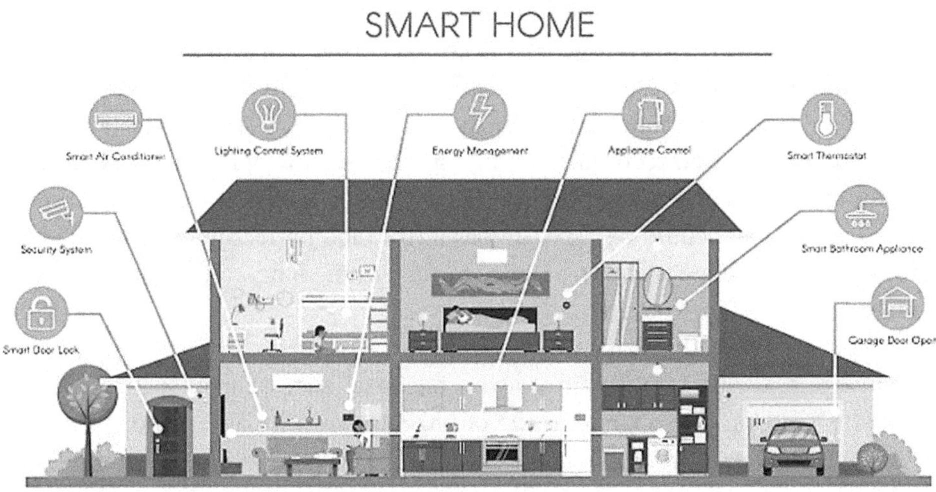

APPLIANCES AND TECHNOLOGY

The appliance themselves constitutes a huge chunk in the "Smart" of Smart Homes and Offices so it is very crucial to inspect and analyze them closely. With time many utility appliances have been adding on to the list giving our experience whole new definition. The amount of appliances exposed to public today was never a case in history and with rise in competition better and affordable ones are easier to access. These appliances ranges from security, ventilation, air conditioning, health and temperature monitoring to virtual assistance. The need for these equipment has also grown with the increase in the knowledge of surrounding and the urge to live life more comfortably than we live earlier. Different part of a smart home has different types of appliances as shown in Fig.1 which shows the variety of appliances in the market today, each appliance has different job to do which when connected with other devices builds a smart ecosystem.

Security Control

Security control and visual surveillance started to gain popularity in late twentieth century and soon became a major player in security related matters in offices and later at Homes, CCTV and access control are real game changers, the development in fingerprint scanner gives genesis to a market of personalized security where later technologies like retina scanner made new strides in the whole security business. Different aspects of security control are discussed in this section with focus on inhouse security.

CCTVs and Access Control

The CCTVs are used to record raw video footage to know about any manoeuvre against the owner or the building, these come handy for analysis of an environment as well as proof against perpetrator. The development of digital multiplexing (Kumar and Svensson, 2015) (Dempsey, 2010) (when multiple digital signals are combined into one signal over a shared medium the process is called digital multiplexing) makes it easier to record with several cameras as well as adds additional features time lapse and motion only recording. Workplaces are perfect for installing security cameras as it is a common experience that Homes and Offices installed with security cameras are less prone to burglary or incidence of violence. A 2009 paper by the researchers of Northeastern University and University of Cambridge showed that crime in general has reduced in the areas with security cameras by 16% (Piza). The new wave in artificial intelligence has affected the video recording but in a good way, now with Machine learning technologies like CNN (convolution neural network) classifiers we are now able to train the computer attached to security systems about what could be classified as a risk, and by recognizing the danger like fire or arms it could promptly alert the concerning figure to take immediate actions and will also contribute in saving man labour.

Access controllers are basically various kind of sensors which can be used to recognize fingerprint and eye retina. Fingerprint scanner are optical scanners or capacitance scanner (Fatima, 2011) which basically takes an inverted and black and white image of one's finger print and compares with prints stored in the system whereas a retina scanner scans the unique pattern of blood vessels on retina. This allows us to secure the premise as well as make a more personalized approach to distribution of the authority. Access controls are now used to mark the presence of employees in industries which makes it easier to maintain the database of attendance and get rid of hectic daily routine of attendance.

Firewall and Device Security

All the connected devises are prone to the viruses and spywares. Since a great amount of people are working from home a security firewall defending the systems and personal files is necessary. According to Pew Research one in every five people in USA has more than 10 connected devices. As the total number of devices are increasing the chance of security breach also increases. Firewall in smartphones, other connected device as well as devices built in SHO prevent such security breaches.

Devices with simple implementation or single purpose are more prone to attack as compared to devices with more functions and much evolved security software, still there many challenges in security of SHO some of major challenges are as follows.

As Security is not a prime concern for device makers, many device makers outsource component and user interface, which makes it difficult to track device security measures in the chain of supply. So external software or even awareness can be enough for security as investing time and money in every device is costly and hectic. Device makers usually have a code base across devices, and it is difficult to make changes and test them.

So the basic solution is to integrate firewall, VPN and antivirus in our Smart Homes and Offices. These are necessary for security within SHO, though these security solutions comes with some problems but are enough to cover your house against ordinary attacks and security breach.

Virtual Private Network (VPN)

A Virtual Private Network (VPN) creates a private network for devices across the internet. A common use is for employees to securely connect to their organization's intranet from home. Many activists also use it to share data that is prohibited by some authority wielding organization. (Mason, 2002)

A VPN creates an encrypted point-to-point tunnel so data can be securely transferred. Whenever one access sensitive information, they would never want that data keeps floating, unencrypted, around the public Wi-Fi near your home or in office, where it could be intercepted by anyone. So, you connect your device over a VPN and it keeps your data safe.

So every device must have a VPN service but that's not possible because not every device support VPN, next frontier to conquer should be making every device digital at low cost so that they can have VPN service inbuilt in it.

A firewall is a layer of security with a set of security rules (Boudriga, 2009) which monitor and control inbound/outbound traffic on a network. Many routers today come with pretty slack firewall settings, and allow all out traffic.

The responsibility of SHO owner increases while maintaining a firewall it has no default or automated setting which can reduce the burden of user, this adds to the problem of slack security of firewall. Logistically, the owner needs to manually make settings in place for connected devices.

Antivirus Software

Antivirus being most popular among the three, can manage many tasks on PC and mobile phones it can protect them from spyware, malware, viruses, worms etc. Antivirus comes with varied protection layer therefore it can even replace firewall in PCs.

In future when attacks on devices like smartwatches will increase antiviruses can counter them too.

Temperature Control

The degree hotness and coldness are controlled as per need in houses and offices to counter extreme weathers, some of these control systems includes centralized temperature control for whole office or non-centralised temperature regulator. These systems can be manual or automated depending upon the use. Refrigerator, Air conditioner, geysers, Heaters are the usual ones in use. 20th Century marks first use of Automatic Temperature Control Systems. Warren S. Johnson was the man behind the idea which occurred to him while he was teaching at a School in Oklahoma. Before then all that was a tedious process where janitors had to switch the dampers in the basement manually. This was how automated temperature control systems were thought off to make our work easier. Now these automated temperature controllers are widely used in cold regions. Since certain levels of temperature can be dangerous especially in extreme whether therefore these systems nowadays come with safety applications like alarm system, child lock, remote controllers and automatic switching etc.

A thermostat is a component which senses the temperature of a physical system or environment and performs actions to maintain the temperature at a set point. (Wlezien, 1995) Users usually demand for ambient temperature after a while when conditions of the surrounding changes so the need for automation of thermostat was felt, and therefore automated smart thermostat were developed which senses the ambient temperature for the user and can provide both heating and cooling without the interference of user.

Smart Appliances

Smart appliances do something what we call "smart", either they communicate with smart phones and tablets, are energy efficient, allow us to control them remotely or perform some task without human intervention while providing insights which can be beneficial to the users. These devices communicate to the same network while responding to the remote controllers, they also have alarms that reminds the consumers to use the devices in off peak time. Smart appliances for Homes and kitchen are becoming part of people's day to day life.

Smart plugs and switches, security cameras, smart locks, automated thermostat, smart leak detector, fire detector, robot vacuum, voice assistance gadget are part of reality that didn't exist 20 year back, it shows us how fast technology is changing and with the level of automation and intelligent gadgets many things are possible now. Smart lighting technologies are often used in buildings and offices and are proving to be cost effective since LED bulbs are lot greener and nowadays are integrated with sensors to detect presence of Humans or animals in the room, it also decreases the headache of switching on and off the lights again and again. Fire detector sense smoke and therefore are beneficial inside Homes and offices but didn't tell us about fire in premises but with security cameras working on neural network technology can sense and set alarm in the premises which can save life and money. Security cameras uses self-learning behaviour analytics and looks at the entire image while using statistical models which helps it to differentiate between normal and abnormal scenarios. (Dufour, 2012) Applications of smart appliances in buildings with security systems integrated within a home automation system is full package. Surveillance through security cameras at different corners of the house can be enhanced by combining with sensors and scanners which helps in locking of doors and windows through access control and central locking systems, it takes the security to another level. Appliances like Leak detector, smoke and carbon ooxide(which is deadly on prolonged inhalation) detectors (Holcombe, 2014) (Geiger and Matko, 2003) are really beneficial for small scale disaster that happens in SHO, these Indoor positioning systems (IPS) detects water leakage which may be symptoms of fouled up boilers or geysers whereas smoke and CO detection can life saver at times. Occupancy-aware control system uses the intelligent meters (Jin and Spanos, 2017) and domain sensors like CO_2 sensors and senses the occupancy inside the Homes and offices, (Jin et. al., 2018) which when integrated with the building automation system triggers an automatic response which is energy efficient and reduces risk.

There are special appliances for Home automation for baby care the and elder people which tracks their movement and alarms the guardians when these people are in vulnerable situation. (Mcdonald, 2010) Air quality controllers are another important device which helps us to map the pollution level inside house and you can be loaded with additional air filters. virtual and voice assistances are the next-gen technologies that assists us and are of great importance, they can be used to control and schedule our instant coffee machines, microwaves, refrigerator and multi cooker, and Instant Pot and Alexa, Google Home are some of the examples and enhances the experience of users. Robotic vacuum cleaner are another major smart appliances which uses some of the machine learning technique (Staugaard, 1987) that is used in self driving cars and yield great result in terms of cleaning a room or kitchen. Motion sensors are used in SHO pretty often and many appliances also uses IFTT (Ovadia, 2014) (it is used to automize web based jobs) applications which is be discussed later. Furnitures with hydraulics to support different movements are also gaining popularity in SHO these furniture increases the comfort and efficiency of employees and are loaded with other appliances for examples a television integrated within a bed in homes. IOT can do a lot as we have the new smart furnitures which would include smart desks in your office. These desks would have the capability to understand the user's preferences and adjust to it automatically.

Entertainment Control

Entertainment Control is something that is contributed to rise of smart phones and smart TVs it comprises all the major streaming services like Amazon prime, Netflix, Hulu, YouTube. These services in company with fast internet connection and enhanced interconnectivity due to Bluetooth and hotspot give a surreal experience. The personalisation of entertainment makes it more focused and intimate to an individual. These services can be provided with USB sticks of service providers and as it is smarter than the average system it can handle more than just entertainment. As Entertainment is matter of luxury therefore a thermostat, a home theatre and security locks can provide an uninterrupted environment for enjoyment. It just requires a strong network which is the foundation of any connected home. Fast speed is much needed because of high demand, it includes Package networking which is purposely built for better performance and incredibly fast network. Wireless devices provide comfort and exclude all the wired mess therefore Wireless devices are more preferred in today's time, every connected device that is brought in SHO are now more assured of their performance because of increase in high speed and robust devices.

IT ALL CONNECTS

Smart phones, smart watches and virtual reality devices are intelligent have data-driven technology, they are responsible for altering many aspects of Human Life. These technologies allows these devices not only to interact with users but also with other devices and this interconnection makes this whole environment much more exciting. You can switch on your fan by giving command to your personal assistance, the Fitbit band on your wrist can interact with your Smartphone via Bluetooth or Wi-Fi connection and can track your heath. The latest Apple Watch enables you to make calls and message back and forth to anyone, and some of the gadgets are really contributing to the genesis of the 'smart' revolution in the IOT and smart device application development sector. The real potential of automobiles and different gadgets lies in the interconnectivity between them. The changing scenario of connected devices and their mobile application is contributing to next big leap that we are witnessing in the first quarter of 21st century, much more evolved interfaces are making things easier to understand and are taking this individual process to the enterprise level. There are several devices like hue light bulbs and Google's Nest thermostat which builds a next-gen ecosystem especially when bulbs are smart enough to sense the brightness inside the office and a smart thermostat that is controlled by a remote control.

IOT is a gateway for smooth integration of several devices together, these could be equipped with *sensing*, *identification*, *processing*, *communication* and *networking* capabilities. It is such an evolved cyberphysical space, it surely opens the window for a whole new environment for learning from such devices and evolving as a tech savvy society. IOT is doing such a great work because of the fast and integrated developments in IOT market and industrial complex that produces these devices with high efficiency while maintaining the international standards. (Yang and Wang, 2018) Network device control and the management of how to manufacture equipment along with controlled assets and handle situations of engineering operations is the key in producing IOT devices. Manufacture Control has brought these devices under the umbrella of global applications and integrated production. The robust systems enable fast paced production of new gadgets, dynamic feedback of products in demand, and real-time optimization in manufacture process and a better demand-production networks, with the network of machineries and robust system. (Severi et. al., 2018)

IOT system architecture and framework comprises 3 different levels

Level 1: devices, Level 2: the Edge Gateway and Level 3: cloud connectivity

Level 1 of the IOT architecture comprises of networked devices, sensors and actuators, these are usual parts which helps in fetching the information about area possible, from the IOT equipment, which use a basic connectivity protocols to connect to an interface that channels the information to an appropriate application. Level 2 includes sensor data clustered into the systems called Edge Gateways that provide components such as pre-processing of the data, establishing connectivity to cloud, using different systems such as WebSocket or event hub etc. Level 3 includes the providing cloud connectivity and similar built platforms for IIOT using other service architectures, which are available in multiple higher coding languages and are inherently secure in nature as they use HTTPS/OAuth request system. This level also stores the sensor data using various database-using, query language system, for instance Cassandra or Postgres but different languages are used in accordance of requirement. This data is helpful for analysis and could be used for predictive requirement with the use of deep learning (Sethi and Sarangi, 2017).

Another environment that takes the advantage of Internet of things and builds on it is the web of things, it is an architecture that takes Internet of things as an application layer while it looks for converging the data from IOT based appliances into Web apps, it creates a creative model that visibly uses IOT. To facilitate programming in the device chip and increase the control over the movement of data in the Internet of things, a sophisticated architectural design is taken for web of things. (Guinard, Mattern, and Wilde, 2011) (Guinard and Trifa, 2016)

Bluetooth mesh networking (Frodigh, Johansson and Larsson, 2000) adopted in July 2017 (Kinney, 2003) is a network-based set of rules that utilizes Bluetooth Low Energy technology that allows for many-to-many(M2M) device communication over Bluetooth radio.

Human-Machine Interface

Home Automation based on IOT allows users to utilize a Smart Home based on interconnected devices in direct interaction with the users. The SHO nowadays are interconnected via internet and homeowners now controls SHO smart appliances with direct or indirect control. The user command over the internet is facilitated over the Wi-Fi modems which regulate tasks by directly interacting with the device Microcontroller. Digitalization has made it very easy to know what device is being used and the small screen in our smart devices usually notifies us all the time about the job being done, all this comprises of what you call a general use IOT based Home Automation system. The SHO market has get its wings since the smart devices are getting cheaper and cheaper and the people are understanding the benefits of a SHO. And from the perspective of smart Homes developer planning it all over the city seems the next logical step but for the smart homes sounds much intimate and risk free, and therefore because of higher level of Human-machine interaction and therefore people are investing to acquire SHO.

The thing that makes Home Automation system more beautiful is that you can configure each and every detail and pretty much personalize and control all your devices from any remote-control device, and this is prime example of how things are getting smarter as they cut the unnecessary human efforts. SHO IOT devices do minimize effective costs and conserve energy. The Home Automation segment includes intelligent lighting, smart TVs and other appliances. Wearable's (Smart Watch, fitness brands) which are always in vicinity of user are prime example of devices with human machine interface. So what's the secret that makes this whole system work? The answer is the applications that provide a suitable interface between human and machines, platforms like IOS and Android are investing a lot to make these experiences as smooth as possible. Today in India, nearly 25 percent uses smart phones with the number is estimated to grow over 60 percent between 2017 and 2022 (Poushter, 2016) which clearly indicates that there is immense opportunity for increased adoption of such technologies. The upgradation of the Home automation market will happen with few key changes in the Automation technology. Lowering prices of Wireless gadget solutions contribute to the acceptance of the Home automation usage in larger volumes. The fast pace evolution of home-based automations and M2M (machine-to-machine) communications is creating hundreds of millions of smart entities in the internet domain. This M2M service layer will provide many services such as services like data transport, devices management, security and device discovery in a coordinated manner. (Abraham, 2016)

Personalisation

The integration of IT-based service with IOT based devices has changed the product interaction with users to a whole new level, the use of IOT is a key factor in smart Home environment. So the scope of the product in context of user experience have been expanding.

It is usually quite hard to fulfil dynamic customer. We are not going to provide all the instantaneous support to the users just by innovating in the device alone, engineers are working from every perspective to provide a support system for devices to connect not just in end to end fashion but in multi-phase manner. As the products are multi-disciplinary in nature, the user can become more involved in understanding the interrelationship between network domains and device domain and it will only help the user to be more efficient in personalizing his curriculum.

Cloud provides very robust system for personalizing the database, as the cloud systems available can carry a lot of memory today, so whatever we do our personal setting are not going anywhere and they will keep benefitting us till the end.

All devices are personalized, they keep our data and settings safe which we store on the internet. Cloud provides personal internet spaces for storing and even executing several applications. Cloud services are provided by the servers carrying a large memory, these servers also do data analytics if one opts for it. These not only helps in personal data storage but also helps small enterprise to run without thinking about the infrastructure requirement. Many small offices buy these server outlets to themselves grow as it can host website, provide powerful systems for computation and platforms required for specific jobs. Cloud services are mid-level services in smart home environment.

These enterprise application will only be useful for uses as they provide services which have more value than cost. This has given rise to more solution catering than real physical services or device manufacturing. The smart service provides combination of hardware and software. For instance, Engineer A is designing a next-gen service powered by smart phones targeted for people using automated gate opening system for vehicles registered in system. He will need an integrated device to capture the plates and physical appearance of the vehicle analyse the data and do the action, now this can be remotely controlled or done with help of scanners and computers. After discussing about cloud we get an idea how smart phones are connected to clouds which is connected to devices in smart homes, Fig.2 depicts a basic connection of clouds enabling IoT and AI to function.

Data analytics and Artificial intelligence is discussed later in the chapter.

Figure 2. Interconnectivity of smart homes devices and cloud support(for personalization)
Source: https://www.appliancedesign.com/articles/94937-consumers-want-more-than-just-connected-appliances-they-want-connected-services

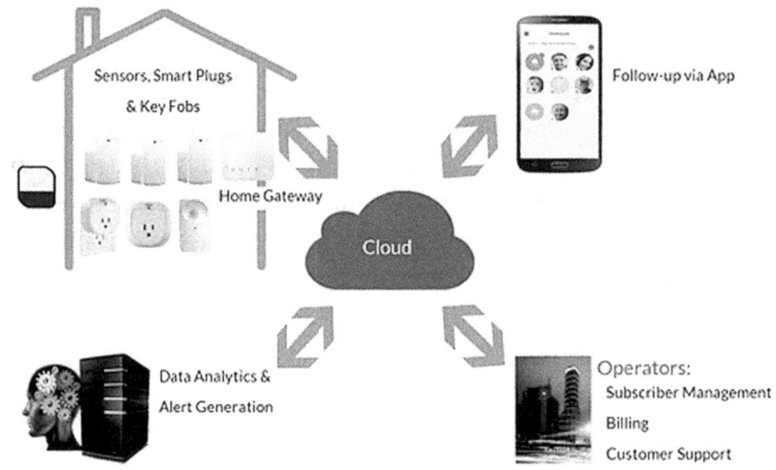

CHANGES AND TECHNOLOGIES

We are facing rapid changes in technology and upgradation of functionality of smart devices coming in market every now and then. The technologies are coming with primary objective of providing comfort for users, they just cut much of our labour and save a lot of time and money. As changes are inevitable some of the changes that are really beneficial to us are discussed below.

Emerging Technologies

The Next-Gen Security

The security in next generation will be only burdened on CCTVs but technologies like smart recognition of face, smoke, locking systems would also accompany the pre-existing system and this will save us from all the security compromises that we have been facing for ages. Smart systems such as child lock are game changer in security especially in automated threat detection. SHO depending on IOT when accompanied with newer technology can be highly beneficial. (Piza, n.d.)

A Greater Role for Artificial Intelligence

AI is revolutionary and it proves to go well with any technology so buying appliances with AI support is certainly beneficial, now dedicated processor chips for AI in phones are improving picture capturing, cleaning and gaming. Play stations also comes with AI acceleration amplifying the gaming experience.

Homeowner Data Sharing

Data sharing with appliances is very useful as this data can be used to track and know your surroundings to a satisfactory level without giving the owner a headache about observing and handing them manually. IOT is a major player for data sharing; low energy Bluetooth sharing makes it possible for us to implement IOT in smaller devices like wrist watches, and then these watches keep the track of information like how much you walked the entire day. The activity within your SHO can be managed by one click of your smart phones and that's exciting to even think about. Cloud is set to revolutionize the home innovation technology to a whole new level. Technology with shared information will become much more efficient, and we will be able to personalize everything with secure cloud technology which gives us better control over targeted applications like increasing television volume from another room. We eventually won't need to manually handle any devices in SHO as they will be able to automatically adjust to our preferences.

Customer Service

The application troubleshooting, appliance management and costumer services can also be automated, they can be inbuilt in devices personal assistance and GPS navigation being the example. Personal assistance when inbuilt in SHO can increase efficiency of user and with proper authority control people can do much of their tasks more comfortably.

Higher Cross-Compatibility Standards

There is high hope for eminent progress on standards. The SHO market has huge potential, but at the moment it's too fragmented. Consumers still have to think them investing in Nest, Amazon's Echo line or products that support Apple's Homekit would yield good result or not. we expect to see greater cross-compatibility and less focus on platform lock-in.

Intelligent Kitchen Gadgets

As more smart kitchen gadgets are coming to market, like multi-cookers which can interacts to virtual assistance, smart crackpots with integrated apps working in kitchen is not that tiring anymore. We can ask Alexa about the status of these devices and can even control them from our smartphones at work. Families benefit a lot with kitchen automation and it is more important for certain group of people than anything else.

Smart Spaces Outside of the Home

Smart app for offices in buildings are future where everything that can be automated or controlled with smartphones will be at our fingertip. Water in tanks can be regulated with sensors and smartphones, air pressure and quality can be regulated in closed offices. Smart buildings are consequence of smarthomes and offices. Parking lot can be replete with sensors alarming for theft and parking related issues. Beacons are installed in free spaces outside SHO to achieve voluntary actions by machines. (Basavaraju, n.d.)

Table 1. Technologies and their impact

On Global Basis	Internet of things (IOT)	AI and cognitive	Augmented and virtual reality.	Hybrid wireless
On Industry Basis	Cloud native application platform.	Real time interaction management.	IOT analytics and spatial analytics	Edge computing.

In 2015 Basavaraju presented a research paper on how smart parking lot can be effective with Internet controlled system.

Upgradation of Existing Applications

New computer language can be designed especially for combination technologies like IOT, AI and many more SHO related concepts to increase efficiency of SHO ecosystems. The updates in existing application to increase efficiency should be welcomed by companies and users. A safer platform with upgraded application will enhance the whole experience of living in smart homes and offices.

Increase in Voice Control Integration

Some technologies are going to change the whole scene and one of these biggest breakthrough is virtual assistance. Sound recognition technologies that have been integrated in our phones, TVs, home audio system and even cars have changed things a lot but now they can be turned into voice control just like remote control. Virtual assistance is going to be the breakthrough advancement that we surely need and is a boon to user's comfort.

Some technologies are implemented on industry level because the industry requires these technologies these are not broad enough to be implemented as for many purposes rather they provide specific functions to be use. Some technologies are broad enough to be used in industries targeting on different fields and application these are used on global basis table.1 shows the technologies used on industry and global level.

Artificial Intelligence

Artificial intelligence is everywhere nowadays and importance of AI is increasing day by day. As the data generated by the users is in huge quantity it could be used for analytics and predictive work, it helps to automize a lot of devices, extract information from raw data and cognitive AI even helps for reasoning and understanding things at higher level.

So, the question is where AI is implemented and what are its effect on our daily life in context of smart Homes and offices.

Few examples of the applications that are given below are the ones that we use nowadays, and they have been selected as the classic examples for broader trajectory of applications of AI. The examples are intended to avoid overly reserved applications of IOT, or that applications that don't include AI in any form of IOT. The following imply how merger of both, AI and IOT are useful in ways to express the smart devices segment with more strong data usage.

Figure 3. Robot vacuum cleaner

1. Automated vacuum cleaners, like iRobot Roomba. IRobot successfully developed automated vacuum machines for commercial uses in 2002. Established by MIT robotics, the organization has made its revolutionary puck shaped vacuum cleaner that can sense whatever ahead of it can clean the by vacuum cleaner technique and is efficient due to its extra ordinary movement pattern, and most amazing thing is that it can dock itself for charging. It learns and maps house with time and makes cleaning more reliable as the time passes by as it gathers the data and process it to find most efficient way to clean the SHO and makes most out of it. As a matter of fact these appliances are not as main-stream or aren't as popular as major player in AI market such as Facebook's facial recognition or Apple's *Siri*. So it is a prime example of AI embedded robot cum vacuum cleaner.
2. Smart thermostat solutions. Smart Homes solutions are major stake holder in AI industry. The big player like Nest and Ecobee4 Alexa are changing the game. The concept of smart Thermostat makes sense as people are required to manually do a lot with a simple thermostat whereas a whole lot of automation is implemented with smart thermostat which is smart enough to know what should be the temperature at a particular time. Thermostat is now smart enough to switch off itself when people are sleeping thus cutting cost to a significant level.

As a smart gadget, Nest's clean digital interface is a refreshing innovation that differs wieldy from the manual dialing and the smartphone integration allows for a remote and robust temperature control structure. This is AI in principle, and it made many companies to follow the success story and integrate AI everywhere it could be implemented.

In terms of artificial intelligence application, these thermostat devices "learn" the default temperature preferences of the users which grows over time, it also figure out the schedule of the users about how and when he uses certain devices, it turns down energy use when it sees necessary, it is possible due to sensors. This AI application is certainly beneficial, but its most important benefit would be the increase in comfort level and cutting of cost by switching off the thermostat when ambient temperature is achieved.

Potential Future applications for AI-Powered IOT Devices include the following examples, these are still in use but the use is still at lower scale.

Today's Smart IOT applications are useful in understanding trends and understanding user demands, and are paving a path to the expanse where big-company and venture money is already moving. ("Dept.

Figure 4. Smart thermostat

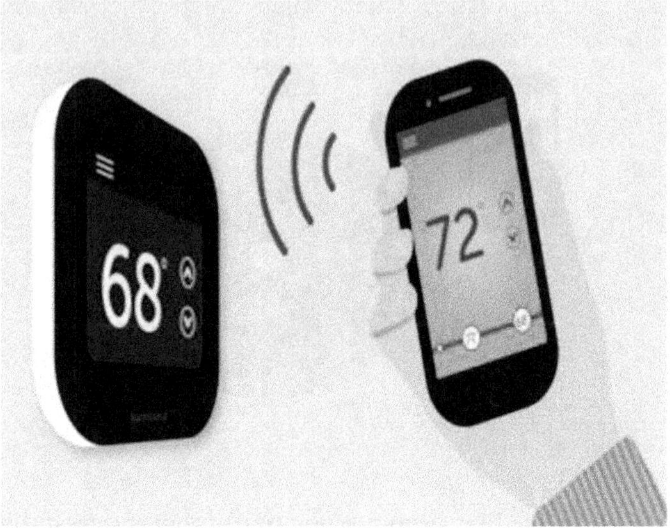

Figure 5. Google Home

of Electronics & Telecommunication", n.d.) However, autos and vacuums are only the tip of iceberg for potential of IOT+ AI applications: Security and access devices in terms of purely AI applications can be a juggernaut in AI market; companies are already making strides in technologies like automated door unlocking by recognising person. Artificial intelligence is kind of technology that can be employ a certain amount of people and they could be well under a thousand people. AI is also useful in detecting fire in the SHO premises where smoke detection is a bit tricky due to open environment.

Emotional analysis, facial recognition are not only valuable assets for Smartphone but also for SHO, these are used for security industry and luxuries like lighting as per mood. (Owayjan et, al., 2015) and are coming out as big game changers from surveillance to marketing, it seems safe to say that its applications haven't been tapped to its fullest. Many Smartphones companies as well as SHO security companies are making efforts to use it to its fullest. With a camera on nearly every computer and smartphone, garnering information from consumer reactions to products and marketing has probably never been easier. Facebook's auto-tagging is probably the most popular example of face detection and recognition while other business models are still to be fleshed out and as per some of the leading magazines and journals it is predicted to be a force to reckon with in the coming future. (Schroeder, 1998) (Mana, et. al., 2016)

With a basic understanding of use-cases, trends, and predictions this intersection of AI and the internet of things is going to be enormous, it'll help for further research. Bear in mind that "IOT" and "AI" are itself very broad sectors of studies and should be considered as concepts rather specific fields. Some general definitions related to topics are described below:

Machine learning: When a Machine learns to perform a task without human intervention from the experiences of humans or machines in the form of data points, this process is called machine learning (Lasserre, Bishop, and Minka, 2006).

Deep learning: when multiple networks share data within themselves each helping other layer to come up with a weighted computational model that can mimic, understand and produce a task it is known as Deep learning, it is a subset of AI just like machine learning (Bengio, Courville and Vincent, 2013).

Smart objects: interaction between humans and virtual entities or between multiple virtual objects are facilitated by computational model known as smart objects (Wikipedia)

STANDARDIZATION

As the IOT market is rapidly expanding it has caused sudden outburst in the IOT sector which is in itself not so surprising because of the technology advancement. Also, big firms are diverting their money and energy to help and push small start-ups in this field to come up with innovative solutions. Hence, the industry mainly focusses on manufacturing universally acceptable hardware to enable the basic solutions these device performs wherever they are being used. Currently, most IOT solution providers just assemble all the components top of the stack from the hardware bunch as they are somehow relevant to the cloud services, they like to call it as "IOT solutions''. Consequently, the grave problem towards this industry used to be presence of irregularity and quality especially in software sector.

Since the market is expanding, the demand for quality in interface and hardware to perform simple IOT backend tasks has risen, and the companies are also catching up for example, processing, storage, and firmware updates, are becoming more common. In that surviving paradigm, it is easy to see various IOT solutions work with the same backend services, and it will achieve certain aspects which are not achievable by current IOT generation such as greater level of interoperability, portability and manageability (Bandyopadhyay and Sen, 2011).

There are challenges that slow the whole process of standardization and implementation of IOT solutions and constant efforts are being made to make a better and more sustainable system.

Table 2. categories in challenges in standardization of IOT related devices and software

Platform	Connectivity	Analytical Applications	Business Model
Platform includes various form and design of the User interface (UI) which provides functions and instruments used to handle the enormous information sharing between all products in a much secure manner, and scalability which means wide adoption of protocols are also provided with some protocols like proIPv6 in all vertical and horizontal markets, which is quite needed for the industry.	The product running over multiple platforms needs similar holding hardware and drivers for connecting network. Therefore universalism in hardware and software is introduced, prime example being USB and Unicode.	The main processes through which every industry have to pass also applies to here, the analytical measures are to **collect, pre-process and analyse** the data. So these IOT analytical application are quite useful as the processed data is always useful to figure out the most efficient way to do certain jobs.	The main objective to start a business is to generate profit and to generate profit you must have a sound model as investing and exploring consumes both time and money so to have a model is quite good, and any model leads to efficient environment without havoc Business model must prefer standardisation but not to a satisfactory extent.

IOT STANDARDIZATION

These standardization challenges can be divided in these **4 categories.** The uncertainty in standardization in IoT is evident due to different platforms, hardwares, applications and business model. This vividness in the approach have to be reduced on industry level first to so that the consumers don't have to think about the type of appliances before buying it. Broad categories of challenges to standardization are provided in Table.2.

These categories are complimentary, one should have implemented all namely *platform, connectivity, Analytical Applications, Business Model* to make it working simultaneously (Ferraro, and Aktihanoglu, 2011). Not implimenting one of the category will affect the whole process adversely. Great amount of efforts are required to get it going, since numerous organizations are trying hard to make all the standards applicable in these categories as a parameter for their work schedule, still it will be quite hard to make it all happen and to accept an unifying model that will be considered always.

IOT IMPLEMENTATION

Another part of the model is implementations of IOT; implementing IOT is not an easy process, there are a lot of variable that comes into play and some of the key components and factors will discussed here. Five components of IOT Implementation are: Sensors, Networks, Standards, Smart Analysis, and Intelligent Actions (Mingjun et. al., 2012).

- **Sensors:** The two types of sensors are: active sensors & passive sensors. Since sensors in IOT today are cheaper, smarter and smaller and can be fitted in lights, fans, taps, doors etc therefore are more efficient. But more work has to be done in areas of security inter-operability (Yan, 2015).
- **Networks:** Next process in the implementation is to collect the signals transmitted by sensors over networks with various types of a standard network hardware including routers, bridges in different topologies of buses. Different technologies can be used for connection of different parts of networks to the sensors which includes Wi-Fi, Bluetooth, Low Power Wi-Fi, Ethernet, Long Term Evolution (LTE) and the recent Chinese technology of Li-Fi (that uses light as a medium to connect to different devices (Patil, 2015).

Figure 6. Different elements of IOT
Source: https://www.bbvaopenmind.com/en/iot-implementation-and-challenges/

- **Standards:** We have discussed standard earlier but this is different from that standardization, third stage in the implementation process includes the sum of all activities of handling, processing and storing the data collected from the sensors(and beacons). With the increase in aggregation level of data the scalability also increases. These are certainly beneficial for predictive analysis and can be successfully implemented in hydraulics like doors and windows of House. This tech is greener as it reduces costs of electricity by switching off lights when no one is there and provides more personalized approach for sensor utilisation.
  Smart Analysis is extraction of insight from the data produced, it is the Last phase in IOT implementation. Utilization of data is driven by *cognitive technologies* and the accompanying systems to facilitate them. As cognitive technology is at height of advancement out capabilities to process and cluster large amount of data has increased. As imagery and sound are also becoming hot market for innovation. Few factors responsible for adoption of intelligent analytics within the IOT are*;* artificial intelligence models, growth in crowd sourcing and open- source analytics software like tensorflow (Google deep learning model), real-time data collection and processing (Gubbi, Buvva, Marusic, and Palaniswami, 2013).
- **Intelligent Actions:** AS discussed earlier intelligent actions are broadly Machine to Machine (M2M) and Machine to Human (M2H) interfaces that are facilitated by smart objects for example with all the advancement in UI and UX technologies certain actions are more plausible as human don't have to invest their time in engaging with the machine and a lot of tasks abstracted from user and are done by the machine itself. These factors contributes to the increase in adaptability of intelligent actions within the SHO*;* lower prices of appliances, improved device operation, profit making as prediction of human behaviour gives advertisers benefits, and free deep Learning tools.

A LOOK INTO THE FUTURE

The use of AI+ IOT along with specialized appliances holds pretty strong future for SHO. The trend of increase in sale of smart appliances and more internet penetration confirms this notion. The culmination of SHO in near future will contribute to HDI of people in developing nations. The question of whether SHO are luxury or necessity also arises as most of the world population is under fed but it should be noted that the increase in efficiency of resources at a large scale can be helpful for diversion of power and resources in appropriate direction. The most connected environment are usually the most efficient ones. SHO being good at data sharing can be beneficial in more ways than just for which people initially started to accept them. SHO itself can act as a smart entity and connect itself to a local monitoring station which can be beneficial in dangerous scenario. From refrigerators to garages in houses, IOT is bringing more and more things into the digital fold every day and it is likely to make SHO a multi-trillion dollar industry in the near future. Possible outcome of successful standardization of IOT is the implementation of "IOT as a Service" technology for global audience. But it is a long way to finally achieve this dream; SHO enthusiasts need to overcome many obstacles and barriers at two fronts, consumers and businesses before we can harvest the fruits of such technology so the future looks good for SHO.

One of the key catalysts to the need for automation is the environmental challenge faced by us today. As climate change becomes a global issue, the wastage of electricity also becomes one, so making our appliances smart enough to adjust to the growing need for electricity and efficient consumption of energy is the need of the hour, and smart homes can be helpful for us to lead in right path. This is counter-intuitive to our assumptions that their embodiment of devices like sensors and monitors makes them less energy efficient. These joined with government home energy incentive programs, make the advantages of smart appliances more credible. Empowering consumers with this technology can encourage sustainable practices that lead to reductions in the national carbon footprint. The use of nonconventional energy methods is a key to this discussion and has to be focused on later.

CONCLUSION

We are facing rapid changes in technology. and we are witnessing upgradation of functionality of smart devices coming in market every day. The technologies are coming and we have to make most out of them as they can cut much of our labour and save a lot of time and money. As changes are inevitable some of the changes that we have discussed are going to change the whole way of living.

So many technologies are considered as boon and bane at the same time for us but SHO are slightly friendly than most, if we can counter the problem discussed earlier and create an ecosystem where smart Homes and offices becomes people's first choice. It can be said that it will lead us for a greener, safer and a more comfortable living condition as of now.

REFERENCES

Abraham, S. C. (2016). *Internet of Things (IoT) with Cloud Computing and Machine-to- Machine (M2M)*. Communication. doi:10.18535/ijetst/v3i09.13

Bandyopadhyay, D., & Sen, J. (2011). Internet of things: Applications and challenges in technology and standardization. *Wireless Personal Communications*, *58*(1), 49–69. doi:10.100711277-011-0288-5

Bengio, Y., Courville, A., & Vincent, P. (2013). Representation learning: A review and new perspectives. *IEEE Transactions on Pattern Analysis and Machine Intelligence*, *35*(8), 1798–1828. doi:10.1109/TPAMI.2013.50 PMID:23787338

Boudriga, N. (2009). Security of mobile communications. Boca Raton, FL: Auerbach Publications. doi:10.1201/9780849379420

Dempsey, J. S. (2010). Introduction to private security. Boston, MA: Cengage Learning.

Dufour, J. Y. (Ed.). (2012). *Intelligent video surveillance systems*. Hoboken, NJ: John Wiley & Sons. doi:10.1002/9781118577851

Fatima, A. (2011). E-Banking Security Issues-Is There A Solution in Biometrics? *Journal of Internet Banking and Commerce*, *16*(2), 1.

Ferraro, R., & Aktihanoglu, M. (2011). *Location-aware applications*. Shelter Island, NY: Manning Publications. Retrieved from https://www.linkedin.com/pulse/what-next-iot-ahmed-banafa?trk=mp-author-card

Frodigh, M., Johansson, P., & Larsson, P. (2000). Wireless ad hoc networking: the art of networking without a network. *Ericsson review, 4*(4), 249.

Geiger, G., Werner, T., & Matko, D. (2003): Leak Detection and Locating – A Survey. *35th Annual PSIG Meeting*, Bern, Switzerland. Pipeline Simulation Interest Group.

Gubbi, J., Buyya, R., Marusic, S., & Palaniswami, M. (2013). Internet of Things (IoT): A vision, architectural elements, and future directions. *Future Generation Computer Systems*, *29*(7), 1645–1660. doi:10.1016/j.future.2013.01.010

Guinard, D., & Trifa, V. (2016). Building the web of things: with examples in node. js and raspberry pi. Shelter Island, NY: Manning Publications

Guinard, D., Trifa, V., Mattern, F., & Wilde, E. (2011). From the internet of things to the web of things: Resource-oriented architecture and best practices. In *Architecting the Internet of things* (pp. 97–129). Berlin, Germany: Springer. doi:10.1007/978-3-642-19157-2_5

Holcombe, W. T. (2014). *U.S. Patent No. 8,754,775*. Washington, DC: U.S. Patent and Trademark Office.

Hukeri, M. P. A., & Ghewari, M. P. (2017). Review paper on IoT based technology. AMGOI, Maharashtra, India.

Jin, M., Bekiaris-Liberis, N., Weekly, K., Spanos, C. J., & Bayen, A. M. (2018). Occupancy detection via environmental sensing. *IEEE Transactions on Automation Science and Engineering*, *15*(2), 443–455. doi:10.1109/TASE.2016.2619720

Jin, M., Jia, R., & Spanos, C. J. (2017). Virtual occupancy sensing: Using smart meters to indicate your presence. *IEEE Transactions on Mobile Computing*, *16*(11), 3264–3277. doi:10.1109/TMC.2017.2684806

Kinney, P. (2003, October). Zigbee technology: Wireless control that simply works. In Communications design conference (Vol. 2, pp. 1-7).

Kumar, V., & Svensson, J. (Eds.). (2015). Promoting Social Change and Democracy Through Information Technology. Hershey, PA: IGI Global. doi:10.4018/978-1-4666-8502-4

Lasserre, J. A., Bishop, C. M., & Minka, T. P. (2006, June). Principled hybrids of generative and discriminative models. In *2006 IEEE Computer Society Conference on Computer Vision and Pattern Recognition,* (Vol. 1, pp. 87-94). IEEE. 10.1109/CVPR.2006.227

Liu, X., Mao, G., Ren, J., Li, R. Y. M., Guo, J., & Zhang, L. (2015). How might China achieve its 2020 emissions target? A scenario analysis of energy consumption and CO2 emissions using the system dynamics model. *Journal of Cleaner Production*, *103*, 401–410. doi:10.1016/j.jclepro.2014.12.080

Mano, L. Y., Faiçal, B. S., Nakamura, L. H., Gomes, P. H., Libralon, G. L., Meneguete, R. I., & Ueyama, J. (2016). Exploiting IoT technologies for enhancing Health Smart Homes through patient identification and emotion recognition. *Computer Communications*, *89*, 178–190. doi:10.1016/j.comcom.2016.03.010

Mason, A. G. (2002). Cisco Secure Virtual Private Network (p. 7). Indianapolis, IN: Cisco Press.

McDonald, I. (2010). *The Dervish House*. Hachette, UK.

Mingjun, W., Zhen, Y., Wei, Z., Xishang, D., Xiaofei, Y., Chenggang, S., . . . Jinghai, H. (2012, October). A research on experimental system for Internet of things major and application project. In *2012 3rd International Conference on System Science, Engineering Design and Manufacturing Informatization (ICSEM),* (Vol. 1, pp. 261-263). IEEE. 10.1109/ICSSEM.2012.6340722

Ovadia, S. (2014). Automate the internet with "if this then that"(IFTTT). *Behavioral & Social Sciences Librarian*, *33*(4), 208–211. doi:10.1080/01639269.2014.964593

Owayjan, M., Dergham, A., Haber, G., Fakih, N., Hamoush, A., & Abdo, E. (2015). Face recognition security system. In *New Trends in Networking, Computing, E-learning, System Sciences, and Engineering* (pp. 343–348). Cham, Switzerland: Springer.

Patil, P. R. (2015). Introduction to li-fi (light fidelity) technology. *IJTRE*, *3*(4), 600–603.

Piza, E. L. (2019). *CCTV and Crime Prevention A New Systematic Review and Meta-Analysis.* John Jay College of Criminal Justice, City University of New York.

Poushter, J. (2016). Smartphone ownership and internet usage continues to climb in emerging economies. *Pew Research Center*, *22*, 1–44.

Schmidhuber, J. (2015). Deep learning in neural networks: An overview. *Neural Networks*, *61*, 85–117. doi:10.1016/j.neunet.2014.09.003 PMID:25462637

Schroeder, C. C. (1998). *U.S. Patent No. 5,787,186*. Washington, DC: U.S. Patent and Trademark Office.

Sethi, P., & Sarangi, S. R. (2017). Internet of things: Architectures, protocols, and applications. *Journal of Electrical and Computer Engineering*, 2017.

Severi, S., Sottile, F., Abreu, G., Pastrone, C., Spirito, M., & Berens, F. (2014, June). M2M technologies: Enablers for a pervasive Internet of Things. In *2014 European Conference on Networks and Communications (EuCNC),* (pp. 1-5). IEEE.

SR, Basavaraju. (2015). Automatic Smart Parking System using Internet of Things (IOT). *International Journal of Scientific and Research Publications, 628.*

Staugaard, A. C. (1987). *Robotics and AI: an introduction to applied machine intelligence* (p. 320). Englewood Cliffs, NJ: Prentice-Hall.

Wlezien, C. (1995). The public as thermostat: Dynamics of preferences for spending. *American Journal of Political Science*, *39*(4), 981–1000. doi:10.2307/2111666

Yan, J. (2015). *A guide to understanding machinery prognostics and prognosis oriented maintenance management*. Hoboken, NJ: Wiley.

Yang, C., Shen, W., & Wang, X. (2018). The internet of things in manufacturing: Key issues and potential applications. *IEEE Systems, Man, and Cybernetics Magazine*, *4*(1), 6–15. doi:10.1109/MSMC.2017.2702391

Yuan, Y., & Zhang, J. J. (2003). Towards an appropriate business model for m-commerce. *International Journal of Mobile Communications*, *1*(1-2), 35–56. doi:10.1504/IJMC.2003.002459

Chapter 7
Wearable Technologies for Glucose Monitoring:
A Systematic Mapping Study of Publication Trends

Gloria Ejehiohen Iyawa
Namibia University of Science and Technology, Namibia

Vijayalakshmi Velusamy
Manchester Metropolitan University, UK

Selvakumar Palanisamy
https://orcid.org/0000-0003-4707-6179
Manchester Metropolitan University, UK

ABSTRACT

With the increasing prevalence of diabetes mellitus, it has become of utmost importance for wearable technologies for glucose monitoring to be introduced in different contexts. While there is a high number of research on noninvasive techniques for glucose monitoring of diabetes mellitus, there is a shortage of studies discussing the publication trend of wearable technologies that support glucose monitoring. The primary purpose of this chapter was to conduct a Systematic Mapping Study of publication trends relating to wearable technologies for glucose monitoring. This study adopted a Systematic Mapping Study approach in identifying relevant papers for analysis. Articles were identified from relevant databases including IEEE Xplore, ACM Digital Library, ScienceDirect and Scopus. A total of 29 papers met the inclusion criteria. The findings of this study are expected to inform health informatics experts and academics on the current research and publication trends in wearable technologies for glucose monitoring.

DOI: 10.4018/978-1-5225-9246-4.ch007

INTRODUCTION

Diabetes mellitus is an ailment that causes an upsurge in glucose levels (Riaz, 2009). According to Aninaw and Seyoum (2017), the estimated number of people living with diabetes mellitus by 2025 will be approximately 300 million. Furthermore, the effects of diabetes mellitus could lead to further complications such as blindness and other health-related problems (Bruen et al., 2017), hence, the increase in technologies to support diabetes mellitus.

Before the use of noninvasive mechanisms for monitoring patients with diabetes mellitus, the finger pricking approach was applied in observing the glucose level of diabetic patients (Nentwich and Ulbig, 2015). This process produces a bit of discomfort to the patient (Bruen et al., 2017), as Heinemann (2008, p. 920) clearly states, the result of repeatedly pricking the finger could lead to "development of massive scarring/callous formation and loss of sensibility/ perception hindrance". Recently, there has been an interest in the use of noninvasive technologies to support patients with diabetes mellitus to facilitate a continuous process of monitoring the glucose level of these patients (Bruen et al., 2017; Nentwich and Ulbig, 2015; Heinemann, 2008; Lin et al., 2018; Pfutzner et al., 2018). There is also an abundance of research elaborating on new advances in noninvasive technologies such as GlucoTrack (Lin et al., 2018), and TensorTip Combo Glucometer (CoG) (Pfutzner et al., 2018). These types of technologies focus on using noninvasive methods to determine the glucose level rather than invasively pricking the finger to draw blood from the patient (Nentwich and Ulbig, 2015).

Recently, there is a rise in the use of noninvasive technologies for glucose monitoring, with systematic literature reviews conducted on wearable technologies (Cappon et al., 2017; Ray, 2018) and studies highlighting wearable and sensor technologies as components of digital health relevant for improving the outcome of healthcare (Iyawa et al., 2016a; Iyawa et al., 2016b). However, there is limited research focusing on the publication trends and technologies on wearable technologies for glucose monitoring. The purpose of this chapter was to provide a Systematic Mapping Study to identify the publication channels that are the main targets for publishing research on wearable technologies for glucose monitoring and identify the publication trends of research focusing on the use of wearable technologies for glucose monitoring. The chapter also identified and examined the different wearable technologies used for glucose monitoring. To the best of the researchers' knowledge, this is the first time a Systematic Mapping Study has been conducted on publication trends of studies on wearable technologies for glucose monitoring. This study, therefore, contributes to the existing body of knowledge on wearable technologies for glucose monitoring. This chapter is structured as follows: Research Methodology, Results, Discussion, and Conclusions.

RESEARCH METHODOLOGY

The review method adopted in this paper is the Systematic Mapping Study. Grant and Booth (2009, p. 94) explain that the purpose of a Systematic Mapping Study is to "map out and categorise existing literature from which to commission further reviews and primary research by identifying gaps in research literature". The above discussion is in line with the scope of this study which is to identify the publication trends of studies on wearable technologies for monitoring diabetes mellitus. Systematic Mapping Study approach has been adopted in a wide range of studies in different fields including enterprise architecture

Table 1. Description of the research questions

Research Question (RQ)	The Aim of this Paper
RQ1: What are the publication channels that are the primary targets for publishing research on wearable technologies for glucose monitoring?	This question aimed to provide an overview of the regular outlets' of academic research that is generally published on wearable technologies for glucose monitoring, databases where these studies are indexed and the countries in which author institutions originate from
RQ2: What are the publication trends of research focusing on the use of wearable technologies for glucose monitoring?	This question aimed to provide an overview of the types of paper, research methods and research focus of academic papers on wearable technologies for glucose monitoring
RQ3: What are the different wearable technologies used for glucose monitoring?	This question aimed to provide an overview of the different types of wearable technologies developed for glucose monitoring that have been discussed in the literature

(Banaeianjahromi and Smolander, 2016), software engineering (Barreiros et al., 2011) and education-related research (Dicheva et al., 2015).

Furthermore, Systematic Mapping Study approach has been applied in healthcare-related studies (Brennan et al., 2013; Liu and Halonen, 2018). These studies have applied Systematic Mapping Study approach in identifying the publication trends in the literature and have provided useful insights into the structure of research in the diverse fields. The findings from these studies propose useful insights into the research and research gaps that need to be addressed.

The review process adopted in this study followed the approach described by Peterson et al. (2008). This process was used as a guideline for conducting the Systematic Mapping Study. Each phase is described in subsequent sections.

Definition of Research Questions

While there is a high number of studies on wearable technologies for glucose monitoring (Lin et al., 2018; Kascheev et al., 2017; Elsherif et al., 2018), it is not clear what the publication channels are for these studies, the publication trend as well as the different wearable technologies that have been developed for the purpose of monitoring the glucose level of diabetic patients. The problem identified in the literature prompted the research questions for the study. The research questions guiding this study are highlighted as follows:

1. What are the publication channels that are the primary targets for publishing research on wearable technologies for glucose monitoring?
2. What are the publication trends of research focusing on the use of wearable technologies for glucose monitoring?
3. What are the different wearable technologies used for glucose monitoring?

Table 1 provides a description to the research questions.

The outcome of defining these questions is that there is a clear definition of what needs to be done in conducting the research.

Search for Relevant Papers

Primary studies related to wearable technologies for diabetes mellitus were identified in academic databases. The first step in conducting the Systematic Mapping Study was to search relevant databases such as IEEE Xplore, ACM Digital Library, ScienceDirect and Scopus. In the second step, the search strings used in identifying relevant papers were: ("wearable" or "technology") and ("glucose" or "diabetes" or "glucose monitoring" or "diabetes mellitus"). The study included papers published between 2010 and 2018. The initial search retrieved 586 articles. The search included three phases; a search was conducted to include keywords in either the title, abstract or content. The second stage retrieved 220 papers which were later reduced to 80 papers based on the examination of the title and abstracts and removal of duplicates. Eighty full-text articles were screened for relevance, and a total of 29 papers met the inclusion and exclusion criteria. The inclusion and exclusion criteria include:

- Papers published in English were included
- Peer-reviewed research papers were included
- Commentaries were excluded
- Systematic reviews were excluded
- Masters and PhD thesis excluded
- Only papers published in journals or conferences were included
- Papers focusing on wearable technologies and glucose monitoring were included
- Papers focusing on diabetes mellitus were included

The search was conducted from October 2018 to November 2018.

Classification Scheme

Following Peterson et al.'s (2008, p. 4) approach for classification, this study adopted the research type facet which is described as follows:

- Validation research: Research with new findings but have not being applied in real life settings
- Evaluation research: Research with new findings which have been applied in real life settings
- Solution proposal: Research with new solution proposed to solve a problem
- Philosophical papers: Research focusing on new findings with the aid of a conceptual framework
- Opinion papers: Opinion of a researcher on the findings of a particular research
- Experience papers: Understanding of the author on how a technique operates

To identify the research method classification, the methodologies in Management Information Systems (MIS) research proposed by Palvia et al. (2004) and research method proposed by (Banaeianjahromi and Smolander, 2016) were adopted. These methodologies include literature analysis, case study, survey, field study, constructive study and discussion.

Mapping Results

This section describes the result of the mapping study. Twenty-nine papers were classified into different categories shown in Table 2.

Table 2. Mapping the result

Publication type	Reference	Paper Type	Database	Research Focus	Author Institution Country	Research Method	Type of Wearable Technology
Conference	Arakawa et al. (2015)	Evaluation	IEEE	Non-invasive	Japan	Case study	Mouth Guard
Journal	Wang et al. (2017)	Evaluation	IEEE	Minimally invasive	Canada	Case study	Wrist band
Conference	Kascheeve et al. (2017)	Validation	IEEE	Non-invasive	Russia	Constructive study	Wearable device
Journal	Lin et al. (2018)	Evaluation	Scopus	Non-invasive	Israel	Case study	GlucoTrack (Ear lobe)
Journal	Pfützner et al. (2018)	Evaluation	Scopus	Non- invasive	Germany	Case study	CnogaTensorTip Combo Glucometer (CoG)
Journal	Favero et al. (2015)	Evaluation	Scopus	Non-invasive	Italy	Case study	Modular model predictive control managed by a wearable artificial pancreas
Journal	Arsand et al. (2015)	Evaluation	Scopus	Non-invasive	Norway	Case study	Smartwatch and smartphone
Journal	Lee et al. (2017)	Solution proposal	Scopus	Non-invasive	Republic of Korea	Discussion	Wearable/disposable sweat glucose monitoring device
Journal	Al-Tamimi et al. (2018)	Philosophical paper	Science Direct	Non-invasive	United Kingdom	Survey	Wearable device
Journal	Zhang et al. (2015)	Philosophical paper	Science Direct	Non- invasive	United States	Discussion	Saliva biosensor
Journal	You et al. (2015)	Validation	Science Direct	Non-invasive	Taiwan, South Korea, United States	Discussion	Wearable technology and smartphone
Journal	Moser et al. (2010)	Experience	Science Direct	Non- invasive	United States	Discussion	Wearable device
Journal	Liu et al. (2016)	Validation	Science Direct	Non- invasive	United States	Discussion	Wearable conductivity sensor
Journal	Mortellaro and DeHennis (2014)	Evaluation	Science Direct	Non- invasive	United States	Case study	Implantable glucose sensor
Journal	Bolinder et al. (2016)	Evaluation	Science Direct	Non-invasive	Netherlands, Sweden, Germany, Austria	Case study	Blinded sensor
Conference	Zheng et al (2018)	Solution proposal	Science Direct	Non- invasive	Australia, China	Discussion	Wearable device
Conference	Vrba et al. (2018)	Validation	Scopus	Non- invasive	Czech Republic	Discussion	Metamaterial and microstrip transmission lines sensors
Journal	Yoon et al. (2018)	Evaluation	Science Direct	Non- invasive	Republic of Korea	Case study	Stainless steel-based semi-implantable and flexible electrochemical sensor
Journal	Vettoretti et al. (2018)	Experience	Scopus	Non- invasive	Italy	Discussion	Wearable sensor data
Journal	Shaker et al. (2018)	Validation	Scopus	Non- invasive	Canada	Discussion	Wearable platform
Journal	Elsherif et al. (2018)	Validation	Scopus	Non- invasive	United Kingdom, Egypt	Discussion	Contact lens
Journal	Carduff et al. (2018)	Evaluation	Scopus	Non- invasive	Switzerland	Discussion	Wearable multisensor
Journal	Zanon et al. (2018)	Evaluation	Scopus	Non- invasive	Switzerland	Discussion	Wearable multisensor
Journal	Schwartz et al (2018)	Experience	Scopus	Non- invasive	United States	Discussion	Wearable physiological sensors
Journal	Chu et al. (2018)	Validation	Scopus	Non- invasive	Japan	Discussion	Wearable amperometric glucose sensor
Conference	Reich and Dunne (2016)	Validation	ACM Digital Library	Non- invasive	United States	Survey	Multi modal wearable ambient
Conference	Rai and Varadan (2012)	Validation	Scopus	Non- invasive	United States	Discussion	Implantable glucose sensor
Journal	Kim et al (2016)	Evaluation	Scopus	Non- invasive	Republic of Korea	Case study	Wearable device
Journal	Kim et al. (2017)	Evaluation	Scopus	Non- invasive	Republic of Korea	Case study	Contact lens

RESULTS

RQ1: What Are the Publication Channels That Are the Primary Targets for Publishing Research on Wearable Technologies for Glucose Monitoring?

To answer the first research question, the selected papers were categorised into publication type, database and authors' institution originating countries. The findings revealed that out of the 29 papers included in this study, 23 papers had been published in journals, and 6 papers have been published in conferences which are highlighted in Figure 1.

Majority of these journals are related to diabetes care and management (Pfützner et al., 2018; Favero et al., 2015; Arsand et al., 2015; Al-Tamimi et al.,2018). Moreover, the others are related to engineering and biosensors (Arakawa et al., 2015; Wang et al., 2017; Zhang et al., 2015; Liu et al., 2016).

In total, Scopus retrieved 16 relevant papers, followed by ScienceDirect which retrieved 9 papers. IEEE Xplore retrieved 3 papers, and ACM Digital Library retrieved 1 paper. This is indicated in Figure 2.

Countries where authors' institutions are linked are mostly developed countries with the United States taking the lead with 8 publications and the Republic of Korea with 4 publications. Majority of the publications emanated from institutions in developed countries, and only three institutions (Egypt, Taiwan and South Korea) from developing countries had publications (Elsherif et al., 2018; You et al., 2015). These papers were published in collaboration with authors from institutions in developed countries. The author institution country is visualised in Figure 3.

Figure 1. Publication outlet

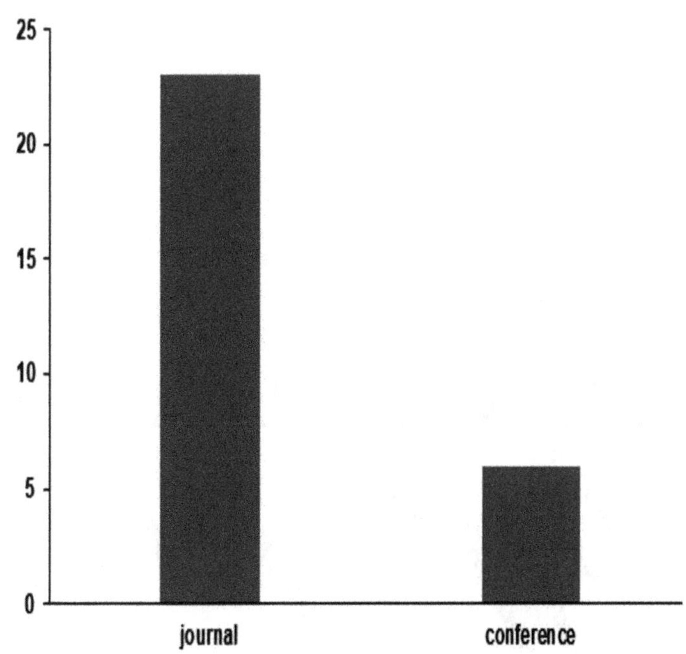

Figure 2. Database search

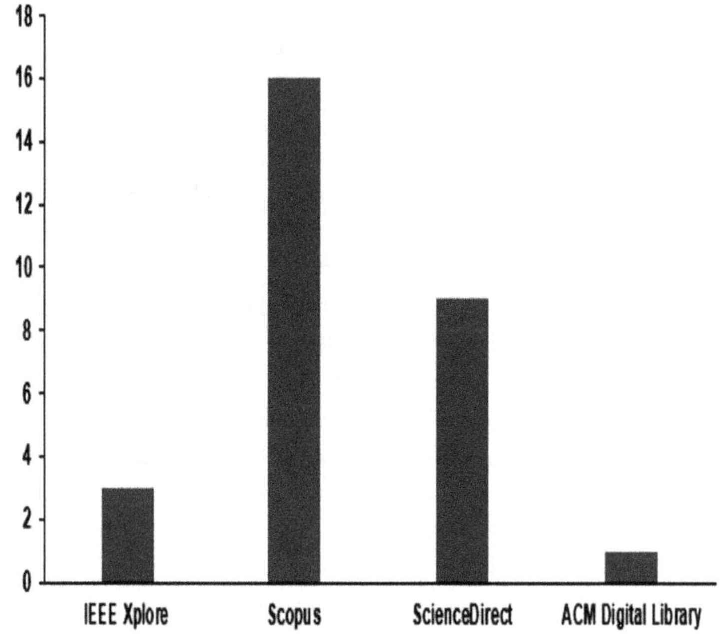

Figure 3. Country of authors' institution

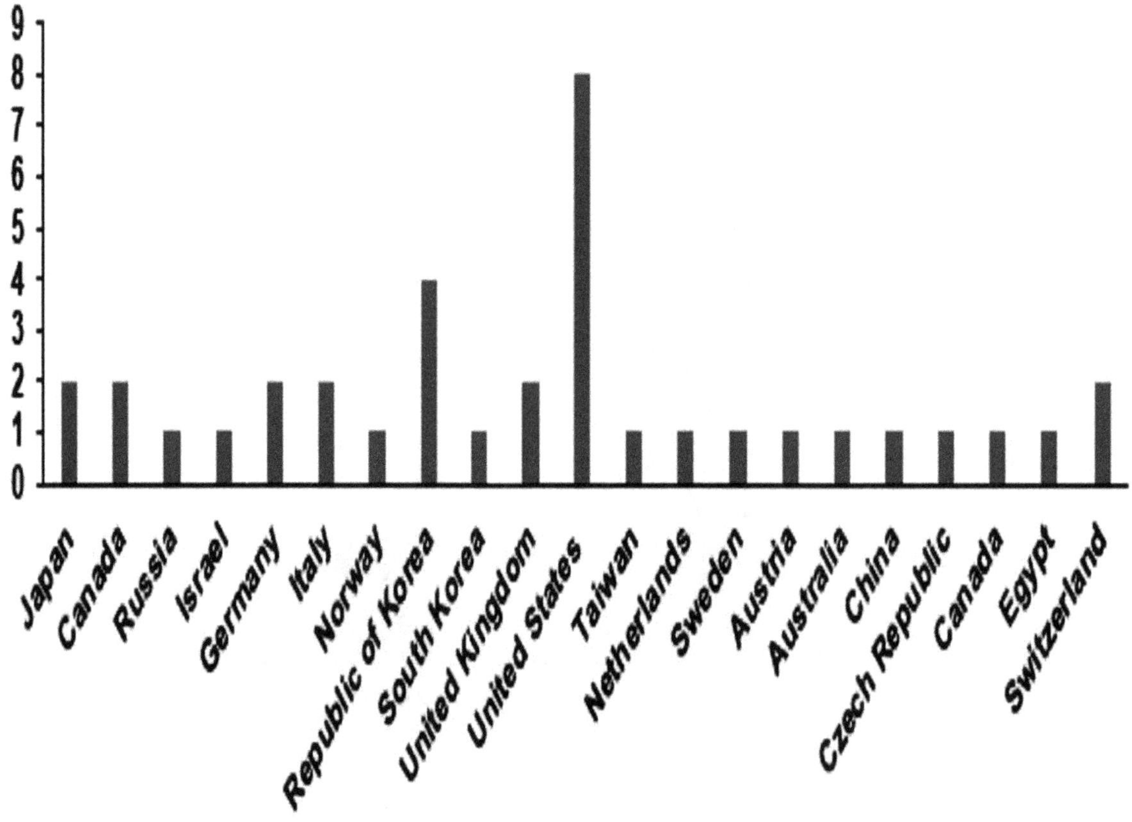

RQ2: What Are the Publication Trends of Research Focusing on the Use of Wearable Technologies for Glucose Monitoring?

The authors followed the categorisation procedure presented by Peterson et al. (2008) and (Banaeian-jahromi and Smolander, 2016). Majority of the publications focused on the evaluation of the technology developed with majority of the discussions revolving around participants either humans or animals testing the technology (Arakawa et al., 2015; Wang et al., 2017; Lin et al., 2018; Favero et al., 2015; Arsand et al., 2015). Other studies presented the technology without experimenting the use of the wearable technologies for glucose monitoring in a practical environment (Liu et al., 2016; You et al., 2015), other studies presented solution proposals (Zheng et al., 2018; Lee et al., 2017). 2 papers were philosophical in nature and presented frameworks and designs. The study paper type is presented in Figure 4.

The search identified two types of research focus found in the papers retrieved, minimally invasive and noninvasive technologies. Minimally invasive techniques generally reduce the occurrence of blood (McAfee et al., 2011). However, noninvasive techniques signify no cutting of the skin to draw blood (Gaillard, 2017). Majority of the studies examined noninvasive wearable technologies as shown in Figure 5.

Different research methods were recorded in the different publications identified in this study with the majority of the studies on Discussion and Case studies as shown in the in Figure 6.

RQ3: What Are the Different Wearable Technologies Used for Glucose Monitoring?

To answer this research question, selected papers were categorised into types of technologies. The findings revealed that different technologies have been used for glucose monitoring which are highlighted in Figure 7. Most of the technologies used for monitoring glucose include a wearable device, wearable conductivity sensor, contact lens and wearable multisensor.

Figure 4. Paper type

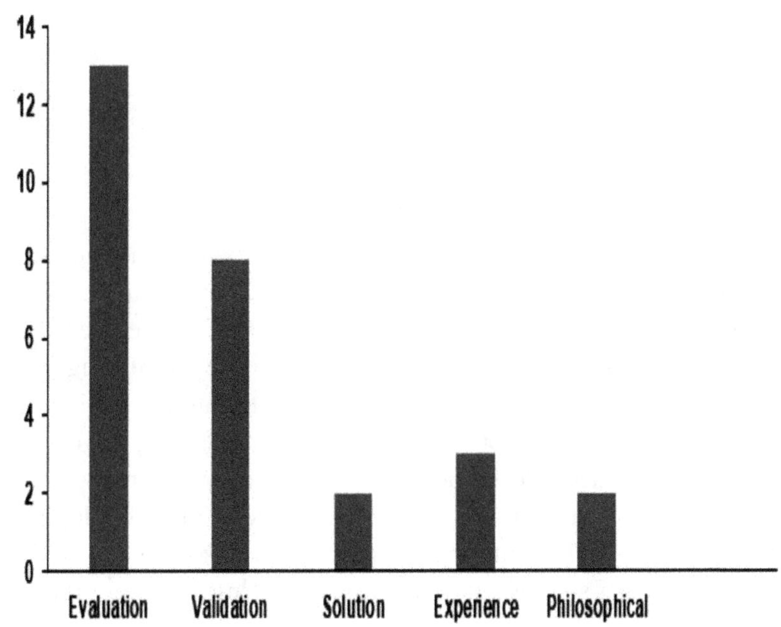

Figure 5. Research focus

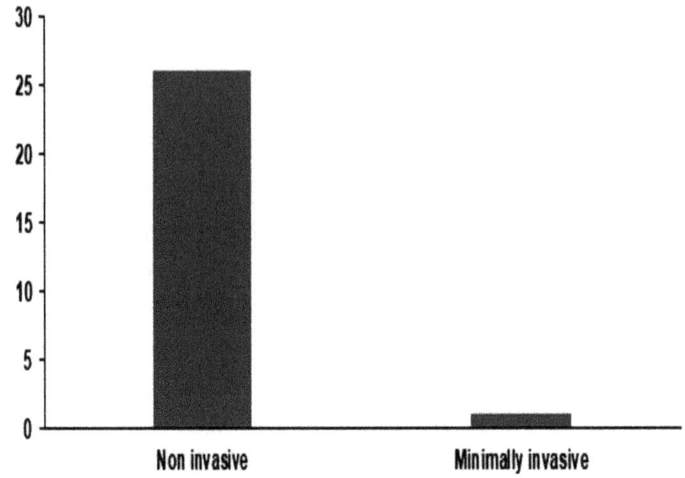

Figure 6. Research method

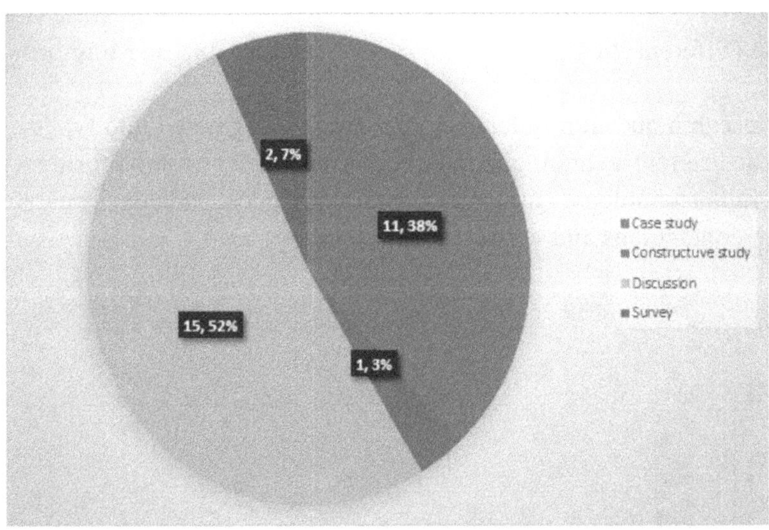

Different technologies for non-invasive glucose monitoring exist in the literature, for example, devices such as GlucoTrack (Lin et al., 2018), Cnoga TensorTip Combo Glucometer (CoG) (Pfützner et al., 2018) and a multisensor-based glucometer (Hewett et al., 1992) have been developed for glucose monitoring. The GlucoTrack device employs a noninvasive glucose monitoring in which an ear lobe is placed around the ear (Lin et al., 2018) and the glucose level presented in the interface. The significant advancement with this technology is that other external factors affecting the diabetes mellitus patient do not affect the reading of the glucose level, hence, can be used by patients who have type 2 diabetes.

Sensors have also played an essential role in the development of technologies for glucose sensor monitoring, for instance, the wearable multi-sensing patch (Anastasova et al., 2017). Sensors have also been applied with other approaches such as tattoos to sense body fluids, namely, the all-printed temporary

Figure 7. Types of technologies

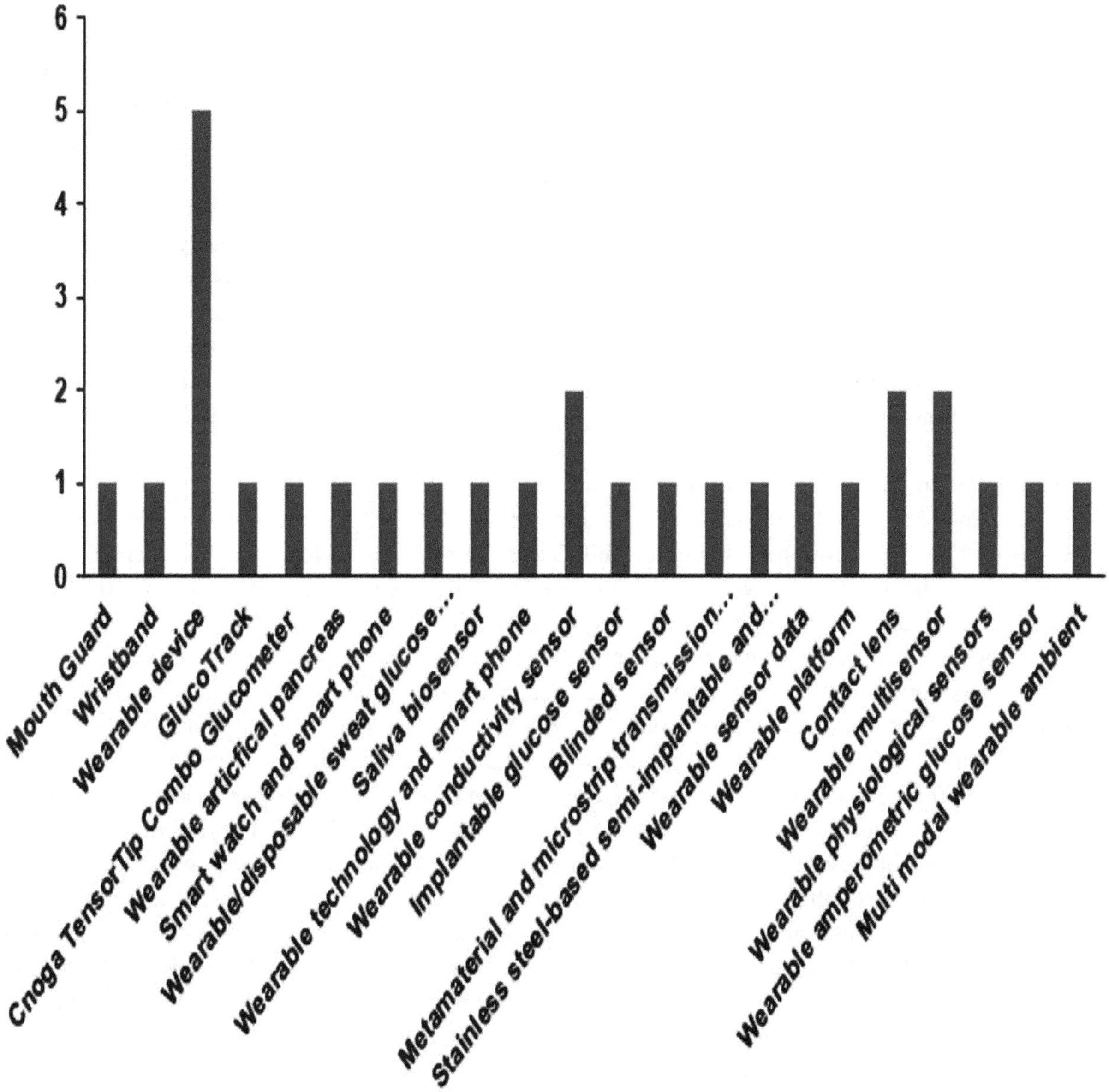

tattoo-based glucose sensor (Bandodkar et al. 2014a), a tattoo-based potentiometric sensor (Bandodkar et al., 2014b) and temporary tattoo-based printable stripping-voltammetric (Kim et al., 2015). Other sensors such as detachable salivary biosensor (Arakawa et al., 2018) can be used to detect glucose level based in saliva. Another dimension in which sensors have taken a new approach include sensing tears for glucose monitoring such as contact lens with an integrated amperometric glucose sensor (Yao et al., 2011) and noninvasive sensor worn under the eyelid (Henning et al., 2014).

DISCUSSION

The purpose of the study was to conduct a Systematic Mapping Study on publication trends on wearable technologies for glucose monitoring. This study identified the publication outlets, types of publications, types of technologies, research focus and research methods of publications on wearable technologies for glucose monitoring. This is the first systematic mapping study that describes the publication trends of studies on wearable technologies for diabetes monitoring.

In total 29 publications were analysed out of the 586 papers that were retrieved from the search. Not all the papers were relevant to the study. Hence, inclusion and exclusion criteria were applied, and relevant papers were analysed. With regards to publication outlets, journal outlets seemed to be a top avenue where research on wearable technologies for glucose monitoring is published. This points out a gap that can be addressed by researchers to ensure that research on wearable technologies for monitoring glucose level is published in conferences. In terms of databases, Scopus topped the list of databases where publications on wearable technologies for glucose monitoring can be found. ACM Digital Library seemed to archive fewer papers in this regard. This could be as a result of the inclusion and exclusion criteria applied in the search for relevant publications.

The findings revealed that institutions publishing on wearable technologies for glucose monitoring were mostly linked to developed countries. This draws the attention of less research emanating from developing countries, where relevant research is needed. This research gap indicates the call for more research to be conducted and implemented in these countries.

The study showed that a significant amount of publications was on evaluating the use of wearable technologies for glucose monitoring. This is a crucial point in this area of research, which indicates that these technologies are not only being designed and developed, but also trialled in different contexts among humans and animals. The essence of this activity is that the trials evaluate how patients react and adapt to these technologies.

Majority of the research focus was on noninvasive technologies. This indicates that the future of wearable technologies for glucose monitoring is focused on noninvasive technologies. The findings were mostly discussed for some of these studies, as 'Discussion' seemed to be a mostly used research method in these studies. It was also noted that studies based on 'Evaluation' mostly adopted the case study approach as the research method. This is in line with another study that reports the same findings (Banaeianjahromi and Smolander, 2016). It was also revealed that there are different wearable technologies and tools that have been adopted for glucose monitoring. However, the contact lens for glucose monitoring seemed to be a unique approach as it is worn directly to retrieve glucose in tears.

It was also revealed that more publications on wearable technologies for glucose monitoring has been published in 2018 compared to other years, highlighting the growing interest of wearable technologies for glucose monitoring.

CONCLUSION

To recapitulate, this book chapter identified the publication trend of wearable technologies for glucose monitoring through a Systematic Mapping Study. The study provided useful insights into what types of publications are available on wearable technologies for glucose monitoring. With this study, it was possible to identify where to locate relevant literature and which countries have conducted research in this area of study.

Despite the effectiveness of the findings of the study, the study was limited in several ways. The first limitation is the use of only four academic databases. This could have had an impact on the search result, hence, a limitation of the findings. Another limitation of this study is the exclusion of publications such as book chapters, Masters and PhD theses. The keywords used might have also put a constrain on the search results. These factors may have had an impact on the findings of the study, and it is possible that we might have missed some important papers on wearable technologies for glucose monitoring.

It was identified that there were fewer publications in conferences on wearable technologies for glucose monitoring; this calls the need for publications in conferences on the subject. As such for future work, more publications in conferences should be encouraged. It was also revealed that the majority of these studies emanated from institutions in developed countries. For future work, the relevant research should be conducted, evaluated and implemented in developing countries.

REFERENCES

Al-Tamimi, N., Slater, N., Kayyali, R., & ElShaer, A. (in press). Perceptions by adult patients with type 1 and 2 diabetes of current and advanced technologies of blood glucose monitoring: A prospective study. *Canadian Journal of Diabetes*. PMID:30026045

Anastasova, S., Crewther, B., Bembnowicz, P., Curto, V., Ip, H. M. D., Rosa, B., & Yang, G. (2017). A wearable sensing patch for continuous sweat monitoring. *Biosensors & Bioelectronics*, *93*, 139–145. doi:10.1016/j.bios.2016.09.038 PMID:27743863

Animaw, W., & Seyoum, Y. (2017). Increasing prevalence of diabetes mellitus in a developing country and its related factors. *PLoS One*, *12*(11), 1–11. doi:10.1371/journal.pone.0187670 PMID:29112962

Arakawa, T., Kuroki, Y., Nitta, H., Toma, K., Mitsubayashi, K., Takeuchi, S., . . . Minakuchi, S. (2015, December). *Mouth guard type biosensor "cavitous sensor" for monitoring of saliva glucose with telemetry system*. Paper presented at 2015 9th International Conference on Sensing Technology, Auckland, New Zealand.

Årsand, E., Muzny, M., Bradway, M., Muzik, J., & Hartvigsen, G. (2015). Performance of the first combined smartwatch and smartphone diabetes diary application study. *Journal of Diabetes Science and Technology*, *9*(3), 556–563. doi:10.1177/1932296814567708 PMID:25591859

Banaeianjahromi, N., & Smolander, K. (2016). What do we know about the role of enterprise architecture in enterprise integration? A systematic mapping study. *Journal of Enterprise Information Management*, *29*(1), 140–164. doi:10.1108/JEIM-12-2014-0114

Bandodkar, A. J., Jia, W., Yardımcı, C., Wang, X., Ramirez, J., & Wang, J. (2014a). Tattoo-based noninvasive glucose monitoring: A proof-of-concept study. *Analytical Chemistry*, *12*(87), 394–398. PMID:25496376

Bandodkar, A. J., Molinnus, D., Mirza, O., Guinovart, T., Windmiller, J. R., Valdes-Ramirez, G., ... Wang, J. (2014b). Epidermal tattoo potentiometric sodium sensors with wireless signal transduction for continuous non-invasive sweat monitoring. *Biosensors & Bioelectronics*, *87*(1), 603–609. doi:10.1016/j.bios.2013.11.039 PMID:24333582

Barreiros, E., Almeida, A., Saraiva, J., & Soares, S. (2011, September). *A systematic mapping study on software engineering testbeds*. Paper presented at 2011 International Symposium on Empirical Software Engineering and Measurement, Alberta, Canada. 10.1109/ESEM.2011.19

Bolinder, J., Antuna, R., Geelhoed-Duijvestijn, P., Kröger, J., & Weitgasser, R. (2016). Novel glucose-sensing technology and hypoglycaemia in type 1 diabetes: A multicentre, non-masked, randomised controlled trial. *Lancet, 388*(10057), 2254–2263. doi:10.1016/S0140-6736(16)31535-5 PMID:27634581

Brennan, N., Barnes, R., Calnan, M., Corrigan, O., Dieppe, P., & Entwistle, V. (2013). Trust in the health-care provider–patient relationship: A systematic mapping review of the evidence base. *International Journal for Quality in Health Care, 25*(6), 682–688. doi:10.1093/intqhc/mzt063 PMID:24068242

Bruen, D., Delaney, C., Florea, L., & Diamond, D. (2017). Glucose sensing for diabetes monitoring: Recent developments. *Sensors (Basel), 17*(8), 1–21. doi:10.339017081866 PMID:28805693

Caduff, A., Zanon, M., Zakharov, P., Mueller, M., Talary, M., Krebs, A., ... Donath, M. (2018). First experiences with a wearable multisensor in an outpatient glucose monitoring study, Part I: The users' view. *Journal of Diabetes Science and Technology, 12*(3), 562–568. doi:10.1177/1932296817750932 PMID:29332423

Cappon, G., Vettoretti, M., Marturano, F., Facchinetti, A., & Sparacino, G. (2018). A neural-network-based approach to personalize insulin bolus calculation using continuous glucose monitoring. *Journal of Diabetes Science and Technology, 12*(2), 265–272. doi:10.1177/1932296818759558 PMID:29493356

Chu, M. K., Iguchi, S., Miyajima, K., Arakawa, T., Kudo, H., & Mitsubayashi, K. (2011). Development of a soft contact-lens biosensor for in-vivo tear glucose monitoring. In *Proceedings of 5th European Conference of the International Federation for Medical and Biological Engineering* (vol 37. pp. 1007-1010). Berlin, Germany: Springer 10.1007/978-3-642-23508-5_262

Del Favero, S., Place, J., Kropff, J., Messori, M., Keith-Hynes, P., Visentin, R., & Di Palma, F. (2015). Multicenter outpatient dinner/overnight reduction of hypoglycemia and increased time of glucose in target with a wearable artificial pancreas using modular model predictive control in adults with type 1 diabetes. *Diabetes, Obesity & Metabolism, 17*(5), 468–476. doi:10.1111/dom.12440 PMID:25600304

Dicheva, D., Dichev, C., Agre, G., & Angelova, G. (2015). Gamification in education: A systematic mapping study. *Journal of Educational Technology & Society, 18*(3), 75–88.

Elsherif, M., Hasan, M. U., Yetisen, A. K., & Butt, H. (2018). Wearable contact lens biosensors for continuous glucose monitoring using smartphones. *ACS Nano, 12*(6), 5452–5462. doi:10.1021/acsnano.8b00829 PMID:29750502

Gaillard, M. (2017). "Invasive" and "Non-invasive" technologies in neuroscience communication. *Bioéthique Online, 6*(11), 1–10.

Grant, M. J., & Booth, A. (2009). A typology of reviews: An analysis of 14 review types and associated methodologies. *Health Information and Libraries Journal, 26*(2), 91–108. doi:10.1111/j.1471-1842.2009.00848.x PMID:19490148

Heinemann, L. (2008). Finger pricking and pain: A never ending story. *Journal of Diabetes Science and Technology*, *2*(5), 919–921. doi:10.1177/193229680800200526 PMID:19885279

Heinemann, L. (2008). Finger pricking and pain: A never ending story. *Journal of Diabetes Science and Technology*, *2*(5), 919–921. doi:10.1177/193229680800200526 PMID:19885279

Henning, A., Lauko, J., Grabmaier, A., & Wilson, C. (2014). Wireless tear glucose sensor. *Procedia Engineering*, *87*, 66–69. doi:10.1016/j.proeng.2014.11.267

Hewett, T., Baecker, R., Card, S., Carey, T., Gasen, J., Mantei, M., ... Verplank, W. (1992). *ACM SIHCHI curricula for human–computer interaction. Association for Computing Machinery*. New York, NY: ACM. doi:10.1145/2594128

Iyawa, G. E., Herselman, M., & Botha, A. (2016a). Digital health innovation ecosystems: From systematic literature to conceptual framework. *Procedia Computer Science*, *100*, 244–252. doi:10.1016/j.procs.2016.09.149

Iyawa, G. E., Herselman, M., & Botha, A. (2016b). Identifying and defining the terms related to a Digital Health Innovation Ecosystem. In M. Herselman, & A. Botha (Eds.), *Strategies, Approaches and Experiences: Towards Building a South African Digital Health Innovation Ecosystem*. Pretoria, South Africa: CSIR.

Kascheev, N., Kozyrev, O., Leykin, M., & Vanyagin, A. (2017, September). *Non-invasive monitoring of blood glucose by means of wearable tracking technology*. Paper presented at 2017 IEEE East-West Design & Test Symposium, Novi Sad, Serbia. 10.1109/EWDTS.2017.8110137

Kim, E. K., Kwak, S. H., Baek, S., Lee, S. L., Jang, H. C., Park, K. S., & Cho, Y. M. (2016). Feasibility of a patient-centered, smartphone-based, diabetes care system: A pilot study. *Diabetes & Metabolism Journal*, *40*(3), 192–201. doi:10.4093/dmj.2016.40.3.192 PMID:27098508

Kim, J., Araujo, W. R., Samek, I. A., Bandodkar, A. J., Jia, W., Brunetti, B., ... Wang, J. (2015). Wearable temporary tattoo sensor for real-time trace metal monitoring in human sweat. *Electrochemistry Communications*, *87*(1), 41–45. doi:10.1016/j.elecom.2014.11.024

Kim, J., Kim, M., Lee, M. S., Kim, K., Ji, S., Kim, Y. T., & Bien, F. (2017). Wearable smart sensor systems integrated on soft contact lenses for wireless ocular diagnostics. *Nature Communications*, *8*(14997), 1–8. PMID:28447604

Lee, H., Song, C., Hong, Y. S., Kim, M. S., Cho, H. R., Kang, T., & Kim, D. H. (2017). Wearable/disposable sweat-based glucose monitoring device with multistage transdermal drug delivery module. *Science Advances*, *3*(3), 1–8. doi:10.1126ciadv.1601314 PMID:28345030

Lin, T., Mayzel, Y., & Bahartan, K. (2018). The accuracy of a non-invasive glucose monitoring device does not depend on clinical characteristics of people with type 2 diabetes mellitus. *Journal of drug assessment, 7*(1), 1-7.

Liu, G., Ho, C., Slappey, N., Zhou, Z., Snelgrove, S. E., Brown, M., & Edwards, J. (2016). A wearable conductivity sensor for wireless real-time sweat monitoring. *Sensors and Actuators. B, Chemical*, *227*, 35–42. doi:10.1016/j.snb.2015.12.034

Liu, Z., & Halonen, Z. (2018). *ICT supporting healthcare for elderly in China: A systematic mapping study*. Paper presented at Bled Econference Digital Transformation: Meeting the Challenges, Bled, Slovenia. 10.18690/978-961-286-170-4.28

McAfee, P. C., Garfin, S. R., Rodgers, W. B., Allen, R. T., Phillips, F., & Kim, C. (2011). An attempt at clinically defining and assessing minimally invasive surgery compared with traditional “open” spinal surgery. *SAS Journal*, *5*(4), 125–130. doi:10.1016/j.esas.2011.06.002 PMID:25802679

Mortellaro, M., & DeHennis, A. (2014). Performance characterization of an abiotic and fluorescent-based continuous glucose monitoring system in patients with type 1 diabetes. *Biosensors & Bioelectronics*, *61*, 227–231. doi:10.1016/j.bios.2014.05.022 PMID:24906080

Moser, E. G., Crew, L. B., & Garg, S. K. (2010). Role of continuous glucose monitoring in diabetes management. *Advances en Diabetologia*, *26*, 73–78.

Nentwich, M. M., & Ulbig, M. W. (2015). Diabetic retinopathy - ocular complications of diabetes mellitus. *World Journal of Diabetes*, *6*(3), 489–499. doi:10.4239/wjd.v6.i3.489 PMID:25897358

Palvia, P., Leary, D., Mao, E., Midha, V., Pinjani, P., & Salam, A. F. (2004). Research methodologies in MIS: An update. *Communications of the Association for Information Systems*, *14*(1), 526–542.

Petersen, K., Feldt, R., Mujtaba, S., & Mattsson, M. (2008). Systematic mapping studies in software engineering. *EASE*, *8*, 68–77.

Pfützner, A., Strobl, S., Demircik, F., Redert, L., Pfützner, J., Pfützner, A. H., & Lier, A. (2018). Evaluation of a new noninvasive glucose monitoring device by means of standardized meal experiments. *Journal of Diabetes Science and Technology*, *12*(6), 1178–1183. doi:10.1177/1932296818758769 PMID:29451016

Rai, P., & Varadan, V. K. (2012, October). *Wireless glucose monitoring watch enabled by an implantable self-sustaining glucose sensor system*. Paper presented at the Nanosystems in Engineering and Medicine, Incheon, Republic of Korea.

Ray, P. P. (2018). Continuous glucose monitoring: A systematic review of sensor systems and prospects. *Sensor Review*, *38*(4), 430–437. doi:10.1108/SR-12-2017-0268

Reich, J., & Dunne, L. E. (2016, September). Multi-modal wearable ambient display: an investigation of continuous glucose monitoring. Paper presented at the ACM International Symposium on Wearable Computers, Heidelberg, Germany. 10.1145/2971763.2971801

Riaz, S. (2009). Diabetes mellitus. *Scientific Research and Essays*, *4*(5), 367–373. PMID:20015604

Schwartz, F. L., Marling, C. R., & Bunescu, R. C. (2018). The Promise and Perils of Wearable Physiological Sensors for Diabetes Management. *Journal of Diabetes Science and Technology*, *12*(3), 587–591. doi:10.1177/1932296818763228 PMID:29542348

Shaker, G., Smith, K., Omer, A. E., Liu, S., Csech, C., Wadhwa, U., & Hughson, R. (2018). Noninvasive monitoring of glucose level changes utilizing a mm-wave radar system. *International Journal of Mobile Human Computer Interaction*, *10*(3), 10–29. doi:10.4018/IJMHCI.2018070102

Vettoretti, M., Cappon, G., Acciaroli, G., Facchinetti, A., & Sparacino, G. (2018). Continuous Glucose Monitoring: Current Use in Diabetes Management and Possible Future Applications. *Journal of Diabetes Science and Technology*, *12*(5), 1064–1071. doi:10.1177/1932296818774078 PMID:29783897

Vrba, J., & Vrba, D., Diaz, l. & Fiser, O. (2018). *Metamaterial sensor for microwave non-invasive blood glucose monitoring*. Paper presented at World Congress on Medical Physics and Biomedical Engineering, Prague, Czech Republic.

Wang, G., Poscente, M. D., Park, S. S., Andrews, C. N., Yadid-Pecht, O., & Mintchev, M. P. (2017). Wearable microsystem for minimally invasive, pseudo-continuous blood glucose monitoring: The e-Mosquito. *IEEE Transactions on Biomedical Circuits and Systems*, *11*(5), 979–987. doi:10.1109/TBCAS.2017.2669440 PMID:28574366

Yao, H., Shum, A. J., Cowan, M., Lahdesmaki, I., & Parviz, B. A. (2011). A contact lens with embedded sensor for monitoring tear glucose level. *Biosensors & Bioelectronics*, *26*(7), 3290–3296. doi:10.1016/j.bios.2010.12.042 PMID:21257302

Yoon, H., Xuan, X., Jeong, S., & Park, J. Y. (2018). Wearable, robust, non-enzymatic continuous glucose monitoring system and its in vivo investigation. *Biosensors & Bioelectronics*, *117*, 267–275. doi:10.1016/j.bios.2018.06.008 PMID:29909198

You, I., Choo, K. K. R., & Ho, C. L. (2018). A smartphone-based wearable sensors for monitoring real-time physiological data. *Computers & Electrical Engineering*, *65*, 376–392. doi:10.1016/j.compeleceng.2017.06.031

Zanon, M., Mueller, M., Zakharov, P., Talary, M. S., Donath, M., Stahel, W. A., & Caduff, A. (2018). First experiences with a wearable multisensor device in a noninvasive continuous glucose monitoring study at home, Part II: The investigators' view. *Journal of Diabetes Science and Technology*, *12*(3), 554–561. doi:10.1177/1932296817740591 PMID:29145749

Zhang, W., Du, Y., & Wang, M. L. (2015). Noninvasive glucose monitoring using saliva nano-biosensor. *Sensing and Bio-Sensing Research*, *4*, 23–29. doi:10.1016/j.sbsr.2015.02.002

Zheng, H., He, J., Li, P., Guo, M., Jin, H., Shen, J., ... Chi, C. (2018). Glucose Screening Measurements and Noninvasive Glucose Monitor Methods. *Procedia Computer Science*, *139*, 613–621. doi:10.1016/j.procs.2018.10.202

Chapter 8
Identifying the Components of a Smart Health Ecosystem for Asthma Patients:
A Systematic Literature Review and Conceptual Framework

Gloria Ejehiohen Iyawa
Namibia University of Science and Technology, Namibia

Asiya Khan
University of Plymouth, UK

Sesinam Dagadu
tinyDavid, Ghana

Kristine Mae Magtubo
Veris Tech, Inc, Philippines

Rupert Calvin Sievert
Veris Tech, Inc, Philippines

ABSTRACT

With asthma being one of the leading causes of death in different countries, the emphasis on improving the health of asthma patients is important. While the use of smart technologies is a good approach for improving the health of asthma patients, technologies need to be connected in such a way that all components of smart health form an ecosystem. However, the components of such an ecosystem have not been identified in the current literature. The purpose of this chapter was to identify the components of a smart health ecosystem for asthma patients through a systematic literature review. A total of 28 articles met the inclusion criteria. This chapter identified the components of a smart health ecosystem for asthma patients and provided a conceptual framework. The findings of the systematic literature review are expected to inform researchers on the components required for building a smart health ecosystem for asthma patients.

DOI: 10.4018/978-1-5225-9246-4.ch008

INTRODUCTION

Asthma is a chronic disease affecting people in different parts of the world (Sinharoy et al. 2018). One of the causes of high prevalence of asthma attacks could be unfavorable weather conditions, lack of medications that have a positive impact on asthma attack in some countries (Sinharoy et al. 2018) and air pollution (Guarnieri and Balmes, 2014). The number of asthma deaths occurring each year is estimated to be about 250,000 (D'Amato et al., 2016). Approximately 300 million people suffer from asthma globally and is a leading cause of disability (Global Initiative for Asthma, 2014; The Global Asthma Report, 2014; World Health Organisation, 2007).

With this growing trend, researchers have developed technologies to improve health conditions for asthma patients. For example, AsthmaGuide, an asthma monitoring system which allows medical practitioners access patient information from a distant location (Ra et al., 2016). Monitoring mechanisms (Kwan et al., 2014, Seto et al., 2009) and prevention mechanisms through technology have been developed (Seto et al., 2009). An emerging term for such applications is "smart health" (Sundaravadivel et al., 2018). According to Sundaravadivel et al. (2018), smart health enables patients to take charge of their health through constant monitoring. Iyawa et al. (2016a) and Iyawa et al. (2016b) suggest that digital health enables patients to keep track of their health by using wearable and wireless technologies and Rahmani et al. (2018) admit that an ecosystem is relevant in providing meaningful care.

Despite the advancement of patient care through wearables and wireless technologies, these technologies need to be connected. These components of smart health should form an ecosystem in which asthma patients, medical practitioners and technologies can communicate at remote locations. Existing literature support the benefits of having a smart health ecosystem. For example, Chang and West (2006) present a digital ecosystem as an ecosystem that enables the interactions of different components in a digital platform. McLaughlin et al. (2009) are of the opinion that components of a digital ecosystem should work together to share knowledge. The literature further suggests that these benefits are not limited by location (Briscoe and De Wilde (2006). The purpose of this chapter was, therefore, to provide a systematic literature review on the components of smart health ecosystem for asthma patients and, using these findings, provide a conceptual framework of a smart health ecosystem for asthma patients. The findings of the study contribute to the growing body of knowledge on smart health specifically for asthma patients. Further, this study could provide a better understanding of what is required in developing a smart health ecosystem for asthma patients. The remainder of this chapter is structured as follows: Research Methodology, Results, Discussion, Conclusion and Future Work.

RESEARCH METHODOLOGY

Research Questions

The methodology applied in the chapter is the systematic literature review. The systematic literature review aimed to answer two research questions (RQ):

1. What is a smart health ecosystem for asthma patients?
2. What are the components of a smart health ecosystem for asthma patients?

Literature Sources

A systematic literature review was conducted on three academic databases: IEEE Xplore, Scopus, and ScienceDirect. In order to identify papers which are not included in the databases, Google Scholar was also used. Google search engine was also used to identify gray literature and non-academic publications relevant to the subject. The following keywords were used to search for relevant papers: ("smart health" AND "asthma") OR ("smart health ecosystem" AND "asthma"). The publication year was between 2013 to 2018.

Inclusion and Exclusion Criteria

The inclusion and exclusion criteria for this systematic literature review were:

1. Only publications written in English were included
2. Studies referring to smart health technologies were included
3. Studies not within smart health domain were excluded
4. Studies describing smart health were included
5. Studies around the concept of technologies and asthma were included
6. Studies within the area of smart health relating to other ailment were excluded

Journal articles, conference papers, non-academic papers and practitioner case reports were screened in three phases: title, abstract and full text.

Data Analysis and Selection

A total of 625 publications were retrieved from the database search, and 125 publications were retrieved from other sources. 430 duplicate records were removed which left 320 publications to be screened. Based on the title and abstract screening, 73 records were excluded. Out of the 247 full-text articles assessed, only 28 papers were analysed based on the inclusion and exclusion criteria.

RESULTS

This chapter analysed 28 papers. Majority of the papers focused on technologies used for smart health. 32 technologies which may be applied in a smart health ecosystem were identified. The findings indicate that smart health has also been discussed around asthma, however, the concept of smart health ecosystem has not been fully determined in the literature; however, this chapter demystified the concept of a smart health ecosystem. The findings helped answer the two research questions which are explained below.

RQ1: What Is a Smart Health Ecosystem for Asthma Patients?

The term *smart health ecosystem* per se was not identified in neither the academic nor non-academic literature. However, discussions on similar concepts were identified in the literature reviewed.

The term *smart health,* also referred to as *s-health,* (Solonas et al., 2014; Zhang et al., 2018) is described as health devices conscious of their environments (Solanas et al., 2014, p.2). *Smart health* is also described as a combination of "m-health and telemedicine to create a novel and richer ubiquitous concept" (Solanas et al., 2014, p. 76). Suzuki et al. (2013), on the other hand, describes *smart health* as the engagement of patients and healthcare practitioners through wireless technology.

Meyer and Boll (2014) defined the three features of smart health systems: a monitoring component, a physical component and a focus on health challenge. Al-Dowaihi et al. (2013) developed an application that illustrates this concept. Their system informs a healthcare practitioner when an asthma patient is having symptoms of asthma attacks. Peak flow meter data is sent to the patient's mobile phone over the Internet to the healthcare practitioner's portal through a server. The study identifies the flow of information between the peak flow meter device, patient, mobile phone, Internet, server, portal and healthcare practitioner with little to no human intervention. This is an example of technology and people working together to solve healthcare problems. Eramo (2018) suggests that technologies should connect asthma patients for integrated care.

Definitions of a smart health were also provided in both gray and the non-academic literature. For example, Blue Stream Consultancy defines *smart health* as "the technology that leads to better diagnostic tools, better treatment for patients, and devices that improve the quality of life for anyone and everyone". Lee (2011, p.1), on the other hand, defines *smart health* as "medical and public health practice supported by smart mobile devices". Lee (2011) also admits that smart health incorporates ubiquitous technologies. Iyawa et al. (2016a, p.247) suggests that participants of a healthcare ecosystem should be "interrelated" and "interconnected" in a digital platform.

RQ2: What Are the Components of a Smart Health Ecosystem for Asthma Patients?

Different components of a smart health ecosystem were identified in the literature. These components explain what is necessary for establishing an ecosystem to support smart health monitoring for asthma patients. Different types of components were identified and have been categorized into human and non-human components. Under human components, two main participants to the system were identified: patients and healthcare practitioners (Al-Dowaihi et al., 2013; Thomson et al., 2017, Kikidis et al., 2016). According to Iyawa et al. (2016a), health institutions and organisations are also important components of healthcare ecosystems.

Technologies relevant to a *smart health ecosystem* include peak flow meter device, mobile phones, portal, Internet and server (Al-Dowaihi et al., 2013; Kassem et al.). Rahmani et al. (2018) also indicates that a smart ecosystem should consist of body area sensor network, Internet-connected gateways and cloud and big data support. Suzuki et al. (2013) explain that sensors and wearable technologies are components of a smart ecosystem. Yung-Cheng et al. (2013) point electronic textiles as relevant in healthcare. Wireless communication is also necessary for this technology (Suzuki et al., 2013). Social media was also pointed out as a component of smart health (Abbasi et al., 2014).

A summary of the result is presented in Table 1.

Table 1. Components of a smart health ecosystem for asthma patients

Publication Type	Author(s)	Components Identified	Year of Publication	Publisher
Conference	Al-Dowaihi et al.	Peak flow meter device, mobile phone, portal, internet, server, patients, healthcare practitioners	2013	IEEE
Conference	Thomson et al.	Patients, healthcare practitioners	2017	IEEE
Journal	Iyawa et al.	Patients, healthcare practitioners, healthcare institutions and organisations	2016	ScienceDirect
Journal	Rahmani et al.	Body area sensor network, internet-connected gateways, cloud and big data support	2018	ScienceDirect
Conference	Suzuki et al.	Sensors, wearable technologies, processor, storage, hardware, software	2013	IEEE
Conference	Kassem et al.	Peak flow meter, mobile phones,	2013	IEEE
Conference	Yung-Cheng et al.	Electronic textiles, cloud computing	2013	IEEE
Conference	Uddin et al.	Mobile phone	2013	IEEE
Journal	Abbasi et al.	Social media	2014	IEEE
Journal	Solonas et al.	Electronic health, Mobile health, Privacy, Security and Professionals	2014	IEEE
Conference	Teixeira and Postolache	Wireless sensor network (sensor nodes, routers and gateways, sensor node firmware, system software)	2014	IEEE
Journal	Kwan et al.	Mobile devices, microcontroller and sensor circuit	2015	IEEE
Conference	Meyer and Boll	Monitoring component, physical devices	2014	IEEE
Conference	Uwaoma and Mansingh	Mobile phones	2015	IEEE
Conference	Sheng et al.	Bluetooth technology, wearable technology, Smartphone sensors	2015	IEEE
Conference	Siddiquee et al.	Internet of Things (IoT)	2016	IEEE
Conference	Wu et al.	Mobile application	2016	IEEE
Conference	Ra et al.	Sensors, mobile phone, cloud web infrastructure, machine learning	2016	IEEE
Journal	Fuentes et al.	Internet of Things	2018	IEEE
Conference	Kokalki et al.	Sensor, wearable device	2017	IEEE
Conference	Kumar et al.	Privacy, security	2018	IEEE
Journal	Zhang et al.	Privacy, security	2018	IEEE
Conference	Do et al.	Deep learning, forecasting	2016	Scopus
Journal	Kikidis et al.	Patients, healthcare practitioners	2016	Google Scholar
Journal	Sampri et al.	Google Trends	2016	ScienceDirect
Journal	Venkatesh et al.	Internet of Things	2018	IEEE
Gray literature	Deloitte	m-health, mobile devices	2018	Google Search Engine
Gray literature	Eramo	Smart inhalers, wearable technologies, mobile applications	2018	Google Search Engine

DISCUSSION

The purpose of the chapter was to conduct a systematic literature review on smart health ecosystems for asthma patients. Specifically, the aims of this chapter were to explain what a smart health ecosystem for asthma patient is and to identify the components of a smart health ecosystem for asthma patients. The chapter also provides a conceptual framework for a smart health ecosystem for asthma patients (Figure 1). In total, 28 studies were included in this systematic literature review, and to the best of the researchers' knowledge, this is the first time such a study is being conducted.

Some of the studies described the concept of *smart health ecosystem*. Based on the discussion in the literature and a combination of definitions provided by different authors, a *smart health ecosystem* for asthma patients can be defined as a set of intelligent devices able to monitor patients before, during and after an asthma attack, and send relevant information to health care practitioners in real time. These intelligent devices should be able to perform functions such as prediction (Ra et al., 2016; Do et al., 2016) and sensing (Suzuki et al., 2013; Kolkalki et al., 2017).

Components of a *smart health ecosystem* include human and non-human components. Human components consist of patients and healthcare practitioners (Al-Dowaihi et al., 2013; Thomson et al., 2017). Human components can also include healthcare institutions which are managed by healthcare stakeholders (Iyawa et al., 2016). The ecosystem does not exist and cannot function without the participation of a human component. The human component of the smart health ecosystem for asthma patients drive the non-human components.

Non-human components of a smart health ecosystem for asthma patients are technological in nature. They are categorized into two types: physical devices and applications. Physical devices may include peak flow meter devices, mobile phones, sensors, wearable technologies, processors, storage solutions, electronic textiles, microcontrollers, smartphone sensors and smart inhalers as indicated in Table 1.

The application component may include portals, the Internet, servers, body area sensor networks, Internet-connected gateways, cloud and big data support, cloud computing, social media, electronic health solutions, mobile health solutions, privacy and security solutions, wireless sensor networks, Bluetooth technology, IoT, mobile applications, machine learning, deep learning, forecasting, and Google Trends as indicated in Table 1. A conceptual framework of how these human and non-human components interact within a smart health ecosystem for asthma patients is presented in Figure 1 below.

Peak flow meter devices are necessary for determining the flow of air in the lungs (Al-Dowaihi et al., 2013), while smart inhalers are able to determine how a patient makes use of the inhaler (Eramo, 2018). The use of smart inhalers (Eramo, 2018) and smartphone sensors (Sheng et al., 2015) is usually facilitated through Bluetooth technology (Sheng et al., 2014). Wearable technologies, wireless sensor networks and sensors can facilitate a body area network to enable the monitoring of asthma patients (Teixeira and Postolache, 2014; Ra et al., 2016). Wireless communication is necessary to ensure the flow of information (Suzuki et al., 2013), which is supported by Bluetooth technology and the Internet. Sensors can be embedded in wearable technologies such as electronic textiles (Yung-Cheng et al., 2013; Kokalki et al., 2017). Microcontrollers such as Arduino devices can be used to facilitate monitoring of patients (Kwan et al., 2015). Fast processors and high storage systems such as cloud computing platforms are necessary in ensuring high-quality processing of patient data and storage. Data storage can likewise be processed more efficiently with big data which can breed other useful applications such as machine

Figure 1. Conceptual framework for smart health ecosystem for asthma patients

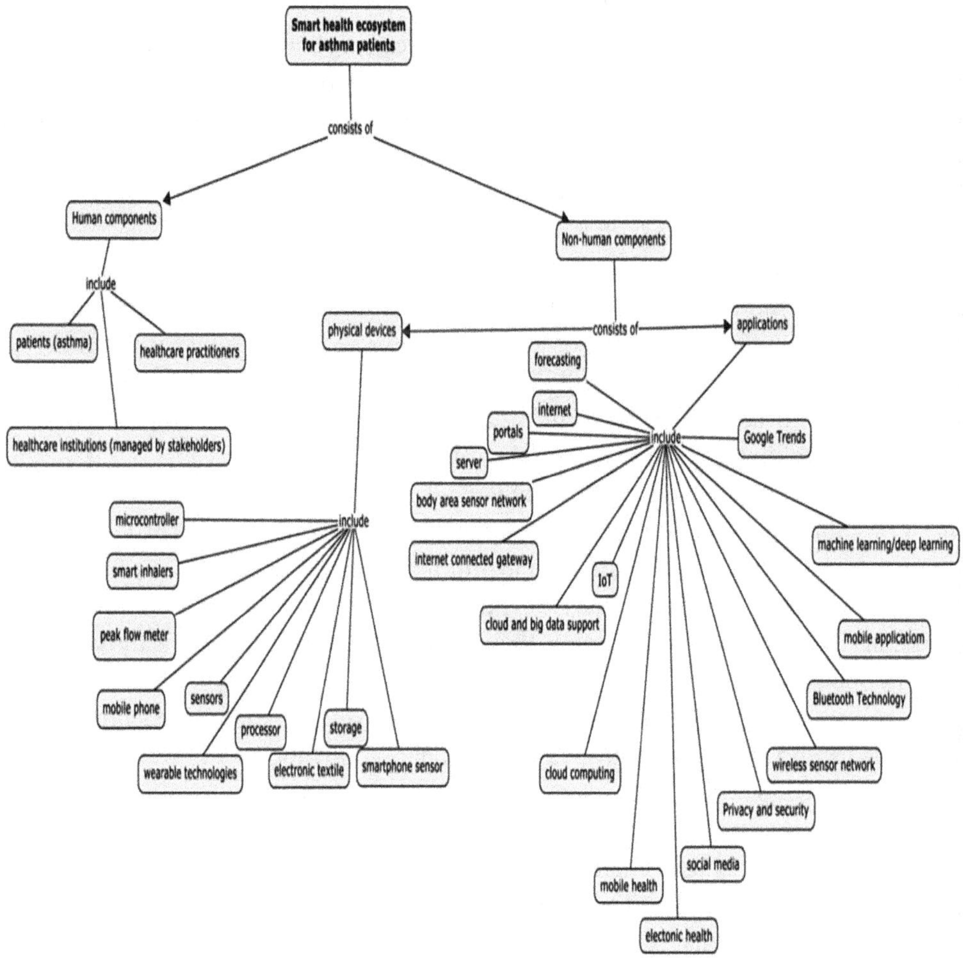

learning and deep learning (Do et al., 2016). Forecasting can be provided through social media analytics (Abbasi et al., 2014) and Google Trends. Mobile phones (Uddin et al., 2013; Uwaoma and Masingh, 2015; Deloitte, 2018) and mobile applications (Kassem et al., 2013; Wu et al., 2016), which are all part of mobile health, are necessary to facilitate deployment and ubiquity. Most of the mobile applications for asthma care are facilitated through the Internet and Internet connected gateways. Smartphone sensors can be used to gather relevant information about environmental information to alert both patients susceptible to potential asthma attacks and healthcare practitioners (Sheng et al., 2015). Portals and servers are also important in viewing patient information and providing services in the ecosystem. Social media can also play a role in ensuring the functionality of the ecosystem as patients and doctors can be linked on social media platforms to facilitate healthcare. IoT cannot be left out in a smart health ecosystem as they facilitate monitoring of patients through sensors and wearables (Fuentes et al., 2018; Siddiquee et al., 2016, Do et al., 2015). Privacy and security in the ecosystem cannot be left out as patients' data have to be protected (Zhang et al., 2018; Solonas et al., 2014).

CONCLUSION

The aim of this chapter to define a smart health ecosystem for asthma patients and describe its components was achieved by developing a conceptual framework. The findings from this study provide useful building blocks for the development of an effective smart health ecosystem for asthma patients by providing a clear set of requirements. It would be interesting to investigate how this ecosystem can be validated by both asthma patients and asthma healthcare providers in different contexts. This first academic study on smart health ecosystems has established the building blocks for future research in this area. Guidelines for implementing such an ecosystem was not provided, as such can be included in future research. For future work, researchers should endeavor to trial some of the technologies proposed as there were less publications in that area.

One major limitation with the study is the use of only three academic databases to search for relevant publications. This might have limited the number of papers that were used in this study. The reason for choosing these databases was because of their reputation for publishing excellent research on the subject. However, Google Scholar and Google Search Engine were used to identify other relevant papers the authors might have missed out. The inclusion and exclusion criteria might have also limited the type of papers which were included in this study. As such, relevant papers might have been excluded in the process. To decrease chances for bias, each paper reviewed was independently reviewed by two other reviewers. The findings were later collated and analyzed comprehensively.

REFERENCES

Abbasi, A., Adjeroh, D., Dredze, M., Paul, M. J., Zahedi, F. M., Zhao, H., & Huesch, M. D. (2014). Social media analytics for smart health. *IEEE Intelligent Systems*, *29*(2), 60–80. doi:10.1109/MIS.2014.29

Al-Dowaihi, D., Al-Ajlan, M., Al-Zahrani, N., Al-Quwayfili, N., al-Jwiser, N., & Kanjo, E. (2013, January). *Mbreath: Asthma monitoring system on the go.* Paper presented at the International Conference on Computer Medical Applications (ICCMA).

Blue Stream Consultancy. (2018). *Smart healthcare.* Available at http://www.bluestream.sg/smart-healthcare

Briscoe, G., & De Wilde, P. (2006). Digital ecosystems: Evolving service-orientated architectures. Paper presented at *1st international conference on Bio inspired models of network, information and computing systems,* Madonna di Campiglio, Trenito, Italy.

Chang, E., & West, M. (2006). Digital Ecosystems a Generation of the Collaborative Environment. Paper presented at Eighth International Conference on Information Integration and Webbased Applications, Yogyakarta, Java, Indonesia.

D'Amato, G., Vitale, C., Molino, A., Stanziola, A., Sanduzzi, A., Vatrella, A., & D'Amato, M. (2016). Asthma-related deaths. *Multidisciplinary Respiratory Medicine*, *11*(1), 37. doi:10.118640248-016-0073-0 PMID:27752310

Deloitte. (2018). *A journey towards smart health. The impact of digitalization on patient experience.* Available at https://www2.deloitte.com

Do, Q., Tran, S., & Robinson, K. (2015, December). Big data and mHealth drive asthma self-management. Paper presented at International Conference on Computational Science and Computational Intelligence, Las Vegas, Nevada. 10.1109/CSCI.2015.129

Eramo, L. (2018). *Three ways asthma treatment is getting connected.* Available at https://www.futurehealthindex.com

Fuentes, J. S., Gonzalez-Manzano, L., Solanas, A., & Veseli, F. (2018). Attribute-Based Credentials for Privacy-Aware Smart Health Services in IoT-Based Smart Cities. *Computer*, *51*(7), 44–53. doi:10.1109/MC.2018.3011042

Global Initiative for Asthma (GINA). (2014). *The Global Strategy for Asthma Management and Prevention.* Available at www.ginasthma.org

Guarnieri, M., & Balmes, J. R. (2014). Outdoor air pollution and asthma. *Lancet*, *383*(9928), 1581–1592. doi:10.1016/S0140-6736(14)60617-6 PMID:24792855

Iyawa, G. E., Herselman, M., & Botha, A. (2016a). Digital health innovation ecosystems: From systematic literature to conceptual framework. *Procedia Computer Science*, *100*, 244–252. doi:10.1016/j.procs.2016.09.149

Iyawa, G. E., Herselman, M., & Botha, A. (2016b). Identifying and defining the terms related to a Digital Health Innovation Ecosystem. In M. Herselman, & A. Botha (Eds.), *Strategies, Approaches and Experiences: Towards Building a South African Digital Health Innovation Ecosystem*. Pretoria, South Africa: CSIR.

Kassem, A., Hamad, M., El-Moucary, C., Neghawi, E., Jaoude, G. B., & Merhej, C. (2013, September). *Asthma Care Apps.* Paper presented in Tripoli, Lebanon.

Kikidis, D., Konstantinos, V., Tzovaras, D., & Usmani, O. S. (2016). The digital asthma patient: The history and future of inhaler based health monitoring devices. *Journal of Aerosol Medicine and Pulmonary Drug Delivery*, *29*(3), 219–232. doi:10.1089/jamp.2015.1267 PMID:26919553

Kokalki, S. A., Mali, A. R., Mundada, P. A., & Sontakke, R. H. (2017). *Smart health band using IoT.* Paper presented at International Conference on Power, Control, Signals and Instrumentation Engineering, Chennai, India.

Kumar, S., Mahapatra, B., Kumar, R., & Turuk, A. K. (2018, March). *Security and privacy solution for I-RFID based smart infrastructure health monitoring*. Paper presented at 2018 Technologies for Smart-City Energy Security and Power. 10.1109/ICSESP.2018.8376733

Kwan, A. M., Fung, A. G., Jansen, P. A., Schivo, M., Kenyon, N. J., Delplanque, J. P., & Davis, C. E. (2015). Personal lung function monitoring devices for asthma patients. *IEEE Sensors Journal*, *15*(4), 2238–2247. doi:10.1109/JSEN.2014.2373134

Lee, J. (2011). Smart health: Concepts and status of ubiquitous health with smartphone. Paper presented at International Conference on Information and Communication Technology Convergence, Seoul, South Korea. 10.1109/ICTC.2011.6082623

McLaughlin, M., Malone, P., & Jennings, B. (2009). A model for identity in digital ecosystems. In *3rd IEEE International Conference on Digital Ecosystems and Technologies,* 295I, Istanbul, Turkey. 10.1109/DEST.2009.5276727

Meyer, J., & Boll, S. (2014, October). Smart health systems for personal health action plans. In *International Conference on e-Health Networking, Applications and Services*, Natal-RN, Brazil. 10.1109/HealthCom.2014.7001877

Moher, D., Liberati, A., Tetzlaff, J., & Altman, D. G. (2009). Preferred reporting items for systematic reviews and meta-analyses: The PRISMA statement. *Annals of Internal Medicine*, *151*(4), 264–269. doi:10.7326/0003-4819-151-4-200908180-00135 PMID:19622511

Ra, H. K., Salekin, A., Yoon, H. J., Kim, J., Nirjon, S. S., Stone, D. J., . . . Stankovic, J. A. (2016). AsthmaGuide: an asthma monitoring and advice ecosystem. In *IEEE Wireless Health,* Bethesda, MD. 10.1109/WH.2016.7764567

Rahmani, A. M., Gia, T. N., Negash, B., Anzanpour, A., Azimi, I., Jiang, M., & Liljeberg, P. (2018). (Article in Press). *Future Generation Computer Systems Journal.*

Sampri, A., Mavragani, A., & Tsagarakis, K. P. (2016). Evaluating Google Trends as a Tool for Integrating the 'Smart Health'Concept in the Smart Cities' Governance in USA. *Procedia Engineering*, *162*, 585–592. doi:10.1016/j.proeng.2016.11.104

Seto, E. Y., Giani, A., Shia, V., Wang, C., Yan, P., Yang, A. Y., ... Bajcsy, R. (2009). A wireless body sensor network for the prevention and management of asthma. In *IEEE International Symposium on Industrial Embedded Systems,* Lausanne, Switzerland. 10.1109/SIES.2009.5196203

Sheng, M., Zhang, Y., Yang, J., Li, C., & Xing, C. (2015, September). Fihuo: A Mobile Smart Health Service Platform. In *Web Information System and Application Conference,* Jinan, China. 10.1109/WISA.2015.26

Siddiquee, J., Roy, A., Datta, A., Sarkar, P., Saha, S., & Biswas, S. S. (2016). Smart asthma attack prediction system using Internet of Things. In *IEEE 7th Annual Information Technology, Electronics and Mobile Communication Conference,* Vancouver, Canada 10.1109/IEMCON.2016.7746252

Sinharoy, A., Mitra, S., & Mondal, P. (2018). Socioeconomic and environmental predictors of asthma-related mortality. *Journal of Environmental and Public Health*, 2018. PMID:29853926

Solanas, A., Patsakis, C., Conti, M., Vlachos, I. S., Ramos, V., Falcone, F., ... Martinez-Balleste, A. (2014). Smart health: A context-aware health paradigm within smart cities. *IEEE Communications Magazine*, *52*(8), 74–81. doi:10.1109/MCOM.2014.6871673

Sundaravadivel, P., Kougianos, E., Mohanty, S. P., & Ganapathiraju, M. K. (2018). Everything You Wanted to Know about Smart Health Care: Evaluating the Different Technologies and Components of the Internet of Things for Better Health. *IEEE Consumer Electronics Magazine*, *7*(1), 1–11. doi:10.1109/MCE.2017.2755378

Suzuki, T., Tanaka, H., Minami, S., Yamada, H., & Miyata, T. (2013, March). Wearable wireless vital monitoring technology for smart health care. In *7th International Symposium Medical Information and Communication Technology,* Tokyo, Japan. 10.1109/ISMICT.2013.6521687

Teixeira, A. F., & Postolache, O. (2014, May). *Wireless sensor network and web based information system for asthma trigger factors monitoring.* Paper presented at IEEE International Instrumentation and Measurement Technology Conference, Montevideo, Uruguay.

The Global Asthma report. (2014). Available at www.globalasthmareport.org

Thomson, J., Hass, C., Horn, I., Kleine, E., Mitchell, S., Gary, K., & Amresh, A. (2017, April). *Aspira: Employing a serious game in an mHealth app to improve asthma outcomes.* Paper presented at Serious Games and Applications for Health, Perth, Australia.

Uddin, M., Gupta, A., Maly, K., Nadeem, T., Godambe, S., & Zaritsky, A. (2013, December). Smart-Spaghetti: Use of smart devices to solve health care problems. In *IEEE International Conference on Bioinformatics and Biomedicine,* Shanghai, China. 10.1109/BIBM.2013.6732598

Uwaoma, C., & Mansingh, G. (2015, January). Towards Real-time Monitoring and Detection of Asthma Symptoms on Resource-constraint Mobile Device. In *12th Annual IEEE Consumer Communications and Networking Conference,* Las Vegas, Nevada. 10.1109/CCNC.2015.7157945

Vankatesh, J., Aksanli, B., Chan, C. S., Akyurek, A. S., & Rosing, T. S. (2018). Modular and personalized smart health application design in a smart city environment. *Internet of Things Journal, 5*(2), 614–623. doi:10.1109/JIOT.2017.2712558

World Health Organization. (2007). *Global surveillance, prevention and control of chronic respiratory diseases: a comprehensive approach.*

Yung-Cheng, M., Chao, Y. P., & Tsai, T. Y. (2013, September). Smart-clothes—Prototyping of a health monitoring platform. Paper presented at IEEE Third International Conference on Consumer Electronics, Berlin, Germany.

Zhang, Y., Zheng, D., & Deng, R. H. (2018). Security and privacy in smart health: Efficient policy-hiding attribute-based access control. *IEEE Internet of Things Journal, 5*(3), 2130–2145. doi:10.1109/JIOT.2018.2825289

Chapter 9
Role of Smart Wearable in Healthcare:
Wearable Internet of Medical Things (WIoMT)

Jana Shafi
Prince Sattam Bin Abdul Aziz University, Saudi Arabia

Amtul Waheed
Prince Sattam Bin Abdulaziz University, Saudi Arabia

ABSTRACT

Remote Health monitoring is encouraged to keep an eye on the patient's health status outside the zone of the clinic. These smart wearable gadgets are now integrated with mobile apps to work efficiently as telemedicine and telehealth to incorporate into the Internet of Medical Things. This chapter introduces WIoMT (Wearable Internet of Medical Things) and reviews smart wearable healthcare devices' scientific terms as well as commercial pains. Internet of Medical Things is displayed through a refined background, wearable computing, wearable technology, cloud frameworks, and architecture design. Included are required hardware and software, body sensors, smartphones, the smart medical application, medical location analyzers for data storage, and finally, a diagnosis. Wearable devices are tested under strict observation fitness, vital signs, and smart environment. Wearable devices are now used for a wide scale of monitoring healthcare.

INTRODUCTION

Today's busy standard of living led people to be less prioritize on their health which is overlapped with another family, work, professional, study issues to take care of it. Health is dynamic on one's part which may be an impact of an environment, outbreak diseases, food intakes etc. in some of the other way. Concerning about health to be monitor is the utmost important issue to be noticed early. Personalize health monitoring system will be an attracting choice in the right place. This system will help to identify critical

DOI: 10.4018/978-1-5225-9246-4.ch009

problems (if there) before getting worsen and ensures about health concerns. It will also prove a good prevention from dreadful diseases which can be detected in an early stage. Thus, It can be understood that the personalized health monitoring system will not be something like a desktop or laptop or even a mobile which accompany all the time with patient or person rather than it will be something which can be comfortably worn and quietly do its job. Here, the concept of wearable devices jumps in, these wearable devices which are monitoring individuals is beyond the fields of health as well these devices are controlled by the Internet. Internet of things connecting wearable device the whole concept now can be called as "Smart Wearable". This chapter will discuss the smart wearable role in healthcare.

The wearable device works primarily for health monitoring on regular basis. The device quickly enables itself to gather the person's data (temperature, heart rate, heat, pulses) including location, date & time. The device is an intelligent enough to sense the abnormal data (crosses the critical level) of person and not wasting a moment it will automatically send notifications(all details) to person's relatives & friends as well location of nearby emergencies. Smart wearable is capable enough to provide details in the form of the test report for the physician as well for the patient which will be highly useful at the instant critical moment. Smart wearable should be thankful for -ever well implemented mobile invention with latest smart technology in everyone's hands. Although smart-phone is quite uncomfortable to be worn all the time due to its confine features (size, weight etc.). This wearable can be trusted enough to capture an individual's live stream data which will enable oneself to have self-knowledge about his/her body's physiology and kinesiology. Smart-phones have limited functionality to help people with health and fitness issues. This can be seen from the popular smart-watches, smart wrist-bands which are able to keep track of physical activities in a whole day, a month etc and also provides with fitness goals and other information which keeps users boost up. Smart self-monitoring wearable is so much helpful in measuring blood pressure, stress, and hypertension which will be able to control a number of diseases which are the root cause of many illnesses (Metcalf et al., 2016). Smart Wearable contribution in health is under research and making stunning progress which will be expecting a lot more in case as in surgery (Atallah et al., 2011), (Cicero et al., 2015). Surgical procedures via smart wearable are a highly convenient way of observing patient's critical signs thus reducing equipment size as well as external devices wiring. This will save the patient from transition and transportation chaos as well keeping them away from deteriorating condition in case of an emergency which may lead to injury complications. These applications contribute to improving medical quality. Leaf Technology is an example of sensor product for the patient which will be able to prevent complications to arise in a patient during a stay in the hospital. Although certain subjects should be considered while installing smart wearable for health care. For instance, People get exhausted of the fitness bands and discard them. They aren't useful; people lose their fitness drive (Wearable and the Internet of Things for Health, 2015).

BACKGROUND AND HISTORY: THE INTERNET OF THINGS (IOT) IN HEALTH

The Internet of Things (IoT) is a promising concept for the variety of new potentials brought about by enveloping connectivity. This concept expands itself with the help of network and computing ability to have connectivity with objects existing with sensor and capable to exchange data useful in daily life without humans help. The IoT principle establishes itself to construct, control, and supervise the corporeal world with the method of enveloping smart-networking, data-gathering, deep optimization, predictive analytics, machine-to-machine means, and other solutions. This is capable enough to make change in one's world of living and work. It is to be predicted nearby in recent years high offices of U.S Defence Department may be expected to add IOT devices to their systems and networks. This perception will provide as a basis of innovative decision making. IOT is another virtual connectivity which will create a different room of never-ending connection prospects to be accomplished. These are many on the scale as, Connected Smart homes, wearable Smart devices (also embeddable), the industrial Smart Internet, connected Smart cities, and connected smart cars and some are still undergoing and can be easily adapted according to the requirements to IOT potentialities. IOT innovations not only work to access data instead it is the whole sole controller of objects connected to internet. IoT development and its growth as well acceptance in world have a major role of well implemented smart phones, tablets and laptops which are known name and aware technology in everyone hands beyond borders. IOT brings in change for healthcare by introducing wearable devices and managing continuous data stream also extract information responsible for changing users outlook. This led to the personalize care with the help of personalize health monitoring systems which will give patient an extra care experience in improving condition. Impact of IoT wearable today is that the individual at the the 3 thousand feet above can also interact with all types of devices, detected, diagnosed just because of the developed algorithms and technology implemented today (Butean et al., 2015), (Lmberis & Dittmar, 2007).

The IoT Origin and Establishment

When one look back to find the origin of IoT it was found that the proposal and idea behind it by the developer of Auto-ID center at the MIT (Massachusetts Institute of Technology) which are Ashton (Ashton, 2009) and Brock (Sarma et al., 2001) .For discovering technologies for applications identification for example reducing errors, efficiency improvement also automation all envelop under Auto-Id term. The relevant Electronic Product Code (EPC) to network was initiated by the center of Auto- ID in the year of 2003 at its executive convention (Roco & Bainbridge, 2002). Meloan mention about the tracking of moving objects from one place to any other place (Meloan, 2003), EPC network launch permit some to imagine emerging IoT on large scale commercially global concept where microchips will be used to form IoT network. The booming advancement of RIFD technology mark that Internet of things (IoT) would be going to set a new era in academy and industry IT (Gershenfeld et al., 2004). National Science Foundation (NSF) in the year of 2002 published a report on convergent technology (Roco & Bainbridge, 2002). In 2005 initial report of the International Telecommunications Union (ITU) it was recommended for IoT to be merged with different technologies to have global objects connectivity. These technologies ready to merge can be object IDs, which supports on large scale networked devices and equipments in communication and interaction (Gershenfeld et al., 2004), (Atzori et al., 2010), (Dinget et al., 2013). IoT-based enterprise devices have been designed for various purposes such as, industrial setting (Sauter & Lobashov, 2011), and community transportations (Akyildiz & Jornet, 2010) and healthcare systems. IoT extends immense interest in developing nations such as in 2009 China established IoT national research centre to encourage the IoT research and development (Mattern & Floerkemeier, 2010), (Gubbi et al., 2013).

WEARABLE COMPUTING

This term denotes to **computer**-powered enabled devices or tools that can be worn by a user, in the form of clothing, watches, glasses, shoes and related items.

Characteristics

- **Connectivity:** The ability to connect devices with Internet of Things (IoT) will entirely utilize networking prospective and use same in the big scale innovations.
- **Interaction:**_A lot of IoT apps will allow users to interact and exchange information with smart environments that offer them.
- **Utilization:** Smart objects can be deployed in minute devices set with components of a microcontroller, a communication (wired or wireless) interface, a power supply, equipped with actuators and sensors that are used to interface with the neighboring environment. Environment: Feasible interconnected daily devices may lead to incompatible solutions, fragmented market demands highly standardized and interoperable environment for IoT (Want et al., 2015).
- **Organizations:** The preventive efforts for IoT fragmentation build up standard organizations which are the Internet Engineering Task Force (IETF), IEEE, also the Internet Protocol for Smart Objects Alliance (IPSO) which consists open principles for communication, which presently comprise IPv6 over Low-Power Wireless Personal Area Networks (6LoWPAN) and still in advancement Constrained Application Protocol (CoAP), (Shelby et al., 2014).
- IoT is now heterogeneous network (Heggestuen, 2013).
- IoT base on Mobile-World is mobile-centric which is portrayed by a symbiosis and interdependence linking of mobiles and users. Smart mobiles are the doorway for IoT objects and its application services (heart rate monitoring, fitness applications)
- **Interaction Defined Pattern:** There should be by default defined patterns for IoT objects and Wearable devices communication which is useful in offering highest usability and the finest user experience.
- **Simplicity:** Wearable computing working with mobile allows people to be always connected no matter which location they are. Still the concept to easily use these devices is required to fill up to generalize IoT users foot.

Communication Standards

- The IoT with existing Internet can be fully integrated and interoperable with the implementation of IPv6 protocol for smart object communication.
- IoT popularity and demand increased because its principle of reusing maximum standard Internet protocols and mechanisms.
- Complaints of low IoT performance devices required new design of data formats, New energy-efficient low- transparency communication protocols considering group communication, mobility, and interface with multiple gadgets.
- The start of 6LoWPAN has solved the crisis of fetching IP to lessen-power devices.

Interaction Patterns

- The smart phones attendance is essential while using wearable devices, now onwards it is the mobile device who takes care of all the tasks including communication and processing of data behind the scenes. Wearable devices will be interacting similar to humans with objects under smart environments such as Google glass enables person to have most natural and easy communication instead of just checking email or text updates (It's a Little Freaky at First, 2013).
- Passive wearable devices doesn't require owner interaction except the smart customize app takes care and have a control on it. It generally works for heart clock monitors and step counter.
- Active wearable devices which in contact with the user connected to smart phones providing information about the data for example of smart watches and glasses. They have the ability to reach out to other remote systems. They certainly able to drive interactions with smart objects.
- The communication patterns that envisage among humans using wearable devices in the smart objects environment. The smartphone identifies the smart objects in its immediacy for which they have access grant. With ZeroConf (Cirani et al., 2014) or low-power radio beacon broadcasting services and CoAP Resource Directory resource identification mechanisms, represented data on the interface.
- Users are able to select any one icon and able to fetch all information for the same including function sets, earlier descriptors that enables interactions with source and user interface (Shelby &Vial, 2014).
- The interaction of user with resources act as connection between smart objects communication can be accomplish with the help of following methods.
- 'Get' allows the users to fetch associated state along resource using CoAP GET request.
- *'Observe' is used when the resource condition changes allowing to* receive asynchronous updates (Heggestuen, 2013).

  One can achieve this by CoAP GET request which contains *Observe* option to send updates with the object permission.
- *'Update'* execute on a resource (depends on the function set) to locate its state with CoAP POST or PUT request.CoAP-unaware clients that may use HTTP for communication.

Implementation

IoT testbed deployed interactive wearable app. (Belli et al., 2015) which consists of 70 IP-addressable gadgets operating CoAP servers that provides resources of sensors and actuators, and also has different type of nodes types of nodes for computations and interfaces for radio access. There are thirty six controlled IoT nodes belongs to class-1 devices[46] established on the Contiki OS and with IEEE 802.15.4 radio six TelosB constrained nodes and thirty four single-board computer (SBC) nodes which are class-2 devices, (Bormann et al., 2014) typically performing on a Linux OS and with several network interfaces of 20 Intel Galileo SBC nodes. The mobile/wearable application was designed on the Android Wear platform with LG G Watches and Android Lollipop smart phones. The mapping of smartwatch screenshots of the running application with flow (Kovatsch et al., 2014).There are chiefly three actors involved in the smart application's life cycle: the smartphone, the wearable device, and smart objects; every process, which can be started by any user, engages in exchange of information that influence all users. Figure 1 displays the interaction with the help of CoAp through the wearable companion application (WA) using the Android Wearable Data Layer application interface among smartphone and the smart objects grounded on Bluetooth for communication (Bojanova et al., 2014), (Bonomi et al., 2012), (Fettweis et al., 2014).

WEARABLE TECHNOLOGY

The popularity based on the portability of the devices which can be easily worn with less visibility in any environment for to access customize data of the owner, it also avail the option of remote observation with integrated communication networks.

Architecture Design

The in general architectural design of the Internet of Things (IoT) system is represented in Figure1. This is a multi-tier architecture that consists of mobile device, middleware, and the cloud-hosted applications and utilities. This system aids sensor data attainment from various sources that is in the users' interests. Therefore, an assortment of user activities including location data can be collected either at home, driving, at work, or vitals. The middleware is light-weight and independent platform since it can be hosted may be on the smart-phone or on a backend server. This is essential because there is variety in the communication practices between sensor devices for object-to object communication, such as, Z-Wave, Bluetooth, ZigBee, RFID, Bluetooth Low Energy (BLE), NFC, and LTE-A. This meant that a protocol coordination service is essential to combine all the diverse protocols for multi-sensor data achievement. The middleware when installed on a smartphone provides a single interface which is use to unite different protocols thus certifying that different supported communication modes. Yet, the other feature of the communication condition is how to thrust sensor data to backend services that do not support small-range practices. For example, it is impractical to send data directly amid a cloud components and Bluetooth enabled device which communicate through HTTP. Furthermore, the cloud services contain many communication protocols such as HTTP, CoAP, MQTT, this can additionally hinder inter-communication with the backend infrastructure. Hence, the middleware installed as a service which ropes the communication among the machines and infrastructure. This is gained via protocol practice composition, sensor data implementation, and direction-finding. Additionally to this it is required the various data to match the devices. In order to make certain derivation or digital audit track, the data produced and collected have to be coordinated to the source sensor device. Moreover, multi-user sight to the architecture has to be modalities in place to coordinate data from different sensor nodes to an exact user.

Figure 1. Wearable Technology IoT Architecture

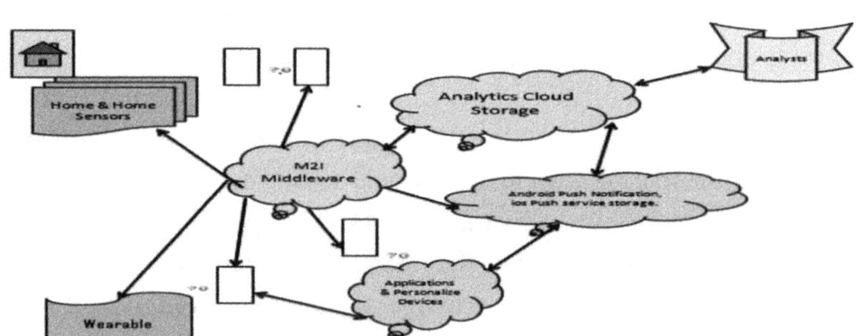

Hardware/Software

Wearable technology based products should have great scale of communication potential which is supported by Bluetooth and WLAN with their respective smart phones (acts as a server).As the sage of the devices increase it led to the necessity of greater bandwidths of bluetooth and WLAN. Several parameters have to balance as acquiring the above goal as many technologies have to be integrated (Bluettoth, Wlan) in wearable device with the consideration of acquiring low transmission rates with less power consumption. The data bus technology use has vast pattern when combining together in platforms Wearable peripherals join in the data bus structure to share data. It is an accurate connection to the way the human being senses effort, where the mind and backbone linking to senses and function control connectivity. This point of view permits the addition of the image highlights like wearables are, as the expansion of the control functions and human senses. The data bus is equivalent to human spine. This is not the new approach. The concept of a 'wearable motherboard' originated for years. The Wearable Motherboard is able to observe the unobtrusive way of user's vital signs. Sensors are related to the motherboard and linked to human body areas that are being checked with the help of Interconnection Technology. An integrated bendable bus transmits data to monitoring devices. It also transmits information to the sensors from exterior sources enabling the 'wearable motherboard' a potentially useful and making the structure for prospect wearable devices.

Cloud Enabled Framework

The cloud framework is now a fundamental service such as electricity, as data now can be accessed from anywhere in the globe with an online (Internet) connection. The collected data by present wearables live mainly in siloed systems or in disparate clouds. This limits analyzing work of the data for possible business intelligence complicated. Whereas the end users still gain profit from the data collected results, the bigger bang is when protected data shared with other data for intelligent relative processing, to which the great tool is a cloud. Open access to data about development in nations around the world is provided by World Bank Open Data (Data.gov, (2014). Data.gov is a important resource from the United States that provides admission to a catalogue of federal government data (World Bank Open Data, 2014).

Body Sensor Networks

In spite of the countless assisting technologies developed for the years, elevated installation costs and inadequate utilities have prohibited their common acceptance. Several new sensing stages have been introduced such as facilitating pervasive blood pressure monitoring, gait, electrocardiogram (EKG), Galvanic skin response, photoplethysmogram (PPG), as well other vital symbols. Widespread research study also has been accomplished on ambient aptitude and its health and welfare. Current developments in low-power wireless connectivity have significantly cut down the installation costs, and permitted large-scale commercial operations. Mutual advancements in low-power wireless network technologies as well the passage from IPv4 to IPv6, each object will shortly be linked to the Internet and happen to part of the Internet of Things (IoT), which was initially anticipated by Kevin Ashton in 1999. Health research broadly adopted IoT technologies as well Wearable, BSN, ambient. For example, a broad study was hold by the U.K. Biobank with wearable sensors to compute actions of a large group of patients (Watts,2012).

Similar wearable device survey for the physical activity, sleep, and strength of a huge population was conducted by The National Health and Nutrition Examination Survey (Troiano et al., 2014).

Movements of Body

Human motion detection can be use in various applications. Recently electronics used by consumers have implied semiconductor-based tracking system which enables its users to control body motions with the help of interface (G &Mahfouz, 2013). Healthcare is an important application area in terms of tracking. This field requires accurate testable devices which limits for the medical application designs. This started with use of uniaxial accelerometer sensors in motion tracking concept by Veltink and Boom in 1996 (Veltink & Boom, 1996). With the series of clinical investigation on position view by accelerometers which introduced the motion tracking concept in the medical application world (Foerster et al., 1999), (Mathie et al., 2003), (Uswatte et al., 2000). Although only an accelerometer is unable to afford accurate information; thus, in 1997, investigators enhanced the tracker's performance and of integrated semiconductor-based gyroscopes to achieve bio-mechanical assessments and gait study (Miyazaki, 1997). After Four years the year of 2002, (Mayagoitia et al., 2002) initiated the initial setup to follow lower edges with a union of gyroscopes and accelerometers in the 2Dimensional sagittal plane. After the successful implementation of reliable integrated gyroscope trackers, the chief center for medical applications using inertial motion tracking was gait analysis (Takeda et al., 2009). It works in exacting gait motion with the help of gravity sensitive accelerometers used to guess the gravity vector and the sensor's axes tilt angles (Miyazaki, 1997). The opening version of recently incorporated sensors (using accelerometer, gyroscope, and magnetometer) which supply very precise data with 9DoF, was proposed in the year 2006 by Roetenberg (Roetenberg et al., 2006). To monitor human motion he anticipated using a set of tri-axial gyroscopes, tri-axial accelerometers also a magnetometer. Body motions are calculated using 9DoF sensors: which are three inertial sensors, an accelerometer, a magnetometer, and a gyroscope. The complete sensors values can be customized by means of definite instructions sent by the microcontroller (Altun et al., 2010) (Kaewkannate & Kim, 2018). The box is of translucent plastic which access the LED indicators observation working on the board. The proposed and implemented device of this research group was intended to be used for three types of monitoring movement. The first and second kinds of monitoring movement are pointed to as short- and long-term monitoring data. In the first form, the device is only coupled to a PC to monitor the outcomes (Bertolotti et al., 2016). Whereas, In the second type, observed data stored locally for long-term monitoring. In the third case, the utmost important element is the body network, where multiple units installed on the human body and wired to a gateway unit, including a local memory or may be a wireless connection to a PC or hand-held gadgets (Misfit: Fitness Trackers & Wearable Technology, 2017).

Diagnosis of Disease

Wearable sensors are a lot of helpful in observing heart rate, activity, skin temperature which expose about human body functioning as well monitors any infection, inflammation, sweat analysis, insulin resistance which helps in diagnosing the user body if any unusual conditions are present.

WEARABLE DEVICES IN HEALTHCARE

Smart phones have constraints in the comfort ability to wear while the body in motion as well by weight, size and aesthetic requirements. Smart phones cannot access inside the health and fitness of human body which is offer by wearable. Mostly wearable are being used for self-monitoring hypertension and stress.

Smart Clinical Devices

- VitalPatch MD: Wireless Biometric Patch
- Proteus Pills: Smart Sensor Pills
- Adheretech: Wireless Smart Pill Bottle

Smart home-care devices

The Table 1 shows the prevailing devices which is used in homes by individuals

Smart Clothing and Footwear

Today footwares also smart and have real time data stream such as smart socks, which permit healthcare experts to distantly keep an eye on patients' vitals counting oxygen levels, heart rate, and body temperature. These Devices are in the form of shoes and soles that compute patient's pace which can help a wounded patient through rehab and diagnose whether a patient is probably undergo from a fall as well check the chances diabetic foot-related wound of a patient. Injury Alerts are sent to the healthcare experts in the occurrence of any variations in the patient's physical condition that need medical aid. Implementation of these alerts from the wearable able to healthcare experts to keep track of the numerous patients health at the same time and guarantee for urgent medical assistance those in need . Family members can also be used these devices. Table 2 Shows below the popular smart textile brands their role and working.

Table 1. Smart Homecare devices

Device	Role	Working
Canary Care: Remote Home Healthcare Monitoring	For the elderly(old) or those with long-term care needs	Sensors network System located around the house to respond to movement, light and temperature, plus a card reader by the front door to check in and out.
Stethee: Bluetooth Stethoscope	Healthy beats for ordinary for the patient.	Heart sounds are wirelessly streamed to a mobile device and can be heard with earbuds.
Insulin Angel: Medication Temperature Sensor	Monitoring of blood sugar levels	Insulin Angel takes readings at five-minute intervals and communicates over Bluetooth Low Energy
MonBaby: Bluetooth Infant Monitor	Wirelessbreathing and movement monitor for infants that sends instant alerts to a parent's smartphone.	Device snaps to a one or other clothing.
Tempdrop: Wearable Fertility Thermometer	Women track their fertility period.	BBT is a common method for figuring out when a woman's body is ovulating.

Table 2. Smart Textiles

Smart Cloth	Role	Working
Clothing+	To track the continuous condition of the patient through the nanotechnology.	Their fabrics are comfortable to wear and accurate for analyzing the data and information.
Camira Fabrics	record the health status of a person .	It helps the healthcare industry to record the health status of a person through the use of smart fabrics.
Hexoskin	assist the healthcare professionals in different therapeutic areas .	Their fabrics are embedded with cardiac, movement and breathing sensors for tracking the patient's data as per the health segment. It uses the smart nanotechnology for the same.
Pireta	For continuous physiological monitoring and emergency care	Develops smart sensor garments using the nanotechnology for assisting the hospitals and nursing homes.
Stitch	streamlined healthcare system by developing yarns and textiles	Best of nanotechnology for manufacturing the smart fabrics that helps the nursing homes and hospitals to have an easy monitoring of patients.
3Tecks	producing non woven fibers assisting the medical care since years	Using the skin as a communication or interface, this technology has brought great solutions for the big medical industries.

PRIVACY AND SAFETY

Mostly smart wearable devices are habitually do not approach with an integrated security system such as user verification or PIN method safety characteristics and they generally accumulate data locally devoid of private encryption (Metcalf et al., 2016), (Wearable Security, 2015).Moreover, smart wearable devices need advanced interactions (Mellisa, 2013) security concerning data integrity, encryption, secrecy and other protection services since it depends on the unrestrained wireless network may be Bluetooth or Wi-Fi wireless connection to transmit data such as HP study (Wearable Technology Devices, 2015) exposed that 30% of the tested smart watches were susceptible to report yielding, which is an attack that get access to the wearable device and to data due to feeble password policy, be short of out account lockout and user details. Additionally, mostly wearable devices do not have integrated security system for example user verification or PIN system security sorts also without encryption they locally store data (Wearable Security, 2015). For example, Smart Apple Watch and Google's Android Wear platform lacking of any security methods defending their pricey wearable from loss or stealing.

WEARABLE SAFETY

Minute particles which are present in the air is a mixture of PM (Particulate matter). These minute particles with a 10 µA diameter or may be less can be breathing in body that origin critical health concerns to lungs. Particles size less then 2.5 µm (PM2.5) diameter able to reach to the lungs depth which pose the greater hazards then the other particles(2.5 and 10 µm (PM10)). These particles are rooted from power plants, automobiles, forest fires, metals and organic compounds (Wu et al., 2018).WSN (Wireless sensor network) is a potential answer for a many monitoring structural health (Torfs et al., 2013), environmental (Wu et al., 2017), internal air quality (Kim et al., 2014), also fire (Silvani et al., 2014). WSN monitoring in indoor air quality (IAQ) is supple in consumption and be able to reduce power wiring and associated infrastructures costs (Jelicic et al., 2013). WSNs typically have strict power necessities to keep uninterrupted monitoring. Thus, a well-organized energy source as well power organization should be elected. Each system measures different kinds of injurious environmental parameters such as WSN system that examines volatile organic compounds (VOCs), and CO in an edifice (Jelicic et al., 2013).

It evaluates methane and temperature in functioning environment (Spiryakin & Baranov, 2016). In Reference (Manes et al., 2012) it measures VOC, dampness, temperature and wind in an industrialized site. Lastly, work in reference (Antolín et al., 2017) measures amount of CO2 concentrations. Wearable sensor nodes are vital fundamentals in wireless body area network in examining the human body (Wu et al., 2017). Medical signals further used to gather environmental circumstances around the individual's body for example in safety applications (Manes et al., 2012) In Reference (Tsow et al., 2009) afford a wearable wireless sensor system for deadly volatile organic compounds monitoring with the help of wireless Bluetooth interface (Wu et al., 2018).

EMERGING TRENDS OF SMART WEARABLE AND MEDICAL INTERNET OF THINGS

Connecting Health Care Provider and Smart Wearable Consumer

Health connected with technology can be understood as tele-care, tele-health, Patients and health care-takers are empowered with the help of existing mobile technologies by handing over the health control by making them independent. Today Digital technology connected patients-healthcare providers which provides convenient personalize quick services resulting in better health which can be achieved by dual remote monitoring method and supporting treatment. To achieve connected health positive outcomes discussed above needs quality connectivity of broadband. Connected technology care use in the growth of healthcare providers as treatment can be communicated and treated as soon as possible without delay to visit the clinics or hospitals which provides care at home itself with the help of digital communication. Technology Development improving the wearable and apps capabilities which helps in remotely monitor the patients and updates the latest status as well alert and remind the patients with the help of digital messages to follow their treatment course well. These all technology driven developments are well promoted in the healthcare industry by the launch of wearable of the world popular companies of Apple and Google .

Wearable Devices In Health Monitoring

We can think consumer wearable as surplus pleasure while medical wearables are mission-critical or life savior (Global wearable technology market, 2012-2018). Popular used Smart Medicinal wearable refers the one which cannot be tackled by apps and are particularly designed to monitor, diagnose and get treatment of the specific situations (Endeavour Partners. (n.d.)) such as asthma, heart arrhythmia, pain management, chronic disruptive pulmonary disease, breast cancer, and a range of other diseases. In spite of tracking information on mobile devices and assemble data by clinicians to provide diagnosis, it helps with patient management and assist studies. We have some examples to understand further available wearable development (Business Insider Intelligence, 2015).

Table 3. Global Market Data

2015	$4.8 billion
2016	$5.5 billion
2021	$19.5 billion
compound annual growth rate (CAGR) expected to increase from 28.8% for 2016-2021	

- **Cardiac Monitoring:** To monitor the heart rate to test for arrhythmias with Smart wearable ECG devices (Wearable Medical Devices Market, 2016).
- **Infant Monitoring:** These wearable devices are able to sense a wet diaper also body movement, as well as blood oxygen levels monitored (Olson & P., 2017).
- **Pain Management**: Wearable devices are equipped to deliver transport pain medication or transcutaneous electrical nerve stimulation (TENS), which is common to use in the medical industry (Olson & P, 2017), (Tech Styles).

Expansion of Healthcare Industry

Global healthcare costs, now anticipated at 6 trillion dollars to 7 trillion dollars, are estimated to attain more than 12 trillion dollars just in seven years. Given this view, essential actions are projected to manage increase expenditure (Reportlinker, 2018).

Highlights

- The wearable medical devices U.S. market is likely to grow from 2.5 billion dollars in 2016 to almost 8.7 billion dollars in the year of 2021 at a CAGR of 27.9% from 2016-2021.
- The wearable medical devices Asian market is likely to raise from 908 million dollars in 2016 to almost 4.0 billion dollars in 2021 at a CAGR of 34.5% from 2016 – 2021 as shown in the Table 3.

Smart Health Applications

Certainly, simply data tracking via gadget is not only sufficient to preserve good quality healthy lasting behaviors, and the usage of software should include motivational implementation for the significant adoption and behaviors training. Conversely, less common amongst these apps were prompts to act and problem-solving chances and action scheduling, which are vital constructs that interpret purpose into long-standing behavior change.

Mobile apps, wearables, and sensors in the clinical setting yet mobile apps, smart wearables, and sensor technologies are hopeful but still needs further research (Yao, & Ho, 2017). Following are the three apps Examples which doctors may choose to recommend to their patients:

- Azumio Instant Heart Rate (Ishikawa et al., 2017) A free available downloadable app on Apple as well on Android operating systems that calculates heart rate and put on view a monitoring band (Ho et al., 2016).

- ◦ Azumio Sleep Time .(Sleep Time by Azumio, 2017) This is a free downloadable app that have the function to monitors sleep patterns, efficiency, and also operates as an alarm clock.
- ◦ RelaxLineBellybio Interactive Breathing (RelaxLine, 2010) A free downloadable app that aids users breathing rate to attain peace. mobile phone Placing the touching the abdomen, users can chase the emitted hum to observe deep, abdominal breathing using this easy yet amazingly helpful tool. In 2014 leading pharmaceutical companies had 63% unique apps in comparison to 2013.

Convergence of Smart Wearable and Mobile for Health Care

Mhealth (mobile health) and digital health have big range on scale uses such as health analysis for complex population, chronic management care. Healthcare transformation remains have m health as a prime partner with the shifting of healthcare as a patient centric, based on delivery model. The partnership will be successful with the stakeholder's eagerness to clinch m-health. To convince about the worthiness of technology success with profit is also a motive in business for the involvement of providers and commissioners. Mobile devices are rapidly increasing in all age groups. Mobile devices in form of tablets, smart phones are rapidly increasing almost in all age populations, although it little bit is behind with the elder population where maximum health care and social services required. Although now the situation started changing with the exposure of smart phones. According to the reports of 2014 (Butean et al., 2015), Smartphones users are motivated with the notifications of losing weight, exercise anf fitness, monitoring and managing health with the help of rising medical apps (Lmberis & Dittmar, 2007). Growth in wearable devices types, Biosensing wearable is the remarkable develop met in wearable technology. For examples Digital hearing aids, fitness bands, smart pills, blood-pressure monitors, smart textiles equipped with sensors. Wearable devices communicate data with the help of smart phone apps requiring internet.

The characteristics are as follows:

- Wearables to work require apps on smart phones.
- Wearable should be comfortable that should not distract with normal movements of human body.
- This is a hot area and market is progressing and demanding for developing wearable.

New Period of Proactive And Preventative Care

Now the period of prevention started which developed healthcare model emphasize on the prevention care which includes prevention of diseases before happening. The best example to sneak is the Case Study of The Future Prediction and Prevention: Chan-Zuckerberg Initiative. In September year of 2016, Mark Zuckerberg (CEO FB) and Priscilla Chan declares to donate 600 million dollars to outline a fresh, sovereign research organization: the Chan Zuckerberg Biohub.

The Stanford Byers Center for Biodesign is developing a numerous predictive devices in perspective to prevent disease prior it hits. These technological innovations comprise a monitoring device that predicts pediatric asthma attacks just before the days they happen, a cost- efficient genetic test to find the heart disease causes also a rapid non-invasive test for likely heart-transplant rejection. From the year of 2011, almost 2 billion dollars has been invested to finance predictive analytics health care companies (The Future of Personalized Health Care, 2014).

Table 4. Commercially available wearable in healthcare

Wearable Company	Porpose	Cost
Apple	ECG function brings fitness trackers	$400, with iPhone cost above than $1,000
Fitbit	identify sleep apnea.	$200
Lifesense	Tracking app to construct a customized exercise program intended to eliminate leaks.	Carin full set €149.00 WIL UNDERWEAR €50,00 WIL SET €289,00
LifeWatch AG	Remote monitoring services for cardiac diagnostic.	14.35$
Philips Electronics	monitor vital signs, body posture and step count, detects falls of the patients to follow up.	Many ranges are present in market.
Samsung	health metrics.	Samsung's Galaxy Watch$299.99, Gear Sport$22.50 or Fit2 Pro $99fitness tracker.
Sotera Wireless	ViSi Mobile® is to observe, doctors and nurses have instant access to vital patient factors in order to intervene and respond rapidly to prevent adverse situations and to progress patient safety	$250
Xiaomi Inc.	The Xiaomi Me Band 2 is the third fitness tracker.	Sold by Second Neutrino*Ltd.* and Fulfilled by Amazon.*Price*: $24.00 & FREE Shipping.
Garmin	use in the community wellness, population health, and patient monitoring markets.	Range from 45$ -999$

Wearable patches are not any kind of removable for clothing, or wraps that enfold the skin, as an alternative they are stick adhesives may or may not be containing a medication or sensor. But they do fit in a means of sticking on the skin for interval of time and all use electrons to work (Wearable Patches, 2017). The receiving end of users for monitoring and analysis are focused in the diseases like cancer, diabetes, cardiac disorders and blood pressure conditions (The age of smart patches, 2018).

Commercially Available Wearable Devices

Wearable Medical Devices are available in market as shown below in Table 4.

SUMMARY AND CONCLUSION

Internet of things emerges as a possible solution to networked inter-connected devices which are better, cheaper, faster, and provisions in health care consumption with customer-driven innovations. These networked smart wearable operate via apps play a major role in laying the foundation to the ever-growing practice of the IoT in healthcare. Smart defensive judgments made for practical individualized healthcare and treatment (Metcalf et al., 2016). In health care provisions, they are enabling to compute vital signs devoid of persistent surgery. Embeddable can be e-tattoos which are sensors equipped that can transmit data by wireless technology. The Wearable Technology power is realistic and clinically evident in patient diagnosis, treatment and health care. This is because of it low-cost facility to collect regular statistics in a distinct style for longitudinal time in any setting. Big data can be easily available with the help of Integration to the cloud, assisting the machine learning application algorithms for new results. The real-time implementation of suitable data directives, validation/verification structures, omnipresent integration to universal network and growing of generations will observe Wearable Technology comprehend its prospective. In this chapter, the key concept of wearable technology are outlined in the terms of healthcare which are in general were examined with its healthcare applications. The study was accomplished including background initial approach of IoT in the area of health. In this chapter we find the smart wearable role in carrying out potential health operations and its effectiveness in the community especially to the elders. In the beginning of chapter IoT is introduced followed by wearable computing and technology in detail with various parameters in the direction of health sector. Middle of the chapter broadly discusses the examples of implemented smart medicinal wearable of different categories which is quite interesting to know which includes smart clinical devices, textiles etc. In all the wearable common technique is to collect data and send to the physician for further diagnosis as well some of the wearable detections are generated on the basis of collected data itself for example stress, energy, blood pressure etc. The chapter chiefly focuses on the smart wearable role in healthcare which is positively accepted by the nation's societies and still encourages to be used. Therefore, expanding the healthcare industry globally due to the growth of connected smart wearable is broadly discussed near the end of chapter. The potentials of wearable device in healthcare is the core point of research question. Here it is concluded that in bid for the information assembled by several wearable devices to work appropriately, companies require to consider organizing methods to store, combine, analyze and distribute this immense quantity of new data. The patients monitor staying in a hospital by wearable technology would help to develop EHR (electronic health records) quality. Elderly people health can be Track with vital signs who are residing lonely through wearable medical technology would improve the reaction time in case of an urgent situation. The wearable available with the essential qualities that add in to better health and fitness monitoring, secure transfer data, user's pains exclusion to care for oneself, pre-alert system to help prevent vital condition.

CHALLENGES AND FUTURE PERSPECTIVES OF SMART HEALTH WEARABLE

To understand and completely realize the advantages of smart wearable technologies in the field of healthcare and fitness research and healthcare providers have to adopt these technologies by studying their requirements and working upon them to develop a broad approach to health and wellness services instead of prevailing apps which checks up to single diseases. Parallel to these researchers and healthcare providers are responsible for providing wide variety healthcare services with the help of available commercial provider sources which are in close contact with users. Communications companies along with Wireless, phone, and cable providers give way in to the home and offer distant caring services billing. A key solution to retain and improve research and development is to build an integrated design of intellectual home services with wearable systems and health and wellness devices for home relieve (Coughlin, 2010). Academic research must undertake issues such as:

- System efficiency, reliability, and unobtrusiveness: (Daniel et al., 2009).
- User requirements and privacy (Mynatt et al., 2004).
- User awareness and approval:
    - Network health data may be unimportant as normally understood (Steele, 2009).
- Legislation (Dickens & Cook, 2006): difficult conditions may arise in tele-health, m-health practice when healthcare experts start to provide services with electronic technology, for instant on Internet; the impression of those services effects the place person resides. The rule affects the healthcare expert and the beneficiaries of the services. The condition becomes more difficult when those persons do not live in the same nation, for case; licensure, documentation, and fortification must be standardized in provisions of laws inside the European societies.
- Interoperability: As healthcare grows from a centered organization through service-centric towards an individual-centric system, information systems concerned must be semantically interoperable, process-associated, decision-compassionate, context perceptive, client - oriented, and truthful (Blobel & G, 2007).
- Services: incorporating prolong and valuable tele-homecare services want key stakeholders concerned in the commencement of the assignment of the delivery services to the elderly people (Essén & Conrick, 2008).
- Compensation or cost: This is the most major determinant persuading elderly approval of technology, for example telemedicine availability for thirty years (Pelletier-Fleury et al., 1997). Personal insurance companies may partly support healthcare expenses. In the case of tele-medicine, monetary assessment of wearable systems still has scope for improvement (Bergmo,2010).
- End-user guidance to use wearable systems: Similar to Telemedicine, tackling technological hurdles is one of the essential settings for its dispersion (Paul, 1999).
- Social enclosure: (Percival &Hanson, 2006) tele-care establishes hazards of enlarged remoteness of users, healthcare experts and healthcare providers through reduce physical facial interaction with healthcaregivers and physicians.
- Moral issues: conquering the obstacles will be complicated and might need financial support and routine incentives by payers that are buyers including doctors, health plans, employers and the government's national social insurance program (JG, 2007).
- Wearable system's Technological potentials: They must convene healthcare expertss and end-users necessities in managing diseases and general remote Healthcare Management.
- Interactive affiliation: focusing Research on relationship amid patients, experts and healthcare providers are required (Daniel et al., 2009).
- Wearable system market diffusion: The government may need to interfere to tackle health information technology market failure (Kleinke, 2005).

REFERENCES

Akyildiz, I., & Jornet, J. (2010). The Internet of nano-things. *IEEE Wireless Communications*, *17*(6), 58–63. doi:10.1109/MWC.2010.5675779

Altun, K., Barshan, B., & Tunçel, O. (2010). Comparative study on classifying human activities with miniature inertial and magnetic sensors. *Pattern Recognition*, *43*(10), 3605–3620. doi:10.1016/j.patcog.2010.04.019

Antolín, D., Medrano, N., Calvo, B., & Pérez, F. (2017). A Wearable Wireless Sensor Network for Indoor Smart Environment Monitoring in Safety Applications. *Sensors (Basel)*, *17*(2), 365. doi:10.339017020365 PMID:28216556

Atallah, L., Jones, G. G., Ali, R., Leong, J. J., Lo, B., & Yang, G. (2011). Observing Recovery from Knee-Replacement Surgery by Using Wearable Sensors. *2011 International Conference on Body Sensor Networks*. 10.1109/BSN.2011.10

Atzori, L., Iera, A., & Morabito, G. (2010). The Internet of Things: A survey. *Computer Networks*, *54*(15), 2787–2805. doi:10.1016/j.comnet.2010.05.010

Belli, L., Cirani, S., Davoli, L., Gorrieri, A., Mancin, M., Picone, M., & Ferrari, G. (2015). Design and Deployment of an IoT Application-Oriented Testbed. *Computer*, *48*(9), 32–40. doi:10.1109/MC.2015.253

Bergmo, T. S. (2010). Economic evaluation in telemedicine – still room for improvement. *Journal of Telemedicine and Telecare*, *16*(5), 229–231. doi:10.1258/jtt.2010.009008 PMID:20501629

Bertolotti, G. M., Cristiani, A. M., Colagiorgio, P., Romano, F., Bassani, E., Caramia, N., & Ramat, S. (2016). A Wearable and Modular Inertial Unit for Measuring Limb Movements and Balance Control Abilities. *IEEE Sensors Journal*, *16*(3), 790–797. doi:10.1109/JSEN.2015.2489381

Blobel, B. G. (2007). Educational Challenge of Health Information Systems' Interoperability. *Methods of Information in Medicine*, *46*(01), 52–56. doi:10.1055-0038-1628132 PMID:17224981

Bojanova, I., Hurlburt, G., & Voas, J. (2014). Imagineering an Internet of Anything. *Computer*, *47*(6), 72–77. doi:10.1109/MC.2014.150

Bonomi, F., Milito, R., Zhu, J., & Addepalli, S. (2012). Fog computing and its role in the internet of things. *Proceedings of the first edition of the MCC workshop on Mobile cloud computing - MCC '12*. 10.1145/2342509.2342513

Bormann, C., Ersue, M., & Keranen, A. (2014). *Terminology for Constrained-Node Networks*. doi:10.17487/rfc7228

Brown, A. (2015). *Technology that is Flexible, Sticky, and Smart = Wearable Patches*. Retrieved from https://www.wearable-technologies.com/2015/08/technology-that-is-flexible-sticky-and-smart-wearable-patches/

Bui, N., & Zorzi, M. (2011). Health care applications. *Proceedings of the 4th International Symposium on Applied Sciences in Biomedical and Communication Technologies - ISABEL '11*. 10.1145/2093698.2093829

Business Insider Intelligence. (2015, May 21). *THE WEARABLES REPORT: Growth trends, consumer attitudes, and why smartwatches will dominate*. Retrieved from http://bit.ly/mpo0617tp3

Butean, A., David, A., Buduleci, C., & Daian, A. (2015). Auxilum Medicine: A Cloud Based Platform for Real-Time Monitoring Medical Devices. *2015 20th International Conference on Control Systems and Computer Science*. doi:10.1109/cscs.2015.135

Cicero, M. X., Walsh, B., Solad, Y., Whitfill, T., Paesano, G., Kim, K., ... Cone, D. C. (2015). Do You See What I See? Insights from Using Google Glass for Disaster Telemedicine Triage. *Prehospital and Disaster Medicine*, *30*(01), 4–8. doi:10.1017/S1049023X1400140X PMID:25571779

Cirani, S., Davoli, L., Ferrari, G., Leone, R., Medagliani, P., Picone, M., & Veltri, L. (2014). A Scalable and Self-Configuring Architecture for Service Discovery in the Internet of Things. *IEEE Internet of Things Journal*, *1*(5), 508–521. doi:10.1109/JIOT.2014.2358296

Coughlin, J. F. (2010). Disruptive Demographics, Design, and the Future of Everyday Environments. *Design Management Review*, *18*(2), 53–59. doi:10.1111/j.1948-7169.2007.tb00083.x

Daniel, K. M., Cason, C. L., & Ferrell, S. (2009). Emerging Technologies to Enhance the Safety of Older People in Their Homes. *Geriatric Nursing*, *30*(6), 384–389. doi:10.1016/j.gerinurse.2009.08.010 PMID:19963147

Data.gov. (n.d.). Retrieved from http://www.data.gov

dataworldbank.org. (n.d.). Retrieved from http://www.dataworldbank.org

Dickens, B., & Cook, R. (2006). Legal and ethical issues in telemedicine and robotics. *International Journal of Gynaecology and Obstetrics: the Official Organ of the International Federation of Gynaecology and Obstetrics*, *94*(1), 73–78. doi:10.1016/j.ijgo.2006.04.023 PMID:16777109

Ding, Y., Jin, Y., Ren, L., & Hao, K. (2013). An Intelligent Self-Organization Scheme for the Internet of Things. *IEEE Computational Intelligence Magazine*, *8*(3), 41–53. doi:10.1109/MCI.2013.2264251

Doukas, C., & Maglogiannis, I. (2012). Bringing IoT and Cloud Computing towards Pervasive Healthcare. *2012 Sixth International Conference on Innovative Mobile and Internet Services in Ubiquitous Computing*. 10.1109/IMIS.2012.26

Endeavour Partners. (n.d.). Retrieved from https://endeavour.partners

Essén, A., & Conrick, M. (2008). New e-service development in the homecare sector: Beyond implementing a radical technology. *International Journal of Medical Informatics*, *77*(10), 679–688. doi:10.1016/j.ijmedinf.2008.02.001 PMID:18514021

Fettweis, G. P. (2014). The Tactile Internet: Applications and Challenges. *IEEE Vehicular Technology Magazine*, *9*(1), 64–70. doi:10.1109/MVT.2013.2295069

Foerster, F., Smeja, M., & Fahrenberg, J. (1999). Detection of posture and motion by accelerometry: A validation study in ambulatory monitoring. *Computers in Human Behavior*, *15*(5), 571–583. doi:10.1016/S0747-5632(99)00037-0

Gershenfeld, N., Krikorian, R., & Cohen, D. (2004). The Internet of Things. *Scientific American*, *291*(4), 76–81. doi:10.1038cientificamerican1004-76 PMID:15487673

Gubbi, J., Buyya, R., Marusic, S., & Palaniswami, M. (2013). Internet of Things (IoT): A vision, architectural elements, and future directions. *Future Generation Computer Systems*, *29*(7), 1645–1660. doi:10.1016/j.future.2013.01.010

Hartke, K. (2015). Observing Resources in the Constrained Application Protocol (CoAP). doi:10.17487/rfc7641

Heggestuen, J. (2013, Dec. 15). One In Every 5 People In The World Own A Smartphone, One In Every 17 Own A Tablet [CHART]. Retrieved from http://www.businessinsider.com/smartphone-and-tablet-penetration-2013-10?IR=T

Ho, K., Borycki, E., Kushniruk, A., Juhra, C., & Hernandez, M. (2016). *The health perspective in using digital media for health and wellness. 2016 Digital Media Industry & Academic Forum*. DMIAF. doi:10.1109/dmiaf.2016.7574895

Ishikawa, T., Hyodo, Y., Miyashita, K., Yoshifuji, K., Komoriya, Y., & Imai, Y. (2017). Wearable Motion Tolerant PPG Sensor for Instant Heart Rate in Daily Activity. *Proceedings of the 10th International Joint Conference on Biomedical Engineering Systems and Technologies*. 10.5220/0006109901260133

"It's a little freaky at first, but you get used to it": Sergey Brin and Google Glass at TED2013. (2014, Jan. 21). Retrieved from http://blog.ted.com/sergey-brin-with-google-glass-at-ted2013/

Anderson, J. G. (n.d.). Social, ethical and legal barriers to e-health. - PubMed - NCBI. Retrieved from https://www.ncbi.nlm.nih.gov/pubmed/17064955

Jara, A. J., Alcolea, A. F., Zamora, M. A., Skarmeta, A. F., & Alsaedy, M. (2010). *Drugs interaction checker based on IoT. 2010 Internet of Things*. IOT. doi:10.1109/iot.2010.5678458

Jelicic, V., Magno, M., Brunelli, D., Paci, G., & Benini, L. (2013). Context-Adaptive Multimodal Wireless Sensor Network for Energy-Efficient Gas Monitoring. *IEEE Sensors Journal*, *13*(1), 328–338. doi:10.1109/JSEN.2012.2215733

Kaewkannate, K., & Kim, S. (2018). *The Comparison of Wearable Fitness Devices*. Wearable Technologies. doi:10.5772/intechopen.76967

Kim, J., Chu, C., & Shin, S. (2014). ISSAQ: An Integrated Sensing Systems for Real-Time Indoor Air Quality Monitoring. *IEEE Sensors Journal*, *14*(12), 4230–4244. doi:10.1109/JSEN.2014.2359832

Kleinke, J. (2005). Dot-Gov: Market Failure And The Creation Of A National Health Information Technology System. *Health Affairs*, *24*(5), 1246–1262. doi:10.1377/hlthaff.24.5.1246 PMID:16162569

Kovatsch, M., Lanter, M., & Shelby, Z. (2014). Californium: Scalable cloud services for the Internet of Things with CoAP. *2014 International Conference on the Internet of Things (IOT)*. 10.1109/IOT.2014.7030106

Lmberis, A., & Dittmar, A. (2007). Advanced Wearable Health Systems and Applications - Research and Development Efforts in the European Union. *IEEE Engineering in Medicine and Biology Magazine*, *26*(3), 29–33. doi:10.1109/MEMB.2007.364926 PMID:17549917

Manes, G., Collodi, G., Fusco, R., Gelpi, L., & Manes, A. (2012). A Wireless Sensor Network for Precise Volatile Organic Compound Monitoring. *International Journal of Distributed Sensor Networks*, *8*(4), 820716. doi:10.1155/2012/820716

Mathie, M. J., Coster, A. C., Lovell, N. H., & Celler, B. G. (2003). Detection of daily physical activities using a triaxial accelerometer. *Medical & Biological Engineering & Computing*, *41*(3), 296–301. doi:10.1007/BF02348434 PMID:12803294

Mattern, F., & Floerkemeier, C. (2010). From the Internet of Computers to the Internet of Things. *Lecture Notes in Computer Science*, *6462*, 242–259. doi:10.1007/978-3-642-17226-7_15

Mayagoitia, R. E., Nene, A. V., & Veltink, P. H. (2002). Accelerometer and rate gyroscope measurement of kinematics: An inexpensive alternative to optical motion analysis systems. *Journal of Biomechanics*, *35*(4), 537–542. doi:10.1016/S0021-9290(01)00231-7 PMID:11934425

Metcalf, D., Milliard, S. T., Gomez, M., & Schwartz, M. (2016). Wearables and the Internet of Things for Health: Wearable, Interconnected Devices Promise More Efficient and Comprehensive Health Care. *IEEE Pulse*, *7*(5), 35–39. doi:10.1109/MPUL.2016.2592260 PMID:28113167

Metcalf, D., Milliard, S. T., Gomez, M., & Schwartz, M. (2016). Wearables and the Internet of Things for Health: Wearable, Interconnected Devices Promise More Efficient and Comprehensive Health Care. *IEEE Pulse*, *7*(5), 35–39. doi:10.1109/MPUL.2016.2592260 PMID:28113167

Misfit.com. (n.d.). *Fitness Trackers & Wearable Technology.* Retrieved from http://misfit.com/products/shine

Miyazaki, S. (1997). Long-term unrestrained measurement of stride length and walking velocity utilizing a piezoelectric gyroscope. *IEEE Transactions on Biomedical Engineering*, *44*(8), 753–759. doi:10.1109/10.605434 PMID:9254988

Mynatt, E., Melenhorst, A., Fisk, A., & Rogers, W. (2004). Aware technologies for aging in place: Understanding user needs and attitudes. *IEEE Pervasive Computing*, *3*(2), 36–41. doi:10.1109/MPRV.2004.1316816

Olson, P. (2017, Feb. 6). *Report: Jawbone Is Jumping Out Of Consumer Wearables.* Retrieved from http://bit.ly/mpo0617tp5

Paul, D., Pearlson, K., & McDaniel, R. (1999). Assessing technological barriers to telemedicine: Technology-management implications. *IEEE Transactions on Engineering Management*, *46*(3), 279–288. doi:10.1109/17.775280

Pelletier-Fleury, N., Fargeon, V., Lanoé, J., & Fardeau, M. (1997). Transaction costs economics as a conceptual framework for the analysis of barriers to the diffusion of telemedicine. *Health Policy (Amsterdam)*, *42*(1), 1–14. doi:10.1016/S0168-8510(97)00038-9 PMID:10173489

Percival, J., & Hanson, J. (2006). Big brother or brave new world? Telecare and its implications for older people's independence and social inclusion. *Critical Social Policy*, *26*(4), 888–909. doi:10.1177/0261018306068480

RelaxLine. (2010, Feb. 20). *BellyBio Interactive Breathing*. Retrieved from https://itunes.apple.com/us/app/bellybio-interactive-breathing/id353763955?mt=8

Reportlinker. (2018, Jan. 15). *The global market for wearable medical devices was valued at $4.8 billion in 2015*. Retrieved from https://www.prnewswire.com/news-releases/the-global-market-for-wearable-medical-devices-was-valued-at-48-billion-in-2015-300582717.html

ABI Research. (2018, Jan. 31). *Smart Footwear to Provide Unique Data Opportunities for Enterprises and Healthcare Providers*. Retrieved from https://www.prnewswire.com/news-releases/smart-footwear-to-provide-unique-data-opportunities-for-enterprises-and-healthcare-providers-300591076.html

Roco, M. C., & Bainbridge, W. S. (2003). Overview Converging Technologies for Improving Human Performance. *Converging Technologies for Improving Human Performance*, 1-27. doi:10.1007/978-94-017-0359-8_1

Roetenberg, D., Luinge, H., & Veltink, P. (2003, October). Inertial and magnetic sensing of human movement near ferromagnetic materials. *The Second IEEE and ACM International Symposium on Mixed and Augmented Reality, 2003. Proceedings*. doi:10.1109/ismar.2003.1240714

Rohokale, V. M., Prasad, N. R., & Prasad, R. (2011). A cooperative Internet of Things (IoT) for rural healthcare monitoring and control. *2011 2nd International Conference on Wireless Communication, Vehicular Technology, Information Theory and Aerospace & Electronic Systems Technology (Wireless VITAE)*. doi:10.1109/wirelessvitae.2011.5940920

Sarma, S., Brock, D., & Engels, D. (2001). Radio frequency identification and the electronic product code. *IEEE Micro*, *21*(6), 50–54. doi:10.1109/40.977758

Sauter, T., & Lobashov, M. (2011). How to Access Factory Floor Information Using Internet Technologies and Gateways. *IEEE Transactions on Industrial Informatics*, *7*(4), 699–712. doi:10.1109/TII.2011.2166788

Science Service Dr. Hempel Digital Health Network. (2018, April 27). *Now, sensors can attach to your skin and internal organs | The age of smart patches*. Retrieved from https://www.dr-hempel-network.com/digital-health-technolgy/wearables-development-smart-patches/

Shelby, Z. (2014). Foreword. From *Machine-To-Machine to the Internet of Things*, xi. doi:10.1016/b978-0-12-407684-6.00020-6

Shelby, Z., Hartke, K., & Bormann, C. (2014). *The Constrained Application Protocol (CoAP)*. doi:10.17487/rfc7252

Silvani, X., Morandini, F., Innocenti, E., & Peres, S. (2014). Evaluation of a Wireless Sensor Network with Low Cost and Low Energy Consumption for Fire Detection and Monitoring. *Fire Technology*, *51*(4), 971–993. doi:10.100710694-014-0439-9

Sleep Time by Azumio. (n.d.). Retrieved from http://www.azumio.com/s/sleeptime/index.html

Spiryakin, D., & Baranov, A. (2016). Uniform Inbuilt Wireless Sensor Node for Working Conditions Monitoring. *Proceedings of the 2016 Federated Conference on Computer Science and Information Systems*. 10.15439/2016F386

Statista. (n.d.). Global wearable technology market 2012-2018 I Statistic. Retrieved from http://bit.ly/mpo0617tp1

Steele, R., Lo, A., Secombe, C., & Wong, Y. K. (2009). Elderly persons' perception and acceptance of using wireless sensor networks to assist healthcare. *International Journal of Medical Informatics*, *78*(12), 788–801. doi:10.1016/j.ijmedinf.2009.08.001 PMID:19717335

Takeda, R., Tadano, S., Todoh, M., Morikawa, M., Nakayasu, M., & Yoshinari, S. (2009). Gait analysis using gravitational acceleration measured by wearable sensors. *Journal of Biomechanics*, *42*(3), 223–233. doi:10.1016/j.jbiomech.2008.10.027 PMID:19121522

Tarouco, L. M., Bertholdo, L. M., Granville, L. Z., Arbiza, L. M., Carbone, F., Marotta, M., & De Santanna, J. J. (2012). Internet of Things in healthcare: Interoperatibility and security issues. *2012 IEEE International Conference on Communications (ICC)*. 10.1109/ICC.2012.6364830

Tech-Styles: Are Consumers Really Interested in Wearing Tech on their Sleeves? (n.d.). Retrieved from http://bit.ly/mpo0617tp2

MEDIA. (2014, Oct. 26). *The Future of Personalized Health Care. Predictive Analytics by @Rock?* Retrieved from https://www.slideshare.net/RockHealth/the-future-of-personalized-health-care-predictive-analytics-press

To, G., & Mahfouz, M. R. (2013). Modular wireless inertial trackers for biomedical applications. *2013 IEEE Topical Conference on Power Amplifiers for Wireless and Radio Applications*. 10.1109/PAWR.2013.6490226

Tolentino, M. (2013, May 30). *4 Security Challenges for Fitbit, Google Glass + Other Wearable Devices*. Available at siliconangle.com/blog/2013/05/30/4-securitychallenges- for-fitbit-google-glass-other-wearable-devices/

Torfs, T., Sterken, T., Brebels, S., Santana, J., Van den Hoven, R., Spiering, V., ... Zonta, D. (2013). Low Power Wireless Sensor Network for Building Monitoring. *IEEE Sensors Journal*, *13*(3), 909–915. doi:10.1109/JSEN.2012.2218680

Troiano, R. P., McClain, J. J., Brychta, R. J., & Chen, K. Y. (2014). Evolution of accelerometer methods for physical activity research. *British Journal of Sports Medicine*, *48*(13), 1019–1023. doi:10.1136/bjsports-2014-093546 PMID:24782483

Tsow, F., Forzani, E., Rai, A., Wang, R., Tsui, R., Mastroianni, S., ... Tao, N. J. (2009). A Wearable and Wireless Sensor System for Real-Time Monitoring of Toxic Environmental Volatile Organic Compounds. *IEEE Sensors Journal*, *9*(12), 1734–1740. doi:10.1109/JSEN.2009.2030747

Uswatte, G., Miltner, W. H., Foo, B., Varma, M., Moran, S., & Taub, E. (2000). Objective Measurement of Functional Upper-Extremity Movement Using Accelerometer Recordings Transformed With a Threshold Filter. *Stroke*, *31*(3), 662–667. doi:10.1161/01.STR.31.3.662 PMID:10700501

Veltink, P. H., & Boom, H. B. (1996). 3D Movement Analysis Using Accelerometry — Theoretical Concepts. *Neuroprosthetics: from Basic Research to Clinical Applications*, 317-325. doi:10.1007/978-3-642-80211-9_39

Want, R., Schilit, B. N., & Jenson, S. (2015). Enabling the Internet of Things. *Computer*, *48*(1), 28–35. doi:10.1109/MC.2015.12

Watts, G. (2012). UK Biobank opens its data vaults to researchers. *BMJ, 344*(mar30 2), e2459-e2459. doi:10.1136/bmj.e2459

Future Market Insights. (n.d.). *Wearable Medical Devices Market - Global Industry Analysis. Size and Forecast, 2016 to 2026.* Retrieved from http://bit.ly/mpo0617tp4

Ching, K. W., & Singh, M. M. WEARABLE TECHNOLOGY DEVICES SECURITY AND PRIVACY VULNERABILITY ANALYSIS. Retrieved from https://www.scribd.com/document/315235407/WEARABLE-TECHNOLOGY-DEVICES-SECURITY-AND-PRIVACY-VULNERABILITY-ANALYSIS

Wearables and the Internet of Things for Health - IEEE PULSE. (2015, May 20). As Fitbit Prepares for IPO, New Consumer Research Reveals Areas of Wearables Market Vulnerability for Fitness Band Leader. Retrieved from http://www.argusinsights.com/fitbit-ipo-release/

Wearables security: Do enterprises need a separate WYOD policy? (n.d.). Retrieved from https://searchsecurity.techtarget.com/answer/Wearables-security-Do-enterprises-need-a-separate-WYOD-policy

Wu, F., Rüdiger, C., Redouté, J., & Rasit Yuce, M. (2018). A Wearable Multi-sensor IoT Network System for Environmental Monitoring. *Internet of Things*, 29-38. doi:10.1007/978-3-030-02819-0_3

Wu, F., Rüdiger, C., & Yuce, M. (2017). Real-Time Performance of a Self-Powered Environmental IoT Sensor Network System. *Sensors (Basel)*, *17*(2), 282. doi:10.339017020282 PMID:28157148

Wu, T., Wu, F., Redoute, J., & Yuce, M. R. (2017). An Autonomous Wireless Body Area Network Implementation Towards IoT Connected Healthcare Applications. *IEEE Access: Practical Innovations, Open Solutions*, *5*, 11413–11422. doi:10.1109/ACCESS.2017.2716344

Yao, C. A., & Ho, K. (2017). Mobile Sensors and Wearable Technology. *Handbook Integrated Care*, 113-119. doi:10.1007/978-3-319-56103-5_7

Chapter 10
Design and Development of Internet of Things-Based Wireless Health Monitoring System

Neetu Marwah
Manipal University Jaipur, India

ABSTRACT

In this world with exponentially increasing fitness risks, the authors are proposing an Internet of Things (IoT) based device named Wireless Health Monitoring System (WHMS), to put patients' thinking at ease. We are dwelling in a world which is no longer viable for a doctor to keep a watch over a patient's indispensable parameters all the time. This device is helpful for aged people staying alone at home, people dwelling on hills, pregnant females anywhere, and for busy people who cannot often contact a doctor. A health practitioner, some distance away from the patient, needs to understand his heart rate and body temperature of the physique to begin preliminary treatment. Keeping this as a preluding landmark, the authors are proposing an embedded gadget which can measure the rate of heart beat and body temperature, and keep the statistics on the cloud server for the doctor to determine the next course of action.

INTRODUCTION

As indicated by the records outfitted by World Health Organization (WHO), around 32% individuals over 18 years old have passed away, throughout the world, and hidden causes are initiated because of cardiovascular maladies, fundamentally an infirmity of heart and veins. ("Cardio Vascular Disease", n.d.) (World Health Organization, n.d.) (Lancet, 2013) Turmoil of the heart and veins incorporates Coronary Heart Disease (CHD), Rheumatic Heart Disease (RHD), expanded circulatory strain (Hypertension), cerebrovascular infection (Stroke), fringe supply route illness, inherent heart affliction and coronary heart disappointment. ("Cardio Vascular Disease", n.d.) Ongoing overviews likewise reason that India will take not long to have the most extreme instances of coronary heart issue on the planet. Various

DOI: 10.4018/978-1-5225-9246-4.ch010

Figure 1. Consequences of poor lifestyle ("Survey on impact of lifestyle on our health", n.d.)

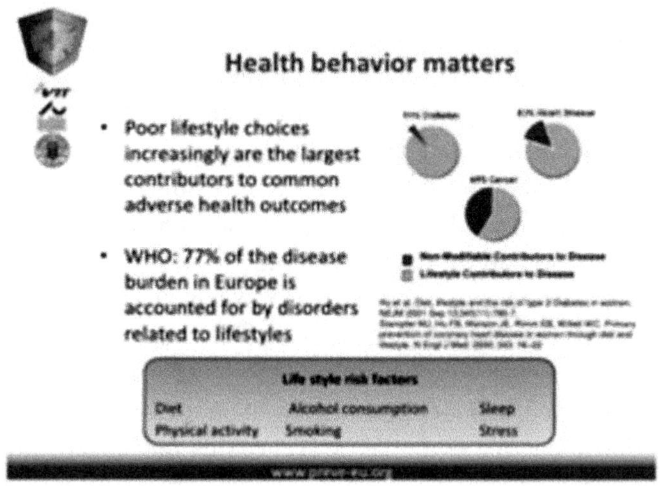

kinds of Cardio Vascular Diseases (CVD) need constant observation of basic body parameters, which may furthermore require protracted stay remains lengthy.

Also, different specialists have demonstrated that low scopes of physical exercises lead to a wide assortment of ailment, for example, abnormal amounts of systolic and diastolic circulatory strain, broadened rates of overweight and stoutness and a lot more to list. Figure 1 expresses the outcomes of sedentary way of life on an individual's wellbeing in agreement to a review led through WHO.

As per WHO studies directed in 2017 in India, there is exclusively one approved allopathic doctor for each 10,189 individuals ("Cardio Vascular Disease", n.d.). This demonstrates a horrifying situation of absence of enough therapeutic specialists and inability to go to every one of the sufferers in the need of hour. Therefore, it is never again possible for a medicinal specialist to take a gander at a patient's heart beat and body temperature for each moment constantly. Once more, a wellbeing specialist far from the patient need to comprehend coronary heart rate and body temperature for fundamental treatment.

In this section, authors present WHMS, a wearable device that will reliably show the wellness of patient. This wearable gadget significantly comprises of temperature sensor and coronary heart rate sensor. WHMS will gauge the coronary pulse and body temperature and aggregate the records fit as a fiddle of bio cautions from sensors and send it to wellbeing specialists control server for capacity and preparing the use of remote correspondence. This information will be accessible to the restorative specialists on server from any region utilizing IoT utility (Gogate, 2018).

Each influenced individual utilizing this gadget will have a unique API key through which all his clinical parameters will be refreshed over constant period of time consequently. Specialist can get directly access information utilizing indistinguishable unique API key, whenever required. Along these lines, doctors know about patient's wellbeing stipulations, excepting the need of being physically present constantly.

A model of WHMS has been made utilizing three-dimensional engineering of Wireless Body Area Network (WBAN). At Level 1, Arduino Nano board dependent on the ATMEGA328P microcontroller, is utilized to gain measurements from sensors. At Level 2, microcontroller will send this data to server for the utilization of ESP8266 Wi-Fi remote transmission and at Level 3, web availability is utilized to exchange data to distant for medicinal applications utilizing IoT utility based ThingSpeak server.

Literature Survey

As a general practice methodology, wellbeing authorities utilize manual system for coronary heart checking to quantify Electro Cardio Gram (ECG) by means of interfacing probes to the chest of patient. The configuration of ECG is checked with the help of unmistakable observing gadgets having cumbersome wire connections. Additionally, as these gadgets don't bolster long separation correspondence, the sufferers need to keep on being in the wellbeing facility for longer time, for authorities to observe and need to hold up under high emergency clinics. Be that as it may, most recent logical advances have prompted robotization in heart and ECG estimating strategies and the utilization of an assortment of techniques. Numerous authors have inspected the state of art frameworks which are open for checking extraordinary indispensable body parameters like temperature, ECG, pulse, beat rate and galvanic pores and skin reaction (GSR) (Alemdar, Hande, and Cem Ersoy, 2010) (Ko, 2010) (Hung and Zhang, 2002). We are exhibiting a short overview on the cutting-edge situation of strategies to develop a WHMS.

A contextual investigation about of a wearable ECG checking machine (Martin and Raskovic, 2008), examined about the issues encompassing wearable PCs, which are utilized as savvy wellbeing screens. Not at all like existing wellness video units (for instance, ECG and EEG holders), that are utilized basically for record securing, the proposed framework gave a continuous remarks to the patient, both as a notice of approaching logical/restorative crisis or as a pending check-up and providing help in the proper frame of time. These logical functionalities are considered novel from use of wearable processing for clinical personnel's like specialists, medical caretakers, and emergency clinical experts, and for restorative observing operability's, in the form of various wearable functionalities in terms of their I/O necessities, sensors, dependability, protection issues, and customer interface. The authors presented a model of wearable ECG showing incredible performance in terms of elite, low power controlled advanced microprocessor with enhancement in basic structure.

A remote advanced stethoscope utilizing bluetooth innovation" was proposed by (Mills and Nketia, 2007). The inventors referenced about stethoscopes which are utilized to focus on acoustic cautions from the inward organs of the human body. Despite the fact that stethoscopes play an extremely vital capacity in the investigation procedure, the chest piece and the interfacing link are known to encourage transmission of pathogens from patient to patient and from patient to the client. They argued that supplanting the interfacing link with a Wi-Fi framework may help in diminishing the sensible dangers and further empowering of broadcasting of the signs to multi-clients for examination. The model incorporated a two-piece Bluetooth-based Wi-Fi device that disposes of the associating links in electronic stethoscopes. The design comprised of a Bluetooth based completely incorporated chest-piece module for catching acoustic sound transmission and a microcontroller-based (MSP430) head-piece recipient module for interpreting the measurements for the three operational methods of the stethoscope.

An easy, convenient, high-throughput Wi-Fi sensor device for phonocardiography applications was proposed by (Sangasoongsong, Kunthong, Sarangan, Cai, and Bukkapatnam, 2007). This paper proposed the design and incorporated a remote sensor gadget created utilizing a Microchip PICDEM developer bundle to gather and screen human heart sounds for phonocardiography applications. This framework has the usefulness to fill in as a financially savvy decision to the most recent advancements in Wi-Fi phonocardiography sensors that have mainly focused on Bluetooth innovation. The other preferred standpoint of this remote sensor gadget was that it has been structured and created in-house utilizing off-the-rack viewpoints and utilized open supply software program for far away and versatile applications. In any case, the device experienced the unwanted commotions of sensor markers in the encompassing condition.

Sensor telemetry utilizing Wi-Fi science was proposed by (Lewalski, 2012). In this paper, a mechatronic insights-based procurement contraption was produced whereby signal molding was completed by sensor and the signal was permitted to go through a Bluetooth transmitter module. Bluetooth recipient was set in exceptionally close distance, identifying with an outer host PC. The Data Acquisition (DAQ) Bluetooth module coordinated with the microcontroller for investigating the molded thermocouple signal and performed the diverse DAQ usefulness as and when required. The microcontroller firmware secured a server that gathers all DAQ measurements for transmission by utilizing Bluetooth to the host PC for showing the outcomes through a customer interface. This Short Distance Telemetry (SDT) framework was considered as a substitute for slip rings and their related cabling.

Potential space applications for body-driven remote and e-material based radio antenna was proposed through (Kennedy, Fink, Chu, and Studor). This paper depicted the great estimated space potential of body-driven Wi-Fi and e-material receiving antenna. Alongside the standard information sources, for example, voice, tracking, and video, new records sources comprising of biotelemetry, high-rate video and information, specific following records and status of handheld devices, units and on-body inventories have been presented. Quickened in general execution from reception apparatus proposes remarkable possible thought for the consequent innovation of room antenna frameworks. Radio antenna designed using fabric materials, anyway ought to be rough and solid for use in space applications.

An aperture coupled patch antenna system for reconciliation into wearable material structures was talked about through (Hertleer, Tronquo, Rogier, and Van Lange hove, 2007). This paper discussed about the rise of wearable fabric frameworks as of late that showed the requirement for Wi-Fi based information exchange, which can be joined into different clothing materials. In literature, various planar receiving antenna dependent on texture materials have been displayed. In any case, other than a custom-made sustaining structure for wearable applications, the remote information transmission would be illogical. An Aperture-Coupled Patch Antenna (ACPA) meets this necessity considering the unbendable coaxial feed used to be changed through a microstrip feed line that coupled its vitality into the reception apparatus through an opening in the ground plane. This model was the first ACPA, completely made out of material. The outcome was an extraordinarily productive, completely adaptable and wearable antenna that was fused into pieces of clothing. In any case, the revealed gain of antenna was quite low.

A wearable, wireless electronic interface for fabric sensors was referenced in (Shu, Tao, and Feng, 2010). This paper depicted the kinds of electronic interface for wearable sensors, protecting wireless connection, suitable and reduced size range, simple hearty structure and immune to external noise and mitigation to wearers in everyday life. A new wearable, wireless computerized interface for resistive material sensors was exhibited, which met the above expressed necessities and had the capacity to work on constant estimations. Top to bottom system arrangement, precision and goals have been investigated and structure techniques have been characterized. Exploratory outcomes demonstrated that this electronic interface uncovered under 1% error in measurement extended for wearable textile resistive sensors. It likewise demonstrated an attractive dependability to control framework-based obstruction. The interface has been effectively used in a foot strain estimation framework.

Advancement of nanoparticle film-based multi hub material sensors for biomedical purposes has been referenced in (Alvares, Wieczorek, Raguse, Ladoucuer, and Lovell, 2013). This paper portrayed the material understanding in biomedical applications, for example prosthetic engineered skin and negligibly obtrusive medical procedure, which requires multi-pivotal identification abilities with high affectability and unwavering quality. This paper featured the fabrication, displaying and test portrayal of a novel nanoparticle resistive-based multi-hub sensor over the scope of 0– 300 mN, for use in material feedback circuit. The fabrication was done using inkjet printing, and smaller scale shaping systems, which were adaptable to sub millimeter measurements.

A wearable yarn-based piezo-resistive sensor was produced by (Huang, Shen, Tang, and Chang, 2008). This paper depicted the smart materials for the utilization of texture-based sensors to show gesture, stance or breath, which have been under assessed in numerous applications. A large portion of textile-based sensors have been manufactured by methods of either covering piezo-resistive materials on a material or by quickly sewing conductive filaments into textures. Clearly, structures of materials, comprehensive yarn material and textile material, will influence the performance characteristics of sensors. Be that as it may, investigate on the consequences of the structures have never again been investigated yet. In this paper, yarn-based sensors have been created by methods for utilizing piezo-resistive filaments, flexible, and standard polyester strands. Single and twofold wrapping strategies had been utilized to manufacture the yarn-based sensors. The exhibitions of the structured yarn-based sensors had been assessed by methods for estimating their resistance change underneath variable load stacking. It is demonstrated that slippage happens between the piezo-resistive filaments and the core elements. The relationship of the resistance versus the strain can't be depicted as a direct component and must be displayed as a second order equation. Because of the symmetric structure, the twofold wrapping yarn ought to withstand the slippage and more noteworthy linearity in the resistance curve can be given. At last, tests had been performed on a respiratory monitoring device to demonstrate the practicality of the yarn-based sensors and the outcomes approved that the yarn-based sensor can follow the respiratory signals accurately.

Microcontroller based temperature indicator and controller was examined by (Jamkar and Chile, 2004). This paper depicted the confined movements of elderly individuals or substantial disabled people or people experiencing some critical illnesses. These people are normally restricted to their homes because of their health conditions. This paper displayed the layout and enhancement of GSM empowered installed gadget for observation of simulated blood strain and body temperature. The framework utilized GSM for conveying the variations from the normal in the simulated biomedical parameters. The indication of peculiar deviation in the estimations of any of these parameters from their set point values was promptly detected and message is transmitted to individual caretakers. In the event that no such help is accessible, this gadget sends SMS specifically to home, doctor or guardian's mobile phone. It is a bidirectional discussion machine in which the guardian/specialist, whenever, can send SMS to know about the present parameter of the influenced individual.

The point of convergence of our work is on relentless checking of patients/individuals at home and the utilization of human services observing contraption dependent on Wi-Fi based sensor network. The work has been done considering various thought processes like decrease in expense of extensive stays at therapeutic establishment, remote observability and simplified access and operability.

Proposed System Architecture

The proposed WHMS depends on three-dimensional design of WBAN. The detailed plan of proposed WHMS is given in Figure 2, which clarifies the development of all the three phases and the signal exchange between imperative parts.

Level 1 comprises of remote sensor hubs connected to the patient's body for detecting predefined body parameters. The information gained by means of the sensors is prepared and transmitted through control unit. Level 2 is the middle of intermediate unit which secures the sent information. It is moreover responsible for storing, processing and presenting the information. Level 3 is related with ready and

Figure 2. Block diagram of WHMS based on 3-level architecture

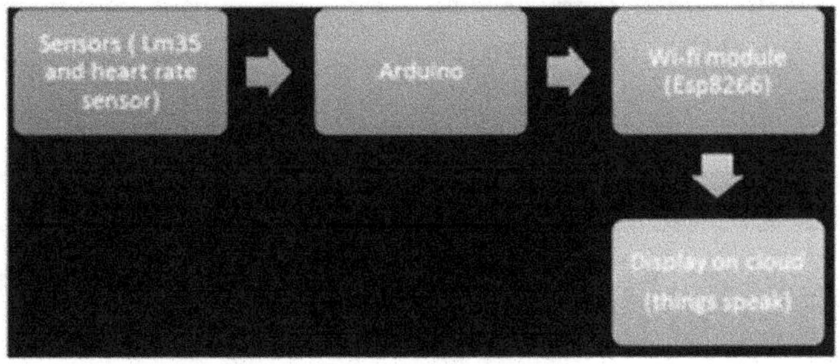

alerted structures and information transmission to longer distance through reliable web administrations. Distinctive remote communication modules, for example, ZigBee, Wi-Fi and Bluetooth are available for level 1, level 2 and level 3 correspondence. The authors have utilized Wi-Fi communication in level 1 and 2 and level 3 for long distance information exchange over IoT.

Proposed Methodology

As apparent from Figure 2, the two sensors-LM35 (which is utilized to gauge body temperature) and MAX30100 (used to quantify pulse and circulatory strain), will quantify client's body temperature, pulse and BP all the while. These sensors not just exclusively measure the quintessential parameters anyway but also exchanging the computerized information to Arduino. Utilizing some predefined capacities, record is prepared utilizing Arduino in such a way, that the favored results can be acquired in a plausible and productive way. When this functionality is over, to information needs to be exchanged to cloud for showing the outcome to the client. This exchange process is completed by utilizing ESP8266, a Wi-Fi module. The need of Wi-Fi module is to set up an invulnerable connection between Arduino and cloud and this information exchange strategy must be quick enough to prevent any loss of information.

The authors have used the cloud administrations of ThingSpeak to present the client's body temperature, coronary pulse and circulatory strain regarding time estimation by means of sensors. Every one of the elucidations are shown continuous with respect to time. The crucial characteristic for the cloud service provider is that it requires a unique API key to give access to the information. It likewise gives arrangement of sharing data to other individuals as per client's decision. Utilizing ThingSpeak server, records are made reachable to affirmed people like specialists, therapeutic professional and overseers who can remotely screen patient's measurements over web and can get ready messages in circumstance of crisis. The flowchart introduced in Figure 3 demonstrates the working system of WHMS.

Working Model

In this research demonstration, Arduino Nano is utilized to amass readings from MAX 30100 pulse sensor and exchange it over Wi-Fi utilizing ESP8266 module. Inbuilt signal molding circuit is utilized for the interfacing of sensor with A to D converter of Arduino Nano. Signals are then send to cloud storage service utilization of ThingSpeak. ThingSpeak is an open source IoT utility and API which is

Figure 3. Work process of proposed WHMS

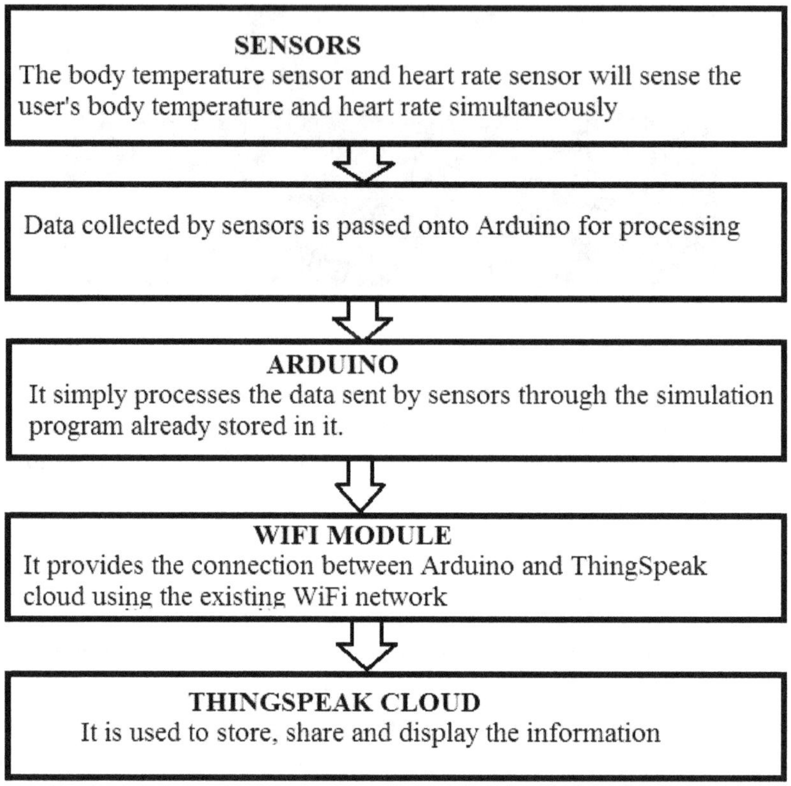

utilized to record and recover data from sensors and makes utilization of the HTTP convention over the web or by utilizing a Local Area Network (LAN). The therapeutic data is made accessible to guardians and restorative professional through web for quick and advantageous reference. The authors have made utilization of web facility accessible at household or medical clinic, to make the correspondence procedure convenient and solid. If there should be an occurrence of crisis, an alarmed message will be send to guardians or family when readings cross the predefined edge values. To develop the viability and precision of the proposed framework, ECG sensor AD 8232 is utilized to uncover the circumstance of coronary heart in agreeable and suitable way.

Cloud Service of ThingSpeak

ThingSpeak is an IoT based analytical stage transporter that grants clients to aggregate, visualize and break down current record of data streams in the cloud. This gives on the spot representations of data uploaded/transferred by different gadgets to server. With the capacity to execute MATLAB code, client can do online examination and preparing of the insights of information as it comes in. This is consistently utilized for prototyping and evidence of ideal IoT structures that require investigation.

IoT depicts a rising system, where tremendous yet evaluated volume of inserted devices (things) are associated with the Internet.

Figure 4. Working of ThingSpeak

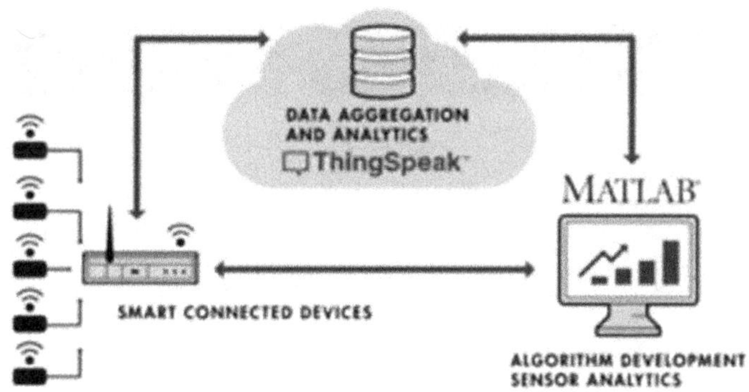

These related gadgets communicate with individuals and with different gadgets and every now and again concede sensors records to centralized storage and distributed computing server for the information to be handled and broke down to procure crucial experiences. Minimal cost and distributed computing power with quick as well as easy framework availability, are reinforcing this pattern. IoT choices are worked for some vertical purposes, for example, ecological checking and control, wellness monitoring, vehicle armada monitoring, industrialization monitoring and control, and domestic robotization.

At a high level, numerous IoT frameworks can be portrayed utilizing Figure 4:

On the left side, the smart contraptions (the "things" in IoT) that stay at the aspect of the system, are put.

These contraptions accumulate insights and comprise of things like wearable gadgets, remote temperatures sensors, coronary pulse screens, and water driven strain sensors, and machines on the assembling unit section. In the center, the cloud is available, where information from numerous sources is accumulated and broke down progressively, utilizing an IoT analytics, which is particularly intended for this reason.

The right side of the layout portrays the algorithm development identified with the IoT application. Here, an engineer or data researcher endeavors to pick up observation from the accumulated insights by performing verifiable assessment on the information. For this situation, the information is pulled from the IoT stage into a registering gadget, utilizing programming program condition, to enable the architect

Figure 5. Hardware setup for WHMS

Figure 6. Test result of variation of temperature on ThingSpeak channel

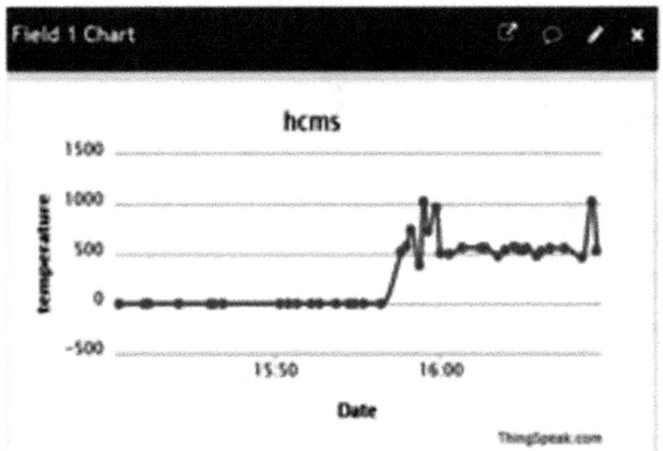

Figure 7. Test result of heart rate on ThingSpeak channel

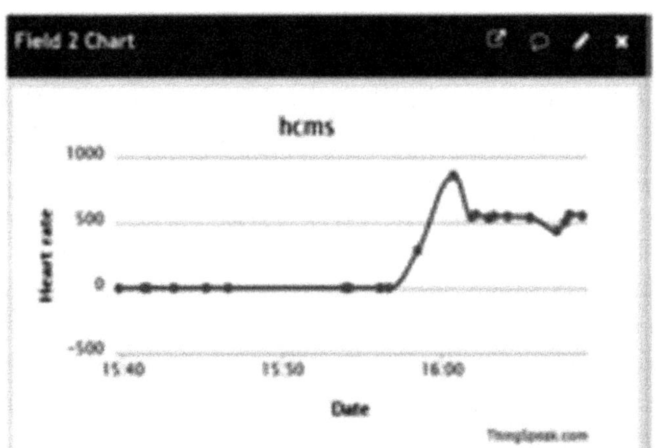

or researcher to plan calculations for the model that may likewise along these lines execute in the cloud or on the keen device itself.

An IoT framework comprises of every one of these components. ThingSpeak suite in the cloud part and offers a platform to assemble information and investigate records from web connected sensors. Information amassed from sensors is exchanged over ThingSpeak channels and only approved clients like overseers and specialists with sign in authority and secret phrase can get in to this information. Figure 5 demonstrates the model of proposed WHMS.

Figure 6 and Figure 7 demonstrates the example of results showed on ThingSpeak channels and Figure 8 indicates outcomes on the screen of screen.

Figure 8. Test results of sensor on serial monitor

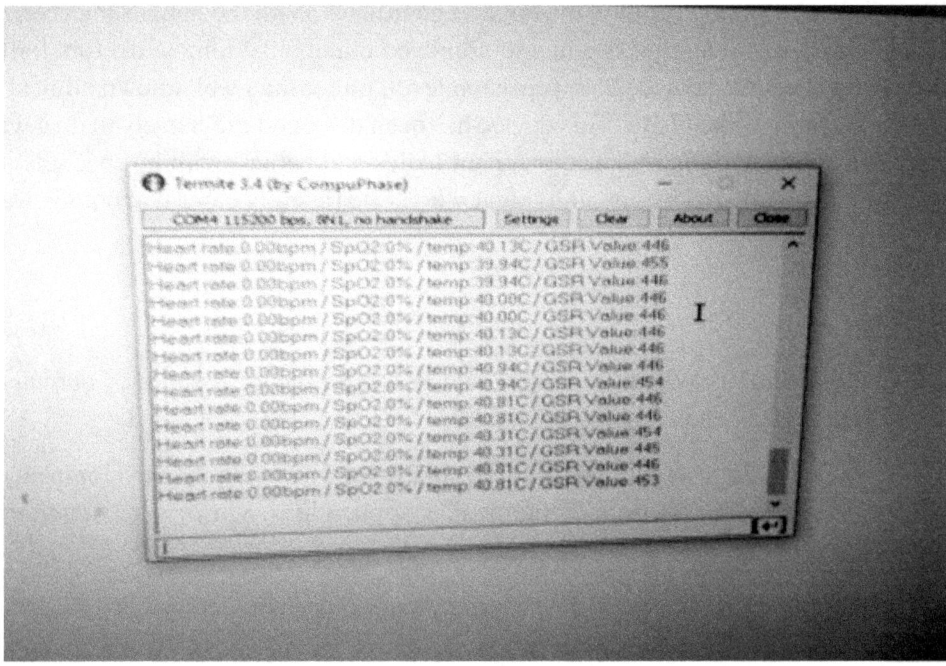

Comparison with Existing Systems

1. In the market, there are wide assortment of hardware's accessible to gauge inner body parameters however they all have a few inadequacies regarding bigger in size, complication in utilization, high maintenance cost, non-compact and so forth. Every one of these constraints are tended to by our WHMS as it is light in weight, easy to utilize, low maintenance cost and versatile.
2. One of the significant preferred standpoint of WHMS is that it sends programmed notice alarms to specialists about the wellbeing status of their patients. Along these lines, specialists are constantly mindful about the state of their different patients. Specialists can even take a gander at past records and dissect the advancement on ThingSpeak. In addition, it likewise refreshes this data in medical clinic records also. While existing frameworks just shows current estimations and not past records and specialists should be physically present for the evaluation.
3. To access any data or records of a patient, the specialist needs a unique API key. This aides in keeping up protection of patient's data and prevents any security rupture in the framework. Along these lines, WHMS is undoubtedly secure, error free and transparent framework when contrasted with existing frameworks.

CONCLUSION

A wireless healthcare monitoring gadget for all people has been proposed, which screens fundamental body parameters like pulse, temperature and circulatory strain. It not just exclusively offers help to guardians and clinic staff for checking and storing patient's vital parameters ceaselessly; yet additionally an easy to understand with ease of access gadget. If there would be an occurrence of deviation from predefined limit, it gives a prompt caution message to guardians. With the help of internet, records can be made accessible to even remote locales and to approved clients like remote doctors for distinct exhortation. The model is able to provide 95% precision level, rather than well known clinical techniques and mechanical contraptions like FitBit. This device has been designed exclusively to deal with matured people remaining in solitude at their homes, pregnant ladies and babies, with no specialist in the sight condition however can be utilized by any individual who is sufficiently careful about his wellbeing.

REFERENCES

Alemdar, H., & Ersoy, C. (2010). Wireless sensor networks for healthcare: A survey. *Computer Networks, 54*(15), 2688–2710. doi:10.1016/j.comnet.2010.05.003

Alvares, D., Wieczorek, L., Raguse, B., Ladouceur, F., & Lovell, N. H. (2013). Development of nanoparticle film-based multiaxial Tactile sensors for biomedical applications. *Sensors and Actuators. A, Physical, 196*(1), 38–47. doi:10.1016/j.sna.2013.03.021

Cardiovascular diseases. (n.d.). Retrieved from http:/ /www.searo.who.int/ india/ topics/cardiovascular_diseases

Gogate, U., & Bakal, J. (2018). Healthcare monitoring system based on wireless sensor network for cardiac patients. *Biomedical & Pharmacology Journal, 11*(3), 1681–1688. doi:10.13005/bpj/1537

Hertleer, C., Tronquo, A., Rogier, H., Vallozzi, L., & Van Langenhove, L. (2007). An aperture-coupled patch antenna for integration into Wearable textile systems. *IEEE Antennas and Wireless Propagation Letters, 6*, 392–395. doi:10.1109/LAWP.2007.903498

Huang, C. T., Shen, C. L., Tang, C. F., & Chang, S. H. (2008). A wearable yarn-based piezo-resistive sensor. *Sensors and Actuators. A, Physical, 141*(2), 396–403. doi:10.1016/j.sna.2007.10.069

Hung, K., & Zhang, Y. T. (2002, October). Usage of Bluetooth/sup TM/in wireless sensors for tele-healthcare. In *Proceedings of the Second Joint 24th Annual Conference and the Annual Fall Meeting of the Biomedical Engineering Society EMBS/BMES Conference, 3*(pp. 1881-1882). IEEE.

Jamkar, R. G., & Chile, R. H. (2004). Microcontroller based Temperature Indicator and Controller. *Journal of Instrument. Society of India, 34*(3), 180–186.

Kennedy, T., Fink, P., Chu, A., & Studor, G. (2007). Potential space applications for body-centric wireless and e-textile antennas. In *Proc. IET Seminar Antennas and Propagation for Body-Centric Wireless Communications*, London, U.K., pp.77–83. 10.1049/ic:20070551

Ko, J. G., Lu, C., Srivastava, M. B., Stankovic, J. A., Terzis, A., & Welsh, M. (2010). Wireless sensor networks for healthcare. *Proceedings of the IEEE, 98*(11), 1947–1960. doi:10.1109/JPROC.2010.2065210

The Lancet. (n.d.). *Global, regional, and national life expectancy, all-cause mortality, and cause specific mortality for 249 causes of death, 1980-2015: a systematic analysis for the Global Burden of Disease Study 2015, 388*(10053):1459–1544. doi:10.1016/S0140-6736(16)31012-1

Lewalski, E. A. (2012). Sensor Telemetry Using WiFi Technology; *International Journal of Electrical & Computer Sciences IJECS-IJENS 5,* pp. 34-42.

Martin, T., & Raskovic, D. (2008). Issues in wearable computing for medical monitoring applications: A case study of a wearable ECG monitoring device. *Sensors and Actuators. A, Physical, 141*(2), 396–403.

Mendis, S., Puska, P., Norrving, B., & World Health Organization. (2011). *Global atlas on cardiovascular disease prevention and control.* Geneva: World Health Organization.

Mills, G. A., & Nketia, T. A. (2007). Wireless Digital Stethoscope Using Bluetooth Technology. *Proc. IET Seminar Antennas and Propagation for Body-Centric Wireless Communications*, London, U.K., pp.77–83.

Sangasoongsong, A., Kunthong, J., Sarangan, V., Cai, X., & Bukkapatnam, S. T. S. (2007). A Low-Cost, Portable, *High-Throughput Wireless Sensor System for Phonocardiography Applications*, *2*, 345–356.

Shu, L., Tao, X. M., & Feng, D. D. (2010). A wearable, wireless electronic interface for textile sensors. In *Proc. IEEE Int. Symp. Circuits Syst. (ISCAS),* Paris, France, pp. 3104– 3107.

Survey on impact of lifestyle on our health. Retrieved from www.preveu-eu.org

The Lancet.(n.d.). *Global, regional, and national age-sex specific all-cause and cause-specific mortality for 240 causes of death, 1990-2013: a systematic analysis for the Global Burden of Disease Study 2013, 385*(9963): 117–71. doi:10.1016/S0140- 6736(14)61682-2.

www.thingspeak.com

Chapter 11
Smart Roads and Parking

Rachna Jain
Bharati Vidyapeeth's College of Engineering, India

Kartik Nagia
Bharati Vidyapeeth's College of Engineering, India

ABSTRACT

The past few years, innovation and technology have reached new heights, much beyond laptops, iPads or smart phones. Innovation and technology is progressing with each passing day. The concepts of smart cities, smart homes, and smart vehicles, are being implemented all over in developed countries. Objects here can get connected to the internet, collect data, and channelize it in order to ease our day-to-day life. The current genre of vehicles is equipped with various kinds of sensors, CPUs, and a software system that has communication capabilities. The increase in population has resulted in more vehicles, congested streets, limited parking spaces, and compromised road safety. Research and industry have proposed many technological advancements, and incorporated a few, in vehicles, but improvements for roads have largely gone unexplored. Smart roads, or Smart highways, are the terms used to define roads having IoT-enabled technologies like smart sensors, wireless connectivity, big data, and cloud computing. They use Solar technology to minimize electricity consumption, making the technology energy efficient. Researchers claim that, in the near future, smart vehicles will not be functional without proper smart roads and smart parking systems. Smart roads and parking contribute to making the drive safe, green, and convenient. They provide real-time information to drivers regarding traffic congestion, weather conditions, natural emergencies, e.g. landslide on the mountains, ice on the roads on high terrain, etc. Smart parking systems help drivers with information regarding available parking spaces as well as warnings about incoming traffic. Smart roads are also conceptualised to be equipped with wireless electric-charging systems and electric-charging stations. Electrical energy generated by the vehicles can be used to light streetlights to provide safe navigation to drivers at night. In this chapter, the concept of Smart Roads and Smart Parking systems is elaborated in a comprehensive manner. Various technologies are highlighted which reduce traffic, limit electricity use, significantly increase safety on the roads, and design a way to use parking spaces more effectively without the need to build new roads/ parking spaces. The chapter elaborates various technologies that will lay a strong foundation for smart roads such as transforming legacy roads to smart roads, solar road highways, etc. The chapter also elaborates on transforming traditional parking systems into smart parking systems.

DOI: 10.4018/978-1-5225-9246-4.ch011

INTRODUCTION

In the era of 21st century the population has crossed almost over 8 million with increase in people's need for transportation. With the staggering rise in the economy and affordable car price, the number of vehicles on the road have shown tremendous growth. People go on outing to movie theatres, shopping malls acquire a lot of parking spaces resulting in shortage of parking spaces. With a lot number of vehicles moving on the road, traffic jams, pollution, congestion on roads and limited spaces for parking often frustrates the commuters. To overcome these problems, the researchers have proposed technological iOT solutions incorporated on roads and parking. Various IT engineers have conducted research and developed significant technological solution in order to make our lives easy. One of the important concern that should be taken into accounting is, regular problem we face every day and how can we overcome it. The chapter discuss basic transportation problems faced by the drivers such as traffic jams, no parking spaces, etc., followed by the existing technologies that are currently used such as traffic lights and unorganised parking systems and their limitations. The next leads to the proposed technology, which include future smart roads such as rethinking roads that can examine the future of highways and transportation looking into the insights of long term drivers and produces a future vision for highways. The chapter the moves ahead with describing the possible future with innovative technologies such as the concept of autonomous cars, electric bicycle schemes, one transport etc. Next section is about various concepts like Smart pavements which include features like radio-connected sensors built on the road to monitor pavement's conditions, solar roadways project which charge the road using photovoltaic cells and applications of smart roads such as Virginia smart roads and honking roads in India. Further we will discuss about the smart parking system using RFID technology, IR Sensors etc. finally the chapter is concluded highlighting the limitations and future research of smart roads and parking system (Wang, Zeng, and Yang, 2006) (Varaiva, 1993).

Current Road Traffic and Parking Conditions

Economic and fast development has promoted rapid urbanization and introduced various changes in the metropolitan cities. Fast motorisation and lack of space are the most highlighted issues amongst the essential variables affecting the mobility and availability within a city. Extreme utilization of the private vehicles has adverse impact over the earth and human mass, in general everywhere, yet private vehicles are being favoured upon and is considered as a helpful method of communication. The present task centres around the normal issues and difficulties faced due to extreme difference in availability of parking spaces in various metropolitan cities of the world.

The vital questions here that needs to be focused are:

- Do highly populated cities of the world face any problems in parking and if so, what are they?
- What are the technological solutions to these problems?

If we carefully examine, we can easily observe that most of the private vehicles are parked for about 95% of the time duration in a day. Parking problems affect, not only vehicle owners but almost everyone. Ranging from fellow drivers to traffic controllers, pedestrians to security guards, everyone is affected from parking problems. Chaotic parking prompts traffic blockage. Disorderly parking can likewise prompt clashes between vehicle proprietors and individuals can get extremely baffled in the event when somebody is utilizing their parking spot. Moreover, vehicles looking around for parking space is a sheer waste of time for vehicle owners, waste of fuel leading to increased emission in a confined space and polluting the whole environment.

A lot of harmful pollutants like heavy metals, grease, sediments etc are aggregated in the parking lots where these toxic substances remain unabsorbed by the impenetrable surfaces, get washed away to the water bodies and pollute them, leading to contamination and degeneration of water table. Parking spaces also increase the temperature of their surroundings by 2 to 3 degree Celsius. As a result, additional cooling mechanisms are needed to be installed around the parking spaces which leads to higher energy consumption and public discomfort.

A way to benefit from excess of heat produced in the parking lots is to use solar panels. These can be utilized as a covering placed on the highest point of the parking slots. Cities with good amount of solar potential can generate a lot of electricity using solar panels over the parking slots considering the huge number of parking spaces available in a city. Buildings increase the demand for parking space especially during the peak hours, thereby increases their demand for electricity. Solar panels installed over parking spaces may help to reduce these demands. However, feasibility of the idea depends upon factors such as, area should be sunny, and parking space should be large enough which is not true for all types of lands.

The interest for parking spots is rising alarmingly and it is vital to understand that our assets are restricted and we can't advantageously design parking spots according to the interest of the mass. Increment in the quantity of our accessible parking spot ought to be the last thought worth considering. This thought of accounting for more private vehicles is disheartened by a recognized academician called Donald Shoup where he stated that, "expanding the current space isn't the most ideal choice for the urban areas, however a superior option would be, putting a maximum cap on the parking requirement" (Karpinski, 2006).

The price of parking space in most parts of the world is not in balance with its high demand. Extremely low parking fees in the parking lots are one of the reasons why people prefer private transport over public one. This leads to chaos on the roads as well as parking areas. Therefore, to push people towards public transport more and more in order to reduce parking problems, increasing the price of parking at public parking spaces can be used as a tool though, this wouldn't be a solution welcomed by public. Thus, technological advancements in the field of smart roads and parking need to be explored [8].

Existing Technology for Smart Roads and Parking

GPS System

Global positioning system is a navigation system which can govern location of communication conveyance from a satellite irrespective of climate and location. GPS system is used for navigating planes, ships and cars or any other mean of transport. GPS tracking mechanism is now highly used in automobile industries for location tracking of vehicle from one place to another. GPS gives a three-dimensional positioning and timing worldwide.

It is divided into three segments:

1. Space segment
2. Control system used by military
3. User segment

Generally, civilians use the user segment of GPS. The user segment comprises of a receiver which receives the signals from satellite and determines its accurate positioning from the satellite. The Global positioning system works on mathematical principle of "trilateration". GPS system is a network of 30 satellites orbiting around earth as a span of almost 20000km. it was first developed by US military but in current is highly used on the roads for navigation purpose. Wherever our position is on the earth you have almost four satellites above covering us. These satellites transmit signal at regular interval of time to the GPS receiver. Position is judged by the time it takes to receive the signal. The trilateration principle can be understood by, imagining we are standing above three satellites and each is calculating their position from us and now this data it transmitted to the same receiver and it calculates the relative position. GPS satellites consist of atomic clocks which gives the accurate time. These clocks comprehend the phenomena of general and space relativity which means times differ in different gravitational forces, and demands certain allowances to be made for that.

Advantage of GPS

1. GPS system is an important navigation tool for military and civilians.
2. Vehicle tracking system can provide us turn by turn direction on the roads and
3. it has very high speed

Disadvantage of GPS

1. GPS signal is too weak compared to phone signals, it doesn't work indoor, underwater etc.
2. The highest accuracy requires line of sight from satellite that is why GPS does work well in urban areas
3. GPS can give error[28][29].

Figure 1. Working of global positioning system

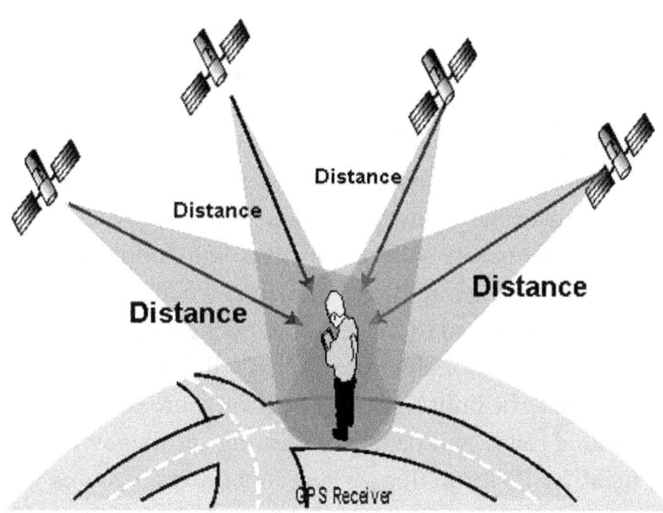

Drawbacks of Existing Technology

The existing road conditions and monitoring technology possess many issues and challenges which needs to be addressed prior to discussion regarding how to enhance them with various advanced mechanism to achieve a Smart Road status.

- **Road Conditions**: The roads have not undergone any change drastically. Though road surface is improved, but new technologies have not yet been developed, while huge progress have been made in our vehicles over time. The newest branch in improvements on the roads is known as *smart roads*. With innovative smart roads technologies, not only function of the roads can be enhanced but it also enables roads to be more energy efficient. Technologies such as solar roads are coming in place. Typically, roads are divided and determined by its use i.e. weather the road is used for driving or it has function to provide comfortable and safe environment. Roads are determined by its characteristic function. Engineers have tried to improve roads safety and comfort by constructing roads in a way to account of human factors. If we can visualise roads as spaces, we will able to raise our horizon and use various technologies. Smart roads prove to be the first step of revolution and have the potential to change the future completely.
- **Innovative Technology**: Smart vehicles are not compatible with existing technologies which create opportunity to change and think about new technologies. Roads are a kind of consumer product, where changes done on its accessories would affect every civilian travelling within the area. Once the new technology is passed by the government it will change the dynamics of the roads completely, roads have to be built in order to prove new facilities to the civilians without harming our environment. To be apt with future scenario, researchers as well as industry have proposed more innovative process based upon the roads and parking system. Few e.g. are, Internet of things smart approach to initiate smart highway innovation program. The highways will not be fully functional if it doesn't have the capability to interact with the vehicles and the ability to adapt the environmental conditions.
- **Smart Highways**: Highways have been tested out by developers and engineers by using various different technologies such as vehicles charging roads, smart pavements & RFID technology and parking sensors in smart parking system. Incorporating IoT network into the roadways is a major challenge since the road system lacks proper internet facility. Current roadways network can be considered as dumb where we have to improvise them which will be safer, reliable and smart. Efforts are being made to make the roads smart and enable IoT network which is equipped with data collecting tools to create data network which will improve safety. Many smart roads are comprised of the sensors which will collect sensitive data and monitor congestion on the roads. The information will allow making intelligent decisions when it comes to maintenance, usage and features. While retrofitting is allowed to make IoT network within the roads system, it also allows incorporate beneficial aspect of this technology which will improve IoT network interconnectivity. Future roads are anticipated to be built with renewable resources such as solar energy will give solar roads, wireless charging roads will be built, charging stations will be built and parking system will be improved by using RFID technology where mobile applications will be availed to book and use our parking spot. Smart roads will also be built by considering in mind the various environmental conditions e.g. heating roads in areas where it snows heavily. Automated weather stations and traffic sensors will make our roads less congested and reusable. Although solar roads are still

far from reality, there are still small efforts required in practice to connect roads everywhere with internet of things technology. Efforts to overhaul roads in order to increase safety are practiced where cost efficiency is the topmost priority for the department of transportation in each country.

- **Weather Conditions**: Among the list of requirements it is a priority to collect the data regarding weather conditions on which smart roads will be built. Weathers will play a crucial role in building of the smart roads, future roads will be equipped with thousands of sensors to collect traffic and weather data. One of the most popular initiative in IoT technology is built a ML system that will predict the weather conditions from the collected data that will help us built roads accordingly to the weather conditions. These system collect atmospheric conditions such as rain, temperature, humidity and monitor water levels on nearby rivers, lakes etc.

Depending on the local and government institutions, change in the dynamics of the roadways technology is expected to be seen in the future. Roads will be built according to the data collected and the needs of the road. Future roads seem to be long way ahead of us but are an achievable aspect in the future.

Proposed System for Smart Roads

Few concepts and technology proposed for achieving the smart roads are:

Rethinking Road

Rethinking roads is a concept which examines the future of highways and transportation by examining the insights of long-term drivers and produces a future vision for highways. It describes the possible future with innovative technologies and new case studies.

Urban population is growing rapidly at two people per second adding roughly 200,000 new city-dwellers out of which 66% will be living in cities by 2050. This indicates there would be a tremendous growth in number of people travelling on the roads, hence it is needed to look for an alternate solution for the existing technology (Juhlin, 1994).

Rethinking roads use nanotechnology, internet of things, biotechnology and cognitive sciences etc. Internet of things provide us with vehicle to vehicle communication using wireless sensors and concept of self-driving cars.

Technologies

1. **Autonomous car**: A self-driving cars or a driver-less vehicle is a vehicle that is capable of navigating without any human interaction by sensing its environment. A self-driving car combine a variety of sensors including a radar, sonar, GPS, computer vision that helps to navigate paths and obstacles in real world.
    a. There are a lot of advantages of using autonomous cars. The concept of autonomous cars increases safety, mobility and reduces infrastructure costs. These benefits include reduction of accidents which would lead to less number of injuries and minimising the need for insurance.
    b. Due to its potential benefits there are some major concerns that include the cost of building up the software and create awareness to increase trust of the consumers to switch existing technology. Security is also a major concern to be considered in this design. It is very vital to

secure the network from hackers. Driver less cars will decrease the need for drivers leading to unemployment problem which is another major concern. Certain aspects of autonomous cars are still in progress and there is no certain prediction of it current developmental status.

2. **One Transport**: One transport is a platform that collects data from mass transportation and give real time datasets and offers intelligent transport solutions. Its typical features are:
    a. More efficient utilisation of transport network: It gives a better solution to the transport services
    b. Customer satisfaction: Improved customer experience
    c. Reduce traffic collision: predict real time parking information
    d. Provides infrastructure planning: Gives us a better long-term investment decisions
3. **Bike sharing schemes**: Bike sharing schemes is an environment friendly solution of transportation which provides user to bike rental options on the roads for shorter travel. Bike sharing scheme are highly used in countries like Germany, Norway, Finland, Sweden etc.
    a. Bikes are available for free for short term travel by the government or some major companies to the registered users.
    b. Website is developed for user registration
    c. Mobile application for giving real time parking information and next available location with the help of GPS
    d. Lessen the air pollution level
    e. Keep the traveller healthy

Smart Pavements

Smart pavement is an emerging concept that can revolutionize the roads everywhere. "Smart pavements" refer to the roadways which have been engineered and build to support a variety of intelligent technologies and hence making it "smart" in the process.

Smart pavements include features like radio-connected sensors which have been built on the road to monitor pavement's conditions. It is also installed with a two-way Wi-Fi transmitter in the roadbed that offers broadband services to the vehicle passing by it. Charging of electric cars can also be embedded onto smart pavement, which will offer a wireless charging to the electric cars passing through leading to reduction of electric charging stops.

Building of smart pavements needs investment therefore, it is being commercialised to the companies as business opportunities for them. Thus smart pavement not only offer travel for the public but also a business for companies who built and maintain them (Lajnef, Chatti, Chakrabartty, Rhimi, and Sarkar, 2013) (Nejad, and Zakeri, 2013).

Pavement Sensor Research

The Federal highway administration (FHWA) conducted research on the sensors that will be installed on smart pavement. It developed and deployed wireless sensors that are capable of loading history of pavement and detecting damages. The sensors were equipped for collecting data which was retrieved by Radio frequency identification reader installed on the vehicles. Each sensor was build using piezo-electric transducer with ultra-power floating gate. The research concluded that sensors were capable of monitoring the roads with minimal power usage and collecting the data which can be beneficial in near future (Nejad, and Zakeri, 2013) (Nassar, Al-Qadi, Flintsch, and Appea, 2000).

Figure 2. Smart pavement integrated system

Benefits of Smart Pavements

There are endless ways that pavements could benefit roads operators and the public. This is always true if the sensors on the pavements are connected to a monitoring roadway centre via wireless communication. For example, remotely accessed roads data would let the engineers make decisions about the maintenance in the future and replacements schedules before the damage can be seen on the surface. This would make it cost efficient and save money due to reductions in equipment's deployment and repair on the roads. Having access to the smart sensors would allow the engineers to make smart decisions which will lead to best production and maintenance of the roads. By linking material data and other information obtained during pavement construction to an agency's pavement management system, they should obtain a greater understanding of why some pavements lasted longer than others. This type of insight could promote a more proactive approach to managing pavements, as an agency may know before the pavement damage becomes visible at the surface whether they may have a potential issue to address. If a high capacity broadband network is installed as part of a smart pavement highway and connected to an asset management operations centre, this network could be used to remotely control and coordinate the entire Wireless network enabled self-driving cars in its coverage area. The ability to coordinate traffic flow on such a large scale would reduce traffic slowdowns by eliminating the stop-and-start behaviour of individually-controlled vehicles. More importantly, taking away control from human drivers could substantially reduce accidents and fatalities; considering that 38,300 people were killed and above 4 million were injured in the U.S. roads in 2015 alone. Beyond the incalculable human cost of these deaths and injuries, the government estimated that they resulted in a $412.1 billion loss to the national economy when related costs such as wage and productivity losses, medical expenses, administrative expenses, employer costs, and property damage are factored in (Nassar, Al-Qadi, Flintsch, and Appea, 2000).

Solar Roadways Project

Almost 0.5% of the land is covered by roads. This percentage will go on to a 60% increase by the year 2050. Using solar energy on smart highways could be the future as it is a renewable source of energy and doesn't create any pollution or harm us. Solar roadways project is a proofed concept to build smart

Figure 3. Solar roadway architecture

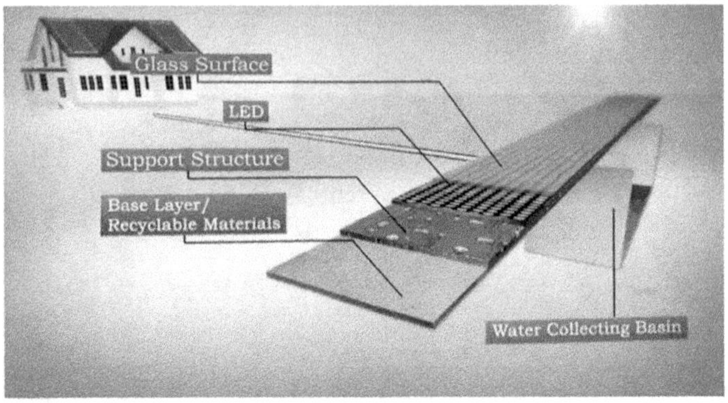

roads. It combines solar cells with a transparent driving surface and electric sensors which acts as a solar array which have programmable activity. The roads are built with recycled material and photovoltaic cells. Solar roadways are designed to melt snow during winter, power the streetlights and even illuminate the pavement yellow and green using LED lights. The solar roadways projects are going on worldwide but are they feasible and produce the future roads that is the bigger concern. There are some of the solar roadways projects (Mehta, Aggrawal, and Tiwari, 2015) (Roadways, 2013).

1. Solar roads in the united states
2. The Dutch solar bicycle path
3. The French wattway
4. The Shandong project in china

Limitations of Solar Roadways Project

Solar roadways project is a slow with respect to making process and roads like Route66 do not offer any benefits that justify its expansion it the entire country. Though, solar roads offer new features but they are not relevant and practical. Solar roads don't offer high safety and the cost of building them are very high. It's unlikely that we will see solar roads in future since many countries have implemented and few have attempted but their results are not very satisfactory (Kulkarni, 2013) (Roadways, 2017) (Xu, 2016).

Wireless Vehicle Charging Roads

Electric vehicles are becoming popular day by day due to awareness to preserve fossil fuels and our environment. People are shifting to electric cars as it is cost efficient and eco-friendly but the charging infrastructure is lacking behind. Most people have to charge their vehicles at home or at a specific location. Wireless charging roads offer a solution by installing inductive charging strips on the roads that charge the vehicle passing through them. A system sets up an electromagnetic field and then an inductive coil harvest power through it. It is the same technology used in our smart phones. The chipmaker Qualcomm has built a 100meter test track in Paris, France to test its version. It is shown that it is possible to charge the vehicles at 20killowatt passing through it and it was operational by charging the vehicle in both direction along the same strip. The benefits of this system is pretty much clear if a vehicle can be charged while moving and then it doesn't need to pull up to electric stations. It also means vehicles with smaller batteries can easily be charged up. Installing inductive hardware on the roads is safer and easy than placing overhead power lines to charge vehicles. Apart from thesebenefits there are certain disadvantages as well. First it requires infrastructure on lay inductive electric strips on the roads which is quite expensive and disruptive. Second, installing the setup on the vehicles is also quite expensive. With the decreasing cost of the batteries installing this setup seems quite a long path. Third it can only be installed on the roads with busy schedule which means selective locations can only be picked up. Wireless vehicle charging cannot the done on the vehicles with higher power batteries such as electric buses and truck (Jia, Hu, Song, and Luo, 2012).

Applications of Smart Roads

The various countries who have constructed smart roads are:

Virginia Smart Road

Virginia's smart road is a highway located in Virginia USA. Its total length is 9.17km and its construction started in July 1997 and accepted completion in 2015. This road was used for testing pavement technologies and as a ground for providing new technologies. This road is operated and maintained by Virginia department of transportation. It is also part of interstate 73 corridor (Al-Qadi, Loulizi, Elseifi, and Lahouar, 2004).

Figure 4. Virginia smart road satellite view

A lot of features are installed in this road such as

1. It has two paved lanes
2. It consists of three bridges including the smart road bridge
3. A 6km controlled access test track built to interstate standards
4. 24/7 Oversight and access control
5. Full time staff that coordinates all road activity
6. Lighting and weather system control
7. Centralized communication
8. Surveillance and safety assurance
9. It consists of 14pavement section including open grade friction course
10. Pavement sensors to detect rain, snow and other weather conditions.
11. A zero crown pavement section to operate Flooded pavement testing
12. An American Association of State Highway and Transportation Officials (AASHTO)-designated surface friction testing facility
13. Seventy-five weather-making towers accessible on crowned and zero-crown pavement sections
14. Artificial snow production of up to four inches per hour (based on suitable weather conditions)
15. Production of differing intensities of rain with varying droplet sizes
16. Fog production
17. Two weather stations with official National Oceanic and Atmospheric Administration (NOAA) weather available within one mile
18. Variable poles spacing designed to replicate 95% of national highway system
19. Multiple illuminating heads, including LED modules
20. A wireless mesh network variable control example- dimming
21. A high-bandwidth network of fiber.
22. A differential GPS base station
23. Remote control signal phase and timing (White et. al., 2003) (Appea, 2003).

System for Smart Parking Using RFID Technology

RFID is an acronym for radio frequency identification. It is a term commonly used in Information and Communication Technologies Engineering. The main purpose of using this technology is to prevent catastrophes and accidents happening on the roads. Thus, this technology helps us to make roads more impregnable and secure.

Radio Frequency Identification (RIFD) is a technology in which digital data are encoded in some tags namely RFID tags which are then captured by a reader via electromagnetic waves (radio waves). Basically RFID technology is used to track objects through the use of electromagnetic waves. The advantage of using these waves is that, the tag which contain the electronically coded information need not be present in front of the user's scanner. This is because electromagnetic waves have a large wavelength and can travel up to large distances without their energy being distorted. In a nutshell Radio frequency Identification (RFID) can said to be a wireless technology which is used to track the objects. The main advantage of using this technology is that the reader and the object are not bound to be in touch. The object can be moving and can still be tracked using this technology. It uses an internal chip and an antenna mounted on that chip. Together they forma tag/inlet. As soon as this tag comes in the range of

Figure 5. RFID Setup architecture

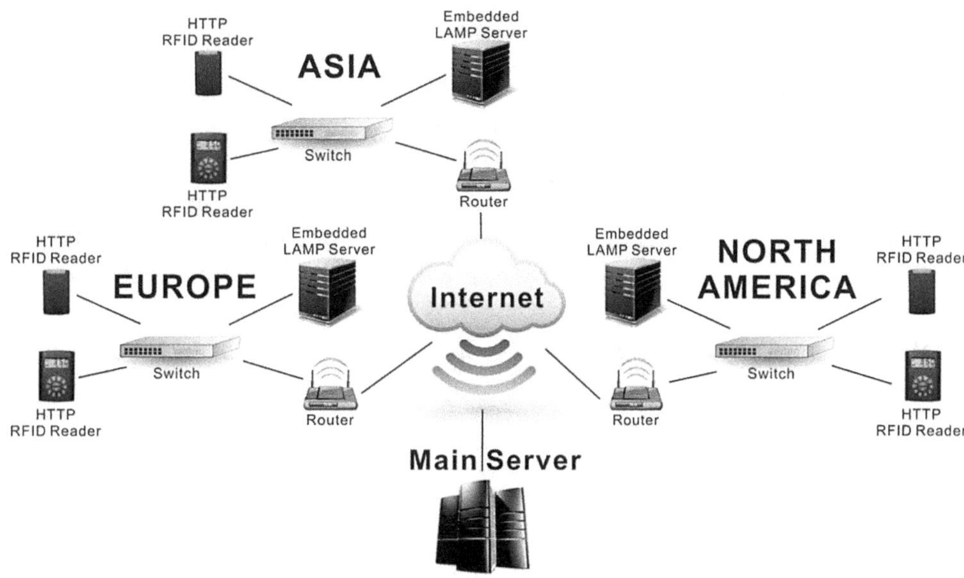

field of reader it transmits information wirelessly in form of electromagnetic waves. These waves carry information from chip to the reader which interprets this information on a computer.

Uses:

1. Tracking of goods/items.
2. Tracking of people.
3. Tracking of airport baggage.
4. Tracking of medicine in warehouse.
5. Smart roads and parking.

Objectives of Using RFID

Since now a days people own private vehicles and prefer to travel by them, they face a lot of traffic problems on a regular basis. This leads to financial as well as temporal loss. Traffic leads to many severe problems that are mentioned below:

- **Vehicle Crashes:** These are very common in many developed countries and can cause severe injuries to the victim of the car crash. It can even lead to death of the victim which causes a lot of pain to the friends and family. Current technologies used to monitor over speeding are not so accurate and can be manipulated with. RFID can monitor speed of the vehicle and even monitor the location of crash which can save lives of hundreds.
- **Traffic Jam:** One of the major problems faced in high populated country is traffic jam. It leads to heavy financial losses to the citizens. People have actually died due to traffic jam since there is no proper technology in in developing countries to ensure that the ambulance reaches the hospital on time. Traffic also leads to pollution and health problems which can lead to death. RFID can be

used to overcome this problem. RFID tags can store information which can ensure proper management of traffic by authorities.

- **Vehicle Theft**: Very common problem which leads to loss of money of the victim. Police in maximum case, fails to recover the car of the victim. RFID can be used in an efficient way to track the stolen car *and* recover it from the thieves. This will save the precious time of the legal bodies as they will not perform random checks to recover the vehicle.
- **Smart Parking**: RFID tags can store the location of parking which can be used by staff members to ensure safe and proper parking of vehicle. Such a vehicle cannot be stolen without the twin tag which is present with the rightful owner. Also it will prevent jams within the parking land.

In a nutshell, RFID tags can store a great deal of information which can be used in an efficient way to prevent mishaps on road and provide a smart road system.

Various Components of RFID system are:

- **Antenna**: It is used to read/ write information on a tag. It is also used to activate a tag in an electromagnetic zone.
- **TAGS:** Tags contain information about the object such as, Identification number. This information can be accessed by the reader using a device and the reader can act accordingly. Tags can be active, passive, and semi-passive based on the source of energy. They can be read only, read-write based upon how information is coded on them.
- **Frequency used RFID technology**: It can be categorized into three zones according to the value of frequency. The types and corresponding values are mentioned below:

LOW: 30 KHz-500 KHz.
MID: 900 KHz-1500 KHz
HIGH: 2.4 GHz to 2.5 GHz

RFID vs. Bar Codes

- No line of sight is required in RFID system. Bar code scanner on the other hand is able to scan a bar code if and only if it is present under line of site. Thus it fails to scan movable objects.
- RFID tags can be rewritten again and again but a bar code and its information cannot be rewritten on the same code.
- RFID can be read easily and with much more speed than bar codes.
- Bar codes have a limited information stored on them unlike RFID chip which has a much greater capacity (Nath, Reynolds, and Want, 2006) (Mainetti, Palano, Patrono, Stefanizzi, and Vergallo, 2014).

Smart Parking System

Internet of things performs a consistent part in joining our bounding substantial conditions through web and also built trivial toward attaining the particular conditions against a part of an isolated area. IoT is a no end in sight network, accessible by computer and people. This project is based on Internet of things and is success. Everywhere now-a-days people are facing problems to park their vehicles in cities. The

Figure 6. Smart parking system

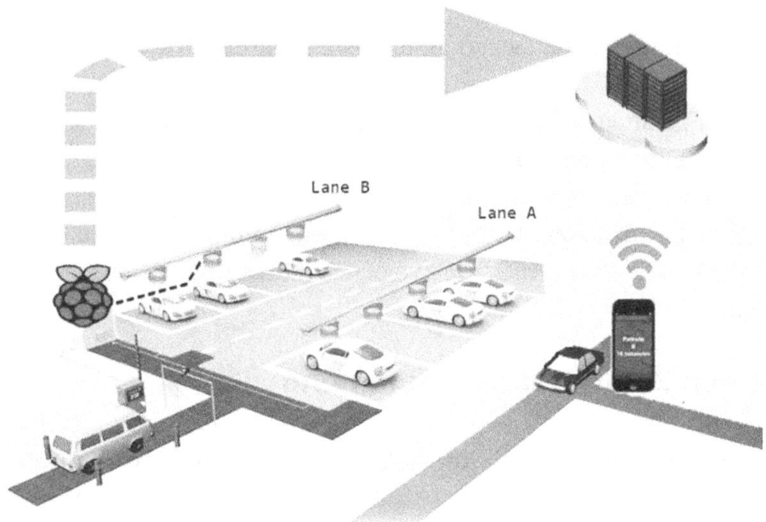

smart parking system makes end user directed toward looking at a nearby place or to look-in parkland space in a peculiar slots. It is especially focusing on minimizing the time and furthermore how to play it close to the vest travelling over filled parking slot (Idris, Leng, Tamil, Noor, and Razak, 2009) (Geng, and Cassandras, 2013) (Kianpisheh, Mustaffa, Limtrairut, and Keikhosrokiani, 2012).

Implementation

The aim of this system is to book a parking slots, within that slots and leaving the parking. It also checks the availability of parking space in order to park a car in a vacant parking places. This is done by implementing the smart parking system.

Working Description

A fundamental aspects-containing proposal must create an app for smart parking, where it will give the information regarding the slot based on its occupancy. If it is occupied, then the app will never allow the user to book that particular slot on the specific time. It will also provide the information of all the slots. Once booking from the app completes successfully it generates the unique booking id for the user (Wang and He, 2011).

Whenever the user enters the main entrance there is a IR sensor will asks about whether the user has already booked the slots or if they have not booked. It will collect user id and will check in the database, once a match, it automatically opens the gate and allow the user to go inside to park the vehicle. In case the user provides the wrong id then it won't allow the user to go inside. Even at the end time also the sensor will ask the user to provide the booking id in order to terminate his/her booking transaction and collects the bid from the user. It consists of a number of components and their functionality is as follows:

- **Centralized Server**: It manages database stored in a cloud server.

- **Raspberry Pi**: This component well-known employ individually devices also connected along a raspberry pi camera.
- **User Device:** This component attached along devices through mobile phones and websites

Now a days, usage of personal vehicles is at increase for commuting rather than being dependent on public transportation. Due to increase of personal vehicles it is a big challenge to find a parking place. Hence is introduced a new concept smart parking system that reduces the risk of finding the parking slots in any parking area and also eliminates unnecessary travelling of vehicles across the filled parking slots in a city. This is easy to access is inexpensive hence reduces both time as well as cost. (Rao, 2017) (Geng and Cassandras, 2011).

Future Research on Smart Roads and Parking

Research on smart roads and parking in on the rising graph. Scientist and engineers are developing and finding new and alternative solutions for smart roads and parking system. Smart roads creates opportunities for new industries such as artificial intelligence, machine learning and deep analytics. Smart roads are able to work within a connected infrastructure network that will improve overall transit of goods and people. Smart roads can provide drivers with warnings and information regarding hazards, such as ice, landslides etc. They can provide lane-flexibility to reduce traffic congestion or provide segmented sections by drive or vehicle type. And smart roads can provide data on driver performance, traffic flow and so much more. Smart roads on the other hand facilitates goods transportation, fleet management and tracking in the short term, albeit we are likely to find other means of goods transportation in the future or maybe the roads will be kept specifically for that. On the actual manufacturing side, there will be emerging opportunities for new industries from major developments on the materials side, AI and sensor technologies, machine learning and deep analytics and network-support products, groundbreaking robotics and autonomous equipment for construction itself, along with new frontiers in BIM and GPS. Future research examines smart roads in future, how we made them, energy and cost efficiency and proper parking management system. In 21st century revolution of roads and parking management is already under action. We have the resources and capabilities to make our roads and parking system smart. Smart roads and parking system incorporate different technologies such as rethinking roads, solar roadways project, smart pavements, RFID technology and many more which can make our life convenient and energy efficient. These concept not only work for particular conditions, it also gives maximum output for many inputs. The potential benefits for having roads and parking smart is huge. It not only make our life less burden but it could regulate speed of vehicles by installing warning system. It transmits real time data and share the information across the network making it simpler and quicker to get around locations. It makes travelling easier and cheaper. New researches are taking place, new developments in roads construction is happening such as in solar roads which uses solar energy to charge road and wireless charging roads which uses electro induction to charge roads, intelligent traffic lights uses improve commute easier. The roads have becomes smarter by integration of technologies, make use of principle and material which are used for making smarter technologies such as solar panels etc. The examples of smart roads technologies are solar roads, smart pavement and charging lanes. Research is being undertaken in order to testify how these technologies can be implemented in the future (Ankrum, 1992) (Wang, Zhang, and Xia, 2010) (Goosem, 2009).

CONCLUSION

In this chapter we have discussed various technologies for smart roads and parking system that can be implemented in the future to make out roads safer and cost efficient and use our parking spot more effectively. Future of the roads lies in the hands of IoT technologies but we have to consider various aspects before implementing these technologies such as cost, weather etc and majorly its need in the future. Many technologies have been tried out in the past, some have shown result some have not but no effort will go in vain. In this chapter we have discussed technologies such as rethinking roads, smart pavements, solar roads, wireless charging roads and smart parking system such as RFID technology, parking sensors and mobile application to book our parking spot. We thus conclude this chapter by saying smart roads and smart parking system is the future of parking and road technologies. Mueller, Loomis, Kalafus, and Sheynblat, 1994) (White, 2003).

REFERENCES

Al-Qadi, I. L., Loulizi, A., Elseifi, M., & Lahouar, S. (2004). The Virginia Smart Road: The impact of pavement instrumentation on understanding pavement performance. *Electronic Journal of the Association of Asphalt Paving Technologists*, *73*, 427–466.

Ankrum, D. R. (1992). Ivhs-smart vehicles, smart roads. Traffic Safety (Chicago), 92(3).

Appea, A. K. (2003). *Validation of FWD testing results at the Virginia Smart Road: Theoretically and by instrument responses* (Doctoral dissertation, Virginia Tech).

Berard, A. J., Mentzer, J. L., & Nixon, D. C. (1996). U.S. Patent No. 5,515,043. Washington, DC: U.S. Patent and Trademark Office.

Geng, Y., & Cassandras, C. G. (2011, October). A new "smart parking" system based on optimal resource allocation and reservations. In *2011 14th International IEEE Conference on Intelligent Transportation Systems (ITSC),* (pp. 979-984). IEEE.

Geng, Y., & Cassandras, C. G. (2013). New" Smart Parking" System Based on Resource Allocation and Reservations. *IEEE Transactions on Intelligent Transportation Systems*, *14*(3), 1129–1139. doi:10.1109/TITS.2013.2252428

Goosem, M. (2009). 36 RETHINKING ROAD ECOLOGY. Living in a Dynamic Tropical Forest Landscape, 445.

Idris, M. Y. I., Leng, Y. Y., Tamil, E. M., Noor, N. M., & Razak, Z. (2009). Car park system: A review of smart parking system and its technology. *Information Technology Journal*, *8*(2), 101–113. doi:10.3923/itj.2009.101.113

Jia, L., Hu, Z., Song, Y., & Luo, Z. (2012, March). Optimal siting and sizing of electric vehicle charging stations. In *2012 IEEE International Electric Vehicle Conference (IEVC),* (pp. 1-6). IEEE. 10.1109/IEVC.2012.6183283

Juhlin, O. (1994). *Information Technology hits the Automobile: rethinking road traffic as social interaction. Changing Large Technical Systems* (J. Summerton, Ed.). San Francisco, CA: Westview Press.

Karpinski, M., Senart, A., & Cahill, V. (2006, March). Sensor networks for smart roads. In *Proceedings of the 4th annual IEEE international conference on Pervasive Computing and Communications Workshops* (pp. 306-310). IEEE.

Kianpisheh, A., Mustaffa, N., Limtrairut, P., & Keikhosrokiani, P. (2012). Smart parking system (SPS) architecture using ultrasonic detector. *International Journal of Software Engineering and Its Applications*, *6*(3), 55–58.

Kulkarni, A. A. (2013). Solar roadways. Rebuilding our Infrastructure and Economy. *International Journal of Engineering Research and Applications*, *3*(3), 1429–1436.

Lajnef, N., Chatti, K., Chakrabartty, S., Rhimi, M., & Sarkar, P. (2013). *Smart pavement monitoring system (No. FHWA-HRT-12-072).* USA: Federal Highway Administration.

Mainetti, L., Palano, L., Patrono, L., Stefanizzi, M. L., & Vergallo, R. (2014, September). Integration of RFID and WSN technologies in a smart parking system. *In 2014 22nd International Conference on Software, Telecommunications and Computer Networks (SoftCOM),* (pp. 104-110). IEEE. 10.1109/SOFTCOM.2014.7039099

Mehta, A., Aggrawal, N., & Tiwari, A. (2015). Solar Roadways-The future of roadways. *International Advanced Research Journal in Science, Engineering and Technology (IARJSET), 2.*

Mueller, K. T., Loomis, P. V., Kalafus, R. M., & Sheynblat, L. (1994). U.S. Patent No. 5,323,322. Washington, DC: U.S. Patent and Trademark Office.

Nassar, W., Al-Qadi, I. L., Flintsch, G. W., & Appea, A. (2000). Evaluation of pavement layer response at the Virginia Smart Road. In *Pavement Subgrade* (pp. 104–118). Unbound Materials, and Nondestructive Testing. doi:10.1061/40509(286)7

Nath, B., Reynolds, F., & Want, R. (2006). RFID technology and applications. *IEEE Pervasive Computing*, *5*(1), 22–24. doi:10.1109/MPRV.2006.13

Nejad, F. M., & Zakeri, H. (2013). The hybrid method and its application to smart pavement management. In *Metaheuristics in Water* (pp. 439–484). Geotechnical and Transport Engineering. doi:10.1016/B978-0-12-398296-4.00019-2

Rao, Y. R. (2017). Automatic smart parking system using Internet of Things (IOT). *Int. J. Eng. Tech. Sci. Res*, *4*, 2394–3386.

Solar Roadways. (2013). *Solar Roadway–A Real Solution.*

Solar Roadways. (2017). *Welcome to Solar Roadways.*

Varaiya, P. (1993). Smart cars on smart roads: Problems of control. *IEEE Transactions on Automatic Control*, *38*(2), 195–207. doi:10.1109/9.250509

Wang, F. Y., Zeng, D., & Yang, L. (2006). Smart cars on smart roads: An IEEE intelligent transportation systems society update. *IEEE Pervasive Computing*, *5*(4), 68–69. doi:10.1109/MPRV.2006.84

Wang, H., & He, W. (2011, April). A reservation-based smart parking system. In 2011 IEEE Conference on Computer Communications Workshops (INFOCOM WKSHPS), (pp. 690-695). IEEE

Wang, H., Huang, Q., Zhang, C., & Xia, A. (2010, December). A novel approach for the layout of electric vehicle charging station. In *2010 International Conference on Apperceiving Computing and Intelligence Analysis (ICACIA),* (pp. 64-70). IEEE

White, S. A., Walley, K. S., Johnston, J. W., Henderson, P. M., Hale, K. H., & Andrews, W. B., Jr., & Siann, J. I. (2003). U.S. Patent No. 6,531,982. Washington, DC: U.S. Patent and Trademark Office.

White, S. A., Walley, K. S., Johnston, J. W., Henderson, P. M., Hale, K. H., & Andrews, W. B., Jr. & Siann, J. I. (2003). U.S. Patent No. 6,531,982. Washington, DC: U.S. Patent and Trademark Office.

Xu, G. (2016). Solar roadways. V sbornike: Rasshirennoye vosproizvodstvo innovatsionnoy ekonomiki i intensifikatsiya sprosa na innovatsii v Rossii Sbornik nauchnykh statey. FGBOU VPO «Rossiyskiy ekonomicheskiy universitet imeni GV Plekhanova, 128-130.

Chapter 12
Exploring IoT-Enabled Smart Transportation System

Sushruta Mishra
Kalinga Institute of Industrial Technology, India

Shikha Patel
Kalinga Institute of Industrial Technology, India

Amiya Ranjan Ranjan Panda
Kalinga Institute of Industrial Technology, India

Brojo Kishore Mishra
https://orcid.org/0000-0002-7836-052X
GIET University, India

ABSTRACT

Internet of Things (IoT) is a platform that makes a device smart such that every day communication becomes more informative. A Smart Transportation system basically consists of three components which include smart roads, smart vehicles and a smart parking system. Smart roads are used to describe roads that use sensors and IoT technology which makes driving safer and greener. Smart parking system involves an automated system model that can assist the drivers in selecting the suitable parking spot for them. The data that the system collects will be sent for some analysis. It provides real time information to drivers about various aspects of transportation like weather conditions, traffic scenario, road safety, parking space, and many other things. A well-built Smart Transportation system reduces the risk of accidents, improves safety, increases capacity, reduces fuel consumption, and enhances overall comfort and performance for drivers. Our chapter deals with the in-depth discussion of these various aspects of a smart transportation system enabled with IoT technology.

DOI: 10.4018/978-1-5225-9246-4.ch012

INTRODUCTION

Internet of Things (IoT) provides an interface for a device to be informative of its surroundings so that data processing and analysis becomes smart. IoT is a recent buzzword that keeps on growing throughout the world. Regularly several platforms and models are being developed in order to benefit mankind. IoT acts as an agent that interconnects things of virtual world to things in reality. It enables anywhere and anytime connectivity of any object with an ON and OFF switch. It consists of an environment where physical things along with living beings with the help of virtual data interact with each other. Massive quantity of data is generated in regular basis since huge number of objects is connected to Internet. Hence these massive data is required to be coordinated and managed to provide relevant knowledge that may be useful in building intelligent automated systems. Applications of IoT range from several domain areas like education, industries, transportation, healthcare and many such sectors. Here our main focus is the application of IOT in transportation zone. IOT can be efficiently deployed to develop intelligent transportation models. These smart transportation systems can be utilized as a supportive framework for smart city architecture. Thus, it may be enabled to employ powerful and modern communication methodologies for effective monitoring of the citizens in a city. A smart transportation system enables futuristic roads and highways in managing congestion of traffic in a better manner than today's scenario. With IoT in center stage, it is estimated that in an interval of around 25 years the present traffic management system will be enhanced to such an extent that vehicles will be able to communicate with each other without any interaction of human beings. In this way, the travelling can be made more secure, simpler and smoother. Vehicles can be embedded with sensors to monitor the present traffic condition of roads and highways. This information can be compiled and collected to a central base station. Based on the status of traffic scenario, essential feedback data can be sent to the vehicles on road. In case of estimated heavy traffic conditions, vehicles can be notified to take alternative traveling route for safety purpose or to restrict their speed limit. Additionally, a parking system can also be enhanced by incorporation of smart intelligent transportation model. Parking becomes a huge headache in urban areas. This can be reduced if the driver can be informed at prior regarding the availability of free parking spaces. In addition to this, the drivers can be made aware of the shortest and feasible route available to reach the destination to control the carbon dioxide emission. It is estimated that by implementing IoT based smart transportation system, the revenue spent on traffic control can be significantly reduced by around 15%. The benefits of IoT is currently being used in all industries and transportation is certainly not an exception. Smart Transportation is the buzzword used to refer the use of IoT technology in transportation domain. By 2025, the smart transportation market is likely to grow to an approximate $130 billion. Some general challenges faced by the use of IoT in transportation sector in modern times are as follows.

- Savings in annual expenditure.
- Enhanced customer service.
- Improved safety of drivers.
- Ensure job satisfaction of drivers.
- Real-time location of vehicles data.
- Vehicle-to-Vehicle (V2V) service.
- Preventive and predictive maintenance.
- Security and safety compliance.

Figure 1. Components of Smart Transportation System

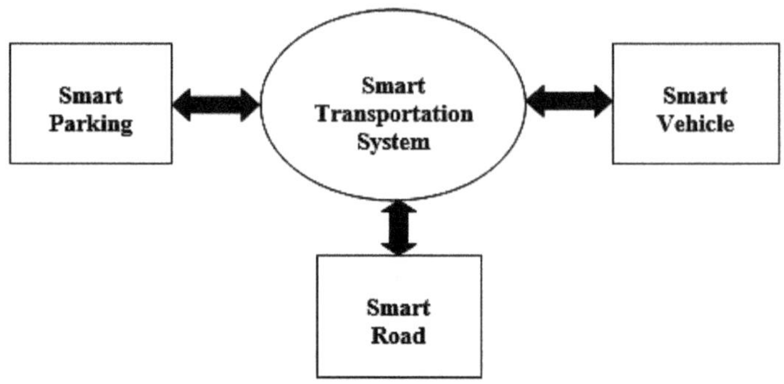

COMPONENTS OF SMART TRANSPORTATION SYSTEM

A smart transportation system can be visualized as comprising of three basic functionalities. This includes smart vehicles, smart parking and smart roads. This is being highlighted in Figure 1. Smart roads are implemented by the usage of various sensors and actuators to make driving safer and green. Smart vehicles are powered with real-time physical devices enabled with computational and communication capabilities to be able to interact with each other for social well-being. Similarly, smart parking deals with an automated parking system to assist drivers in choosing a suitable parking region for their vehicle. The raw data gathered from these components are transmitted to base station for further processing and analysis and thereby it helps in effective decision making on road. Hence these three components taken together form a precise and intelligent transportation model. Such a system can be of immense help to drivers and other associated customers in providing real-time information regarding several aspects of transportation such as traffic scenario, weather conditions, parking space availability and overall road safety. An efficient smart transportation model controls the accident risk, enhance safety, reduce consumption of fuel and increase overall comfort etc.

LITERATURE SURVEY

Comfortable parking facility, effective transportation and coordination are the three critical criteria in the emerging smart city scenario (Zhou, & Li, 2014). In (Panchal, 2017), a wireless sensor network based on IEEE ZigBee concept was designed which responded more to emergent situations and notifies right people in right time. In another work by In (Ferris, Watkins, & Borning, 2010), an analytical survey was undertaken where iit was discussed the importance of public transport system and its significance in handling traffic congestion and minimizing carbon emissions. In (Hari Baabu, Senthil Kumar, Deb, & Rai, 2016), a smart automated parking system model was developed with the help of IR sensors. It takes help of a feedback mechanism to determine availability of parking areas in nearby regions. Special infrared sensors and actuators are embedded in monitoring suitable parking regions available. A smart large-scale parking model with the use of RFID was discussed in (Ganesan, & Vignesh, 2007) to perform matching of unique RFID tag of vehicle with the database value when the RFID reader reads the value in the entrance of parking slot. (Nasir, & Mansor, 2011) built up a programmed key for the two-wheeled vehicle by utilizing a micro-controller. (Tang et al., 2006) created security frameworks for four-wheeled vehicles utilizing remote sensor arrange innovation, and (Yihua, 2010) prevailing with regards to building up a security framework utilizing SMS to track missing two-wheel vehicles. In (Kenmotsu, 2012), a navigation technique is proposed to reduce congestion period by the use of real time data regarding parking slots. It applied various algorithms to allocate vehicles to available parking slots and subsequently evaluation of the model is simulated to determine its effectiveness. In (Suhr, & Jung, 2014), parking slots are identified and tracked by the fusion of sensors of an Around View Monitor (AVM) model which uses an ultrasonic sensor oriented parking model. It facilitates drivers in selecting suitable parking slots which are available and thereby supporting parking coordination by updating of data regarding parking regions. An efficient and real time time coordination system for parking space control had been proposed in (Hong, Soh, Jaafar, & Ishak, 2013) that provides data to user of available parking slots and support the parking coordination through updates about parking slots. In (Wei et al., 2012), a smart parking control model base don RFID tags have been proposed, designed and implemented in the local area by the use of global uninterrupted internet connection. A study in (Nalawade, & Devrukhkar, 2016) presented a bus tracking model with the usage of Raspberry Pi to track the bus on Google map with the help of an android app service. Its primary job was to provide extra information of bus arrival and departure time, location of bus and it uses variety of sensors to enhance security in campus. (Almishari et al., 2017) developed an automated IoT based tracking model which is based on GPS. It is also an energy conserving model used for real time tracking of vehicles. It was seen that the energy consumption is reduced by 17% as compared to traditional systems. (Shinde, & Mane, 2015) presented and implemented a model to enhance the security of school buses through automated follow up and coordination. The system model made use of linux based GPS receiver, GSM, GPRS and Raspberry Pi.

IMPLEMENTATION OF IoT BASED SMART TRANSPORTATION MODEL

The transportation system can be made intelligent by the efficient use of IoT technology. Actually, the three components described in the previous section constitute the backbone of smart transportation. In this section, an example case study showing the framework for each one of the components is described in detail.

Smart Parking System with IoT

Recently the notion of smart city has seen a great amount of value. A crucial criterion of a smart region is facilitating Smart Parking. Trying to find some specific parking area for a vehicle can become an infuriating task. It is observed that drivers find it difficult to get a stopping limit without thinking the place of open parking area. The wait due to unavailability of parking slot causes not only traffic congestion but also higher fuel consumption. To produce an optimize solution for this crisis, a lot of technology has evolved but it wasn't able to benefit all mainly because of high cost, effectiveness, energy, preciseness with some more miscellaneous criteria. In this study, a framework of IoT based parking architecture for a domain application Internet of Things (IoT) by usage of Raspberry Pi is presented. Originally, the study provides a specific idea of smart parking model and the requirement for smart objects to be embedded within cloud. Here the prototype as well as the architectural functioning of the model is studied, framework using Optical Character Recognition and Facial Recognition for providing two-way privacy using Raspberry Pi is presented. By showing these vital attributes of this work, its convenience and

Figure 2. Design of Smart Parking system

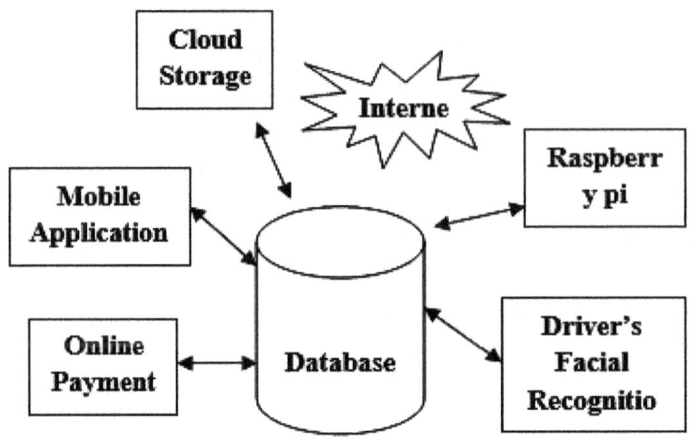

benefits are shown. At the end it is shown that the framework on the basis of smart parking model with use of IoT has found an alternative for the huge traffic and simplify the way of obtaining a parking area.

Working Principle of the Model

The architectural framework of the Smart Parking System is displayed in Figure 2. At start, the user has to create his account by entering his personal information in the smart application to supply the related data integrated with the server database. When his details are entered, he is permitted to log in to his account, make a booking for a slot for his car in the parking area. If booking has already been done, a slip time limit of 15 minutes is allotted and during this time he must get in the parking area designated to him. Hence when the vehicle arrives at the parking area, the number plate of the vehicle is confirmed with the number plate inserted when parking section is booked. This is done with the use of a Raspberry pi3 camera. Not only is that but the driver's identity authenticated by clicking a photo using a Raspberry pi3 camera. It helps in preventing vehicular theft and for other privacy purposes. When these procedures are finished then the vehicle is allowed entry into the parking area and then a slot number is provided analyzing the closest parking slot available. A soon as the parking of vehicle is done in the slot, parking time is started. So, when the driver comes back to take his vehicle, the parking time is terminated and the fee token is mailed to the smart phone. One can then pay the receipt with the usage of his e-wallet or using online banking. Eventually during exit, face of driver is verified again to identify a match with the previous scan with the use of facial recognition and after confirming the identity he is permitted to go out of the parking lot.

Figure 3. Steps of Working Model

Step 1: Install the application in the cell.
Step 2: Create an account in the app using the user's information.
Step 3: Log-in to the system.
Step 4: Choose the location where you need to book the parking slot.
Step 5: Choose the sub-location and the slot for booking.
Step 6: If it is empty, the application asks for the License plate number.
Step 7: Within the time duration, car is to reach the parking slot to complete verification process.
Step 8: Car is parked and removed from parking slot; parking cost are deducted from e-wallet.

Implementation and Working

We have presented a system model using a small toy car and mobile application in Figure 3. The overall workflow of the system is highlighted in Figure 4. This system can be used in malls, multi-storied buildings and any other place with large parking. Above are the steps involved in booking a slot for our parking system model. The steps above are completely explained using the below screenshots. Figure 5 shows the registration and the login details for the parking slot application. Figure 6 shows the location details and sub location details. As shown in Figure 7, you can get a slot and if it is empty, then you can confirm your tenancy by typing the license plate number. But if the driver isn't able to park the car in the certain time limit, the connection will be severed and he will have to book again.

Figure 4. Flowchart of the Smart Parking model

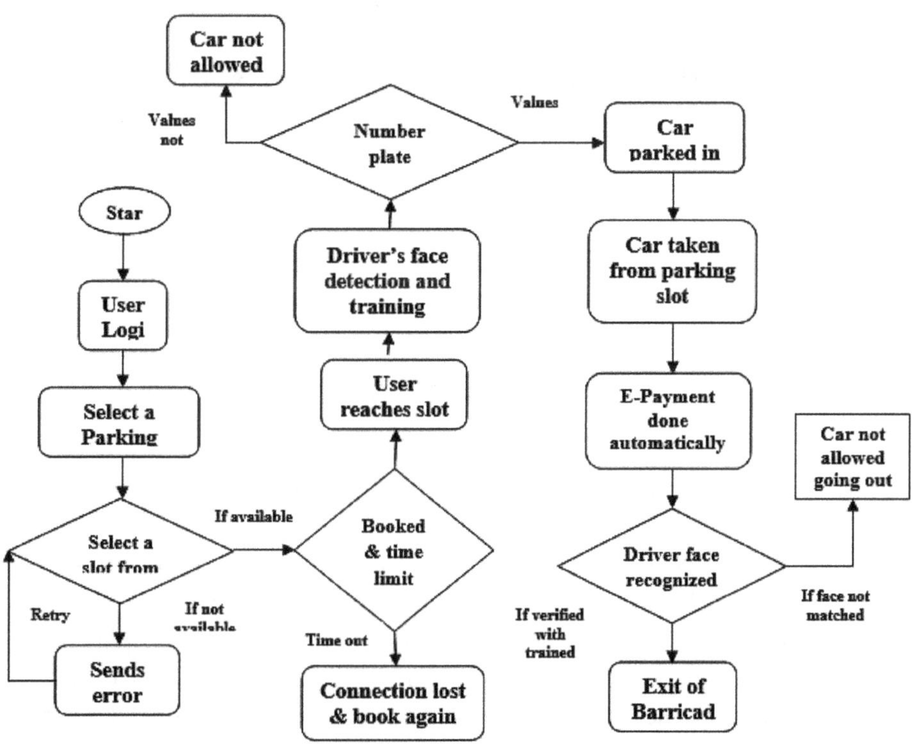

Figure 5. Registration and Login details of the Smart Parking model

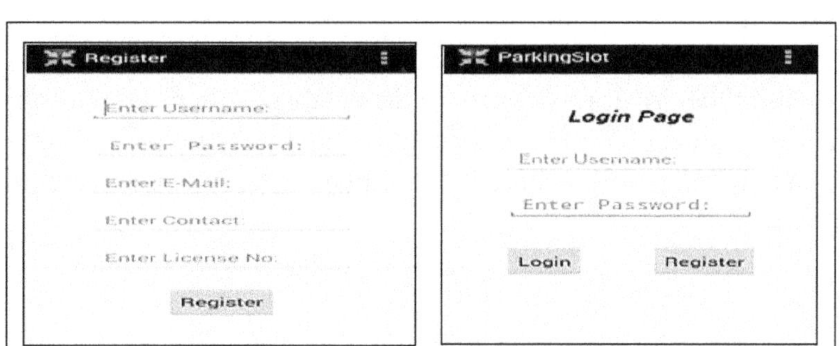

Figure 6. Location details of the Smart Parking model

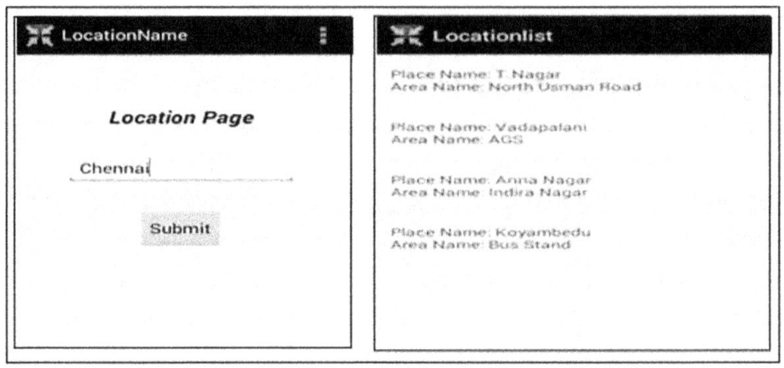

Figure 7. Vehicle booking details of the Smart Parking model

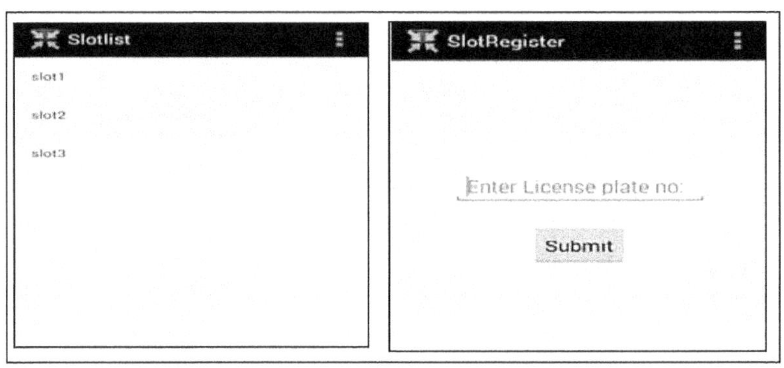

Internal Mechanism and Results Obtained Facial Recognition

Driver's face identification is performed as soon as the vehicle enters the parking area. OpenCV is used by Raspberry pi camera to develop the data samples of the face matrix of driver. Subsequently it is trained to detect the identity of driver. His face scan is again done when he exits the parking area in order to provide proper security. He is permitted to move out only if the image of his face finds a perfect match with the trained samples. Eigen faces technique with Support Vector Machine algorithm and Principal Component Analysis are used in this process. Figure 8 depicts the facial recognition sample scan at the entry and exit point.

Automatic License Plate Recognition

After clicking the photo of the driver it is trained in the database. Then raspberry pi receives the signal and the license plate gets scanned. This process works on the basis of Optical Character Recognition. It comprises three crucial phases which include License plate detection, Character Segmentation, Character Recognition.

Figure 8. Facial Recognition at the entry and exit of the parking slot

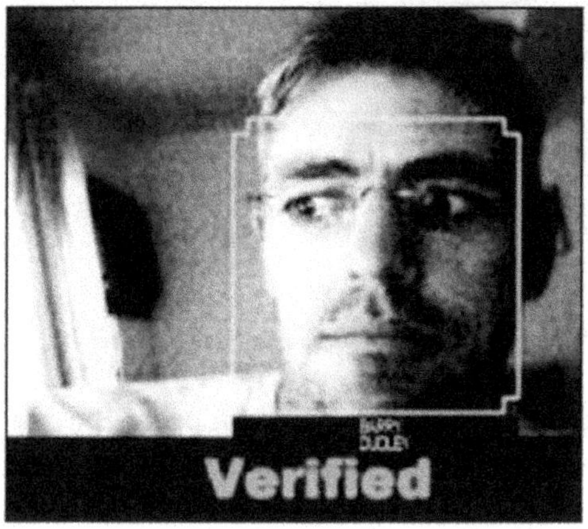

LICENSE POSITIONING

License Plate detection is initiated when the graying of the image is pre-processed. It is done since occasionally it is seen that the color of license plate as well as car is the same. After this, morphological tasks like expansion and corrosion takes place to transform it into binary image. Edges of license plate are detected on the basis of aspect ratio with the frame.

CHARACTER SEGMENTATION

Initially tilt correction is undertaken by placing a horizontal line along with characters and then pinpointing a identification line with the total character. It is followed by noise identification and on the basis of character width; it is further segmented into sub parts. It accurately determines the segmentation line related to the character central position. Hence the characters are extracted by segmentation. Figure 9 shows the license plate recognition image.

CHARACTER RECOGNITION

This model discussed a new technique of character recognition based on KNN algorithm. It basically determines the training points which are close to the required character. Eventually it is compared with the characters present in the database. Hence we extract the characters as denoted in Figure 9. KNN algorithm is used to train the character image along with all English letters and numeric literals of varying size.

A simple and user-friendly application may be developed in Android as well as iOS platform. It makes the driver's work simple by keeping it simple since it is of great help by removing pressure of locating a suitable parking location at a close area and moving towards it. He can get the credential data and the method of payment that is used regularly. This information is to be stored on cloud. The basic advantage is that parking areas may be booked much before entering into the parking slot. It can be of great help

Figure 9. Original image, License positioning image and character Segmentation

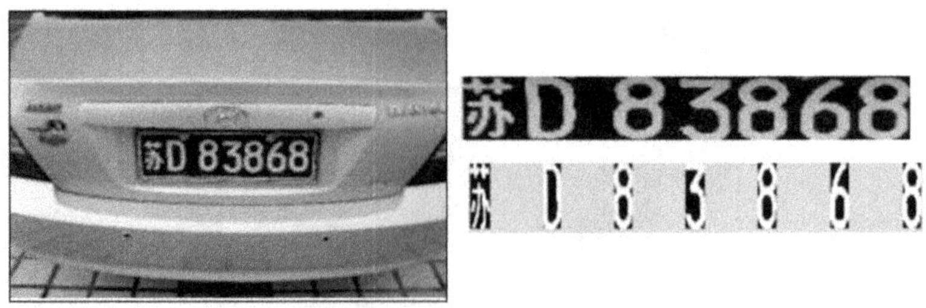

for senior citizens and handicapped people. They also receive the receipt in the application when the car gets exit from the parking area and the driver might make payment at any time using e-wallet or mobile banking for the exact time period that his vehicle was in the slot. Once the vehicle makes enter the parking structure, identity of driver should be checked. If a slot is not filled, it is to be shown using a green signal while a full slot will be shown using a red light. This makes the entry into the parking slot quiet efficient and it makes the customer's life easy. During exit, the requirement for pending queues is removed by complete automation of the payment which is helpful for driver to save time. Also, it strengthens the security by recognizing driver's faced when entering and also while exiting.

Smart Road Model with IoT

Traffic congestion is a crucial problem worldwide causing time, fuel wastage also pollution is increasing day by day. In Indian traffic-system, the problems like overcrowding, incalculable travel time are taking a serious shape which is also very confusing and noisy. In order to minimize the wastage of electricity in street lights which is unnecessarily used, we used light sensors and motion sensors. Various steps have been taken to reduce traffic jams. Recently, researchers have initiated to employ connected vehicle technology which is quite difficult to implement on roads. In this project, we present a low cost innovative and new technology for smart roads. We are executing "Smart traffic" by the use of motion sensors, light sensors, ultrasonic sensors with camera and other IoT things.

Reason for heavy traffic is due to presence of number of vehicles on busy hours. Actually, there are a lot less technology which can be implemented in roads. Roads constitute asphalt or at times concrete, that

Figure 10. Block Diagram for Smart Roads using Raspberry Pi 2

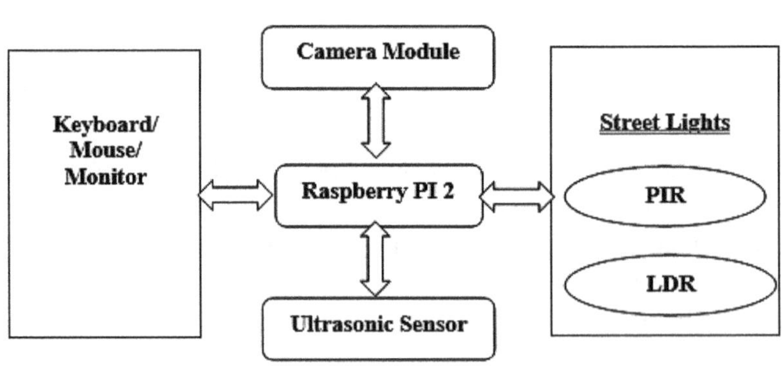

is condensed into a solid and smooth surface which is painted to show certain constraints and information. The automated highway system is specifically designed for well-furnished cars and related vehicles so that it could move efficiently by its own. It is a feature of intelligent transportation systems (ITS), that use nodes with different automated mechanism which were proposed to run air craft, the aerial events and for the betterment of roads and general transportation. Traffic-signal control systems are used in coordinating every traffic signal so to get traffic operations of different networks. They constitute various traffic signals, a transportation medium to hold them altogether, and a central node or a network to deal with the overall nodes. Cooperation among them may be achieved using many techniques such as time-base and hard-wired communication techniques. Cooperation of traffic signals among various organizations needs the creation of information exchange as well as traffic signal coordination. Therefore, an important institutional part of Traffic Signal Control is combination of formal and informal arrangements to

Figure 11. Flowchart for Smart Road model

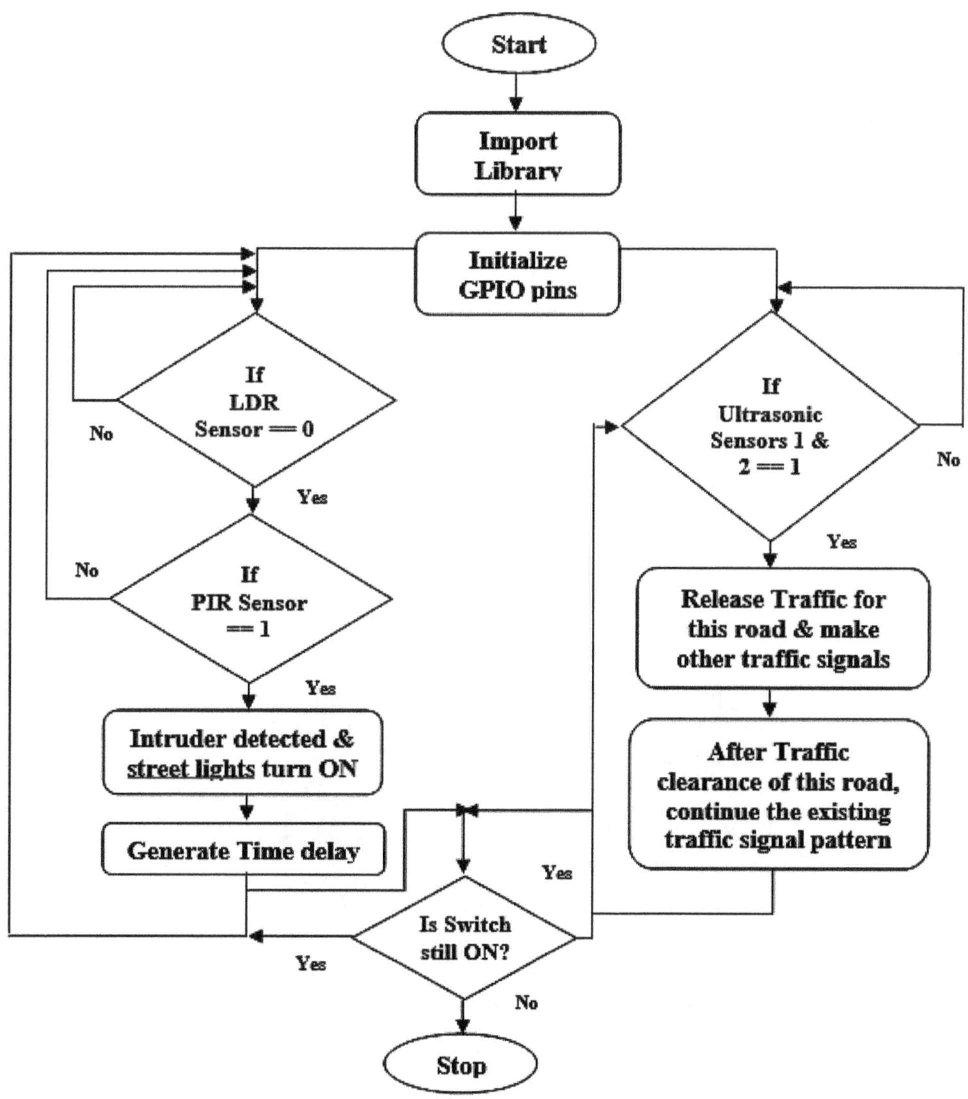

exchange traffic coordination data along with real coordination of traffic signal operation over various authorities. Signal cooperation systems can be created to give access. A well-developed traffic-signal system has no other motive than to provide right signals to drivers. This provides efficient functions that enhance the traffic engineer's power to achieve this goal. They are basically access features. They give control to the intersection signal controller for maintenance, development and various operations. More complete and suitable the access, more systematic the operator will be and the more efficient the whole system will be. Along with the control of traffic signals, new technology gives various surveillance capabilities, including different kinds of traffic observation and video surveillance. When there is massive traffic on one side of the traffic signal, it should be cleared as soon as possible. But in our existing traffic system, one have to wait till their turn comes even if there is a heavy traffic in that particular lane. To avoid this problem we can use a sensor which is capable of analyzing the intensity of traffic. This can be the efficient way to clear the traffic as soon as possible before causing a traffic jam.

Architectural Framework

In this system, sensors are used to avoid energy loss due to irrelevant utilization of street bulbs at late night. Sometimes at night street lights are switched ON even if there are no travelling vehicles. So, in order to overcome this, sensors are used for street lights and when a vehicle crosses the sensor, the street light is switched 'ON' and when the vehicle passes the particular street light, the next street light will be switched 'ON' automatically and the previous lights are switched 'OFF'. This process is continued throughout the street thereby saving lots of energy. Ultrasonic sensors are mainly applied in detection of the traffic volume which signals Raspberry PI 2 of the intensity of traffic and this will be able to clear the traffic. Camera module identifies the vehicles which are breaking the traffic rules.

Experimental Results

Figure 10 shows the top view of the model with its components such as Raspberry Pi 2, Ultrasonic Sensors, LDR sensors, PIR sensors, LED lights. Figure 11, shows the working of traffic signals and ultrasonic

Figure 12. Top view of Model with Raspberry Pi 2 and ultrasonic sensors

Figure 13. Interfacing of Raspberry Pi 2 with Traffic signals

sensors with Raspberry Pi 2. If there is traffic overcrowding in a particular road it will be detected by the two ultrasonic sensors and the signal will be sent to the Raspberry Pi. The Raspberry Pi will turn this signal green thereby releasing the traffic overcrowding. And the PIR sensors present in street lights detects the motion based on which the street lights will be turned ON or OFF when necessary. Figure 12 shows the top view of system model with ultrasonic sensors. Figure 13 presents the interfacing of Raspberry pi with traffic signals.

Road over-crowding is a challenging issue in most of these developing countries. Many cities lack well managed and controlled traffic systems with a lot of traffic crowded points. Here we discussed the nuisance of road traffic congestion in highly congested areas in the developing regions. Ultrasonic sensors are utilized in detection of the number of vehicles sent as a signal to Raspberry Pi 2 to know regarding congestion in traffic. Smooth traffic flow mechanisms are likely easy to run in real-world settings and can increase the flow of traffic at important areas in road traffic systems. This shows only some of the first step in the creation of cheap, efficient and deployable strategies for removing congestion from developing regions. The next scope of work lies on creating electricity by using smart speed breakers. We can even create charging system for electric vehicles along traffic signals by using induction coil. Here a method was discussed to spot and remove over-crowding in a small area. This solution is now only used for small area traffic; it is unable to eliminate congestion ranging for miles and miles because of the small area of attention of the method.

Smart Vehicle Model with IoT

In a Smart City, all things present has to be smart too, it must contain integrated processing units and intercommunication ability using wireless or wired networks. These smart entities would not only help in convenient and riskless surrounding through increasing interoperability and interconnection, called as Internet of Things (IoT). Here, the main role is to connect real-life entities with communication and computing capabilities in order to enhance its performance through interaction with each other. Goal of IoT also contain vehicle to smart object communications. Automobiles also act as a major component in safe and convenient travel. As being a crucial part of IoT, Internet of Vehicles (IoV) has witnessed swift development in intercommunication process. At present every vehicle can efficiently communicate private and effective based information with each other using vehicular ad hoc networks (VANETs). Automobiles are the most crucial components in machine-to-machine vehicular social networks. Real-time applications provide comfortable and safe travel for all vehicle users and all data collected make sure vehicles perform smart actions and analysis of data for the transport authorities. VANETs constituents have been integrated into IoT framework reference model to give optimal integration of IoV with other IoT domains. In the end, we give execution information and experimental study to show the effectiveness of the submitted model as well as contain many other practical deployment of the submitted model, more application scenarios for all the different user-groups, and utilizing it to gather real time multi-modal sensory data in cities. We imagine IoV will be a very important component in smart automation transport in future smart cities framework. In today's time, wireless communication technologies are used in many different region of day to day life. Automobiles are getting furnished with wireless communication devices, allowing inter-vehicular communication and with the centralized system by utilizing road-side infrastructure nodes. These communications provide more chance for creating new devices for automobiles. Utilizing this technology, the automobile industry can enhance transportation systems effectively. In automobiles, wireless technologies allows peer-to-peer mobile communications among vehicles (V2V), along with communication among vehicles and infrastructures (V2I). Hence, vehicular networks are used much more in ITS.

Working of the Model

So, in our examination, A* algorithm is used to find the shortest path that vehicles can choose from the correct paths to avoid traffic as depicted in figure 14. In today's time, DSRC built upon the Wi-Fi standard are largely utilized in VANETs to join infrastructure-to-vehicles and vehicles-to-vehicles using two way short range radios that are inexpensive in relation to other wireless standards present. The DSRC/WAVE systems fill a nook in wireless infrastructure by making high data rate, high mobility communications, low latency, and geographically local easier.

Figure.15 shows the concept of VANET. In VANETs two types of links are present: vehicular-to-vehicular communication (V2V), created on an Ad-hoc architecture, vehicles communicates without a central coordinator; and vehicle-to-infrastructure or infrastructure-to-vehicle (V2I or I2V), here communication happens among vehicles and the RSUs. VANETs were created for a big span of cooperative applications that are assistance to supply details to the driver using to the data shared among automobiles over the internet. It can ensure safe as well as non-safety implementations that may permit many more services as information, traffic management, auto toll collection, geo location based services and many more.

Figure 14. Use of A algorithm in Smart vehicle model*

- A* resembles Dijkstra algorithm and is used to determine the shortest path.
- A* make use of heuristic and is like Greedy Best-First-Search to guide itself.
- In general scenario, it is identical to Greedy Best-First-Search with faster execution.
- In A*, g(n) represents the accurate cost of path from the initial point to any vertex n, and h(n) represents the estimated heuristic cost from vertex n to the goal.

  $f(n)=g(n)+h(n)$.
- A* uses a heuristic to limit the distance calculations and thus results in a much faster routing compared with Dijkstra.

Figure 15. Vehicular Ad hoc Network Scenario

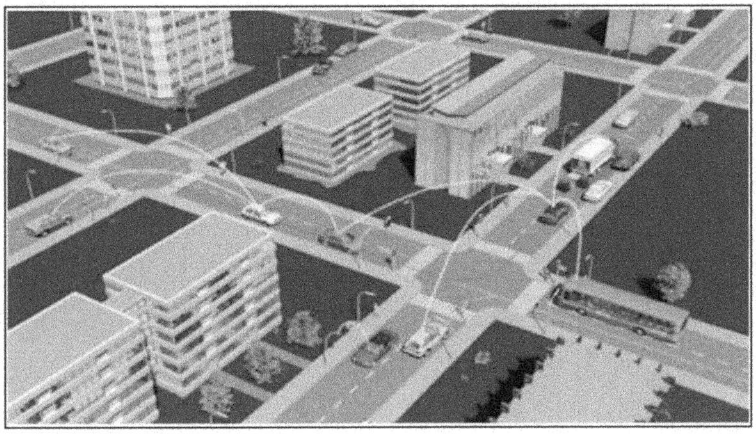

Figure 16. Vehicle to Vehicle Communication using Simulator

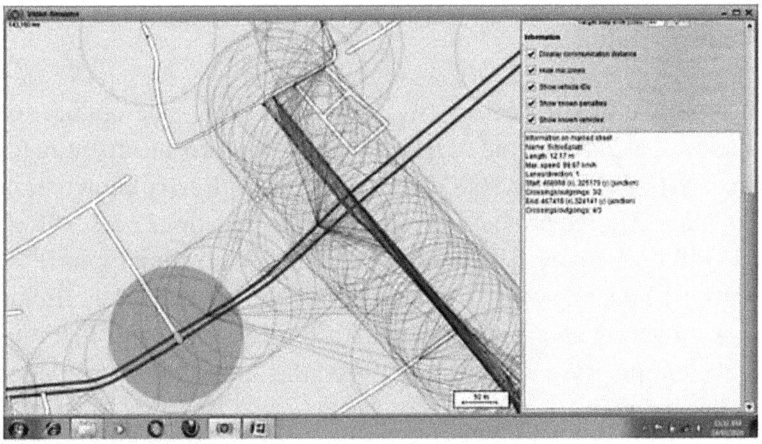

The suggested plan initiates an Intelligent Vehicular network, in which every vehicle acts as a sender, a receiver as well as a router simultaneously and thus information can be successfully broadcasted to the VANET, which utilizes this data to give instructions to the driver. OBU is the hardware responsible to manage this data and even activate such small range ad hoc wireless networks (Range is 200 meters) and it has to discard other nodes that uses reporting location data such as like Global Positioning System (GPS) to get correct position detail if needed. These details are very significant as most of the services of VANET depend upon the geological location of sender and receiver.

To implement the VANET we used Java Simulator to show the idea of our project. In this we have to run the major project, start the simulator and load the map. Then we need to load the map scenario, for map we used OSM (Open Street Map). After loading map and scenario we can see the dot, i.e., vehicle and range of vehicles as well as RSU (Road side units), figure16 represents the actual execution of project. We used A* Algorithm to find the optimal path for vehicle. Wi-Fi is used to connect the vehicle or to start the communication between vehicles and VANET. When vehicles come in the scope

Figure 17. Comparison between Path-finding Algorithms

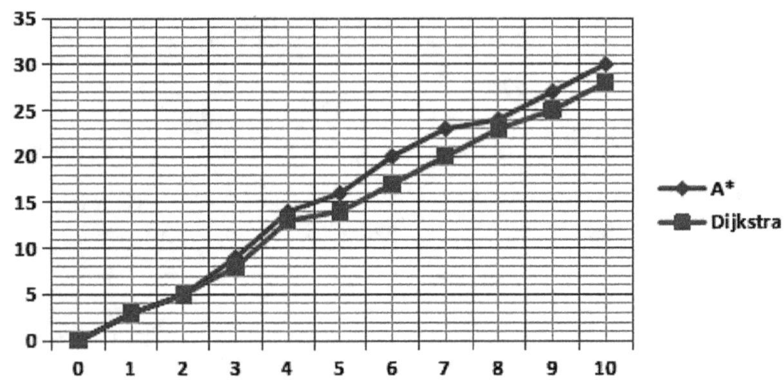

of wireless network, i.e., range of 100 meters, we can observe the line between or connection between Known Vehicles that mean communication between VANET. Here we used DSRC (Dedicated Short Range Communication) to establish the connection between vehicles.

Experimental Result

For implementation project in above section, we used Wi-Fi to create connection between VANET. Normally Wi-Fi works well at 2.4GHz and 5GHz. Due to recent advancement in technology, few firms are even able to extract Wi-Fi to work at 60GHz frequency. ZigBee is known to operate at a speed of 900-928 MHz and 2.4 GHz. ZigBee protocol-oriented communications possess a channel bandwidth of 1 MHz whereas Wi-Fi links generate a bandwidth of around 2 MHz limitations of ZigBee is up to Wireless Personal Area Networks ranging 10 – 30 m approximately. Wi-Fi can be efficiently used for WLAN and PAN area networks covering an average range of 100 m. So, we used Wi-Fi instead of ZigBee. As shown in Figure 17, the Graph gives a contrast between the A* and Dijkstra Algorithm w.r.t. the time and the number of paths and nodes.

CONCLUSION

Internet of Things (IoT) is a platform that makes a device smart such that every day communication becomes more informative. This technology can be effectively used in transportation field. In this chapter we have discussed three major constituents of smart transportation system model which include smart roads, smart vehicles and smart parking system. Our chapter deals with the in-depth discussion of these various aspects of smart transportation system enabled with IoT technology. IoT based automated models are presented and explored that deals with these aspects and its implemented is also highlighted. A case study illustrating individual component is presented in this study. Its implementation is also highlighted here. A well-built Smart Transportation system reduces the possible chances of any accident, enhance safety, improve capacity, minimize consumption of fuel and improve general performance of drivers.

REFERENCES

Almishari, S., Ababtein, N., Dash, P., & Naik, K. (2017). An Energy Efficient Real-time Vehicle Tracking System. *IEEE Pacific Rim Conference on Communications, Computers and Signal Processing*. 10.1109/PACRIM.2017.8121884

Ferris, B., Watkins, K., & Borning, A. (2010, April). OneBusAway: results from providing real-time arrival information for public transit. *Proc. SIGCHI Conf. Hum. Comput. Interact.,* (pp. 1807–1816). ACM. 10.1145/1753326.1753597

Ganesan, K., & Vignesh, K. (2007). Automated parking slot allocation using RFID technology. (pp. 1-4). *9th International Symposium on Signal Processing and Its Applications*. 10.1109/ISSPA.2007.4555296

Hari Baabu, V., Senthil Kumar, G., Deb, P., & Rai, A. (2016). Smart Parking Assist System using Internet of Things (2016). *International Journal of Control Theory and Applications, 9*(40), 2016.

Hong, T. P., & Soh, C. Jaafar, & Ishak (2013). Real Time Monitoring System for Parking Space Management Services. System, Process & Control (ICSPC), IEEE.

Kenmotsu, M., Sun, W., Shibata, N., Yasumoto, K., & Mi-noru, I. (2012). Parking Navigation for alleviating Congestion in Multilevel Parking Facility. *Vehicular Technology Conference (VTC Fall). IEEE*. 10.1109/VTCFall.2012.6398911

Nalawade, S. R., & Devrukhkar, S. (2016). Bus Tracking by Computing Cell Tower Information on Raspberry Pi. (pp. 87–90) *International Conference on Global Trends in Signal Processing, Information Computing and Communication (ICGTSPICC)*. 10.1109/ICGTSPICC.2016.7955275

Nasir, M. M., & Mansor, W. (2011). GSM based Motorcycle Security System, (pp. 129–134), *Control and System Graduate Research Colloquium (ICSGRC) IEEE*.

Panchal, J. R. (2017). Energy Efficient Wireless Sensor Network System for Transportation. International Journal of Engineering Technology. *Management and Applied Sciences, 5*(5), 663–667.

Shinde, P. A., & Mane, Y. B. (2015). Advanced vehicle monitoring and tracking system based on Raspberry Pi. *9th International Conference on Intelligent Systems and Control*. 10.1109/ISCO.2015.7282250

Suhr, J. K., & Jung, H. G. (2014). *Sensor Fusion- Based Vacant Parking Slot Detection and Tracking. Intelligent Transportation Systems*. IEEE.

Tang, V. W., Zheng, Y., & Cao, J. (2006). An intelligent car park management system based on wireless sensor networks. Pervasive Computing and Applications, (pp. 65–70), *1st International Symposium on IEEE*. 10.1109/SPCA.2006.297498

Wei, L., Wu, Q., Yang, M., Ding, W., Li, B., & Gao, R. (2012). Design and Implementation of Smart Parking Management System Based on RFID and Internet. *International Conference on Control Engineering and communication Technology (ICCECT)*. 10.1109/ICCECT.2012.12

Yihua, Z. (2010). VIP customer segmentation based on data mining in mobile-communications industry in Computer Science and Education, (pp. 156–159), 5th *International Conference on IEEE*.

Zhou, F., & Li, Q. (2014), Parking Guidance System Based on ZigBee and Geomagnetic Sensor Technology. *13th International Conference on Distributed Computing and Applications to Business, Engineering and Science (DCABES)*, pp. 268-271. 10.1109/DCABES.2014.58

Chapter 13
A Neural Network-Based Automatic Crop Monitoring Robot for Agriculture

E. Udayakumar
Kalaignarkarunanidhi Institute of Technology, India

S. Balamurugan
https://orcid.org/0000-0003-3428-3849
QUANTS IS & Consultancy Services, India

P. Vetrivelan
PSG Institute of Technology and Applied Research, India

ABSTRACT

The economy, being highly based on agriculture, demands innovative and reliable methods of irrigation. In this paper, an idea of automatic irrigation method is proposed. Automatic irrigation is done using a soil moisture sensor. The manual method of irrigation is done by using automated process. In this proposed method, apart from a moisture sensor, other sensors like PIR sensor, ultrasonic sensor, humidity, temperature sensor, and water level sensors are used. This method has additional features like GSM. In wireless systems, electricity will be provided through solar panels. Whenever the moisture content of the soil reaches its maximum threshold value, the system sends a signal to the motor and it turns ON. The robot can do its work automatically through artificial neural network. Every time the motor starts or stops, the user will get the status of the motor's operation through SMS. The robot will continuously monitor the crop field using wireless camera. This provides security for the agriculture land. The main advantages of this system include minimization of water wastage, & error reduction

DOI: 10.4018/978-1-5225-9246-4.ch013

INTRODUCTION

In some countries, agriculture is considered as one of the major sources of economic progress. The income of many countries depends directly on agricultural advancement. Moreover, continuous increase in the population of a country demands more innovation in food production technology. The factors affecting agricultural progress must be studied thoroughly to obtain maximum results. The significant building block of agriculture is the irrigation system. In other words, the efficiency of irrigation system may induce ample effects on agriculture (Pathan, & Hate, 2016).

Irrigation process should provide water to soil consistently when it is required and as well stop water flow when the soil has soaked enough water. The excess water in the crop is of no good. It not only leads to wastage of but also destroys crops. In India, the country's economy is mainly based on agriculture. It requires efficient and modern methods for water provision in the crops fields (Bircher et. al., 2012). Problems caused through manual methods of irrigation have let us think about some advance methods which could be relied upon. Any method which is cost effective, reducing manpower and energy saving is considered efficient. Hence, in this proposed system a method which uses very less or no labour (runs on its own) has been recommended, saves electricity and is easy to use.

The proposed system is an automatic irrigation system. Here, automaticity means it turns itself on and off depending on the soil moisture requirement. This automatic irrigation is achieved by using different sensors that sense and notify the user if the soil requires water or not. It also tells the user how much water would be sufficient for the soil, so that water wastage can also be avoided. The errors which may arise when manual irrigation is used would also be rectified to the maximum using this method. The major source of electricity in India is hydroelectric power but this source doesn't pay the country with requisite amount of electrical power. Hence, there is shortage of electricity which is not good for the process of irrigation as motors need uninterrupted supply of electricity (Alsayid et. al., 2013). As deficiency of electricity is a major problem in India, this system is made more feasible by using solar energy. The system is independent of any labour but the status of undergoing process will be received by the user through GPRS (Santhi, 2019) (Nallani, Sandeep, & Hency, 2015). The main objective of this paper is to reduce labour dependency and human effort. Advancement is made by modifying the mechanism in farming, which works automatically, to automate the agro sector. Progressive innovation becomes necessary to meet the raising demands on agro product quality, to reduce the human effort and also to increase crop yield. Solar power is used for operating the robot which is controlled by an obstacle avoidance sensor. This robot is used for monitoring and maintaining crop field using ANN.

SYSTEM DESIGN

The countries where agriculture has a big impact on economy demand a highly efficient way of irrigation. A timely and consistent irrigation is need of the hour in such countries. In places where lack of water is not tolerated by the soil during irrigation, the excess of water provision is also not recommended for crops' flourishment. Hence, a feasible irrigation for any land requires suitable amount of water with minimum amount of delays. Today's world demands improved methods as compared to older ones to carry out processes faster. The world is moving towards automation of every process. In this proposed system, automatic irrigation system has been suggested which detects the soil moisture level and is programmed in a way such that if the water level in soil goes below a particular threshold value, it automatically switches ON the pumps to supply water. In this way, maximum results are obtained out of the fields and water wastage is also reduced to significant level (Udayakumar et. al., 2016) (Nallani, Sandeep, & Hency, 2015) .

In traditional agriculture, monitoring was based on kit and wiring system. Humans were used for these processes which consumed a lot of time.

In the proposed approach, artificial neural networks (machine learning algorithms) are used to teach the robot to automatically work over the field. Wi-Fi cameras are used for consistent crop field surveillance (Agrawal, & Singhal, 2015).

The assembly of the robotic system is built using DC motors, solar panels, LCDs, Wi-Fi cameras and sensors. The direction of DC motor is controlled by using Atmega 2560 microcontroller.

The block diagram of robot is shown in Fig. It includes a solar Panel, DC motors, LCD, Wi-Fi camera, sensors and microcontroller assembly.

The basic working principle of this system is easy to understand. The system is divided into smaller circuitries. First one is solar circuit. It provides DC power to the components whenever needed. Second circuit is the sensor network. Moisture sensors are submersed into soil and connected back with the main system. The sensors give values of moisture content of the soil and these values can be read on the LCD. Another circuit is the GSM module. This is also connected with Arduino and is responsible for sending information to the user about all the operation taking (Alsayid et. al., 2015).

Arduino Mega 2560

Arduino is the genesis of this proposed system. It's the centre of all operations taking place in the system. The components are connected to Arduino through different ports and are dependent on its instructions. Arduino Mega 2560 has been used because of its versatility. It has 54 digital I/O ports. There are 16 analog inputs, 4 UARTs, 16 MHz crystal oscillators, a USB port, power ports, reset button and an ICSP header. The flash memory is 256 Kb and EPROM memory is 4 Kb (Arduino.cc). All the data from sensors come directly to the Arduino which processes it and sends the signal further. The Arduino commands the motor either to start or stop. The code that is fed into the Arduino will judge the moisture condition of the soil and decide if the motor has to be turned on or off. The code is written on Arduino software and is transferred to the device using an USB cable. C language is used to code and the upper and lower thresholds are defined in the code. Basically, the code tends to keep the water content in between the threshold levels. If the water level in the soil crosses either values, status of the motor will be changed (Rani and Kamalesh, 2014).

Sensors

The participation of sensors in automatic irrigation is very important. They play a vital role in making the system automatic. Without them, the process cannot be stated as automatic. Three different sensors have been used for measuring three different parameters.

YL 69 is the soil moisture sensor. It senses the water content in the soil. Its role is the most important as the information forwarded by it is most relative regarding water requirement. The sensor has two prongs which are submersed in the soil. It has 4 ports- GND, VCC and outputs for analog and digital values (Udayakumar et. al., 2016).

DHT11 is the humidity sensor. It detects the water content in atmosphere. High atmospheric humidity might increase soil dampness. LM35 is the temperature sensor. It measures the temperature of environment. The advantage of LM35 is its significant feature that always gives temperature in Celsius. This avoids further calculations required to convert the output to get temperature in Celsius. A passive infrared sensor (PIR sensor) is most often used in PIR-based motion detectors. An Ultrasonic sensor is a device that can measure the distance of an object by using sound waves.

The latter two sensors are used to make the system more reliable. These two sensors are left in open environment. They constantly give the value of temperature and humidity. When the temperature or humidity level of environment gets altered, it may affect the moisture level of soil. So, to eradicate any changes that may fluctuate the process of irrigation these sensors send signal to Arduino to take some action. In the design, if humidity level goes above our defined value then the Arduino sends signal to DC fans located near the sensors to mild its effect. The DC fans automatically turn on and keep running till normal conditions are achieved (Aarthi, & Khadir, 2015) (Pathan, & Hate, 2016).

GSM Module

SIM900D has been used for GPS purpose. This module makes the system wireless. The status of undergoing process will be updated to the user via SMS. This module is connected to the Arduino board.

F.GPS Module

A GPS navigation device is a device that is capable of receiving information from GPS satellite and then calculates its geographical position. Using suitable software, the device may display the position on a map, and it may offer direction (Rane, et. al., 2019).

G. Photo Voltaic Panels

Solar panels are used to liberate irrigation from the shackles of load shedding. The requirement of water is judged and information is then transmitted to the solar circuit which modifies its configuration such that it provides enough DC power to drive the pumps and fulfil the assigned task. This method is not only power efficient but also proves to be cost effective when considered in long run. The solar irrigation process proves to be of great worth to the irrigation cites that are far from power grid stations (Alsayid, 2013) (Rajpal et. al., 2011).

Artificial Neural Networks

Artificial neural networks are forecasting methods that are based on simple mathematical models of the brain. They allow complex nonlinear relationships between the response variable and its predictors. In between the input units and output units are one or more layers of hidden units, which, together, form the majority of the artificial brain. Most neural networks are fully connected, which means each hidden unit and each output unit is connected to every unit in the layers either side (Rane, et. al., 2019) .

Figure 1 depicts the block diagram of automatic irrigation process. The diagram includes all basic components used in the system.

Figure 1. Block diagram of system

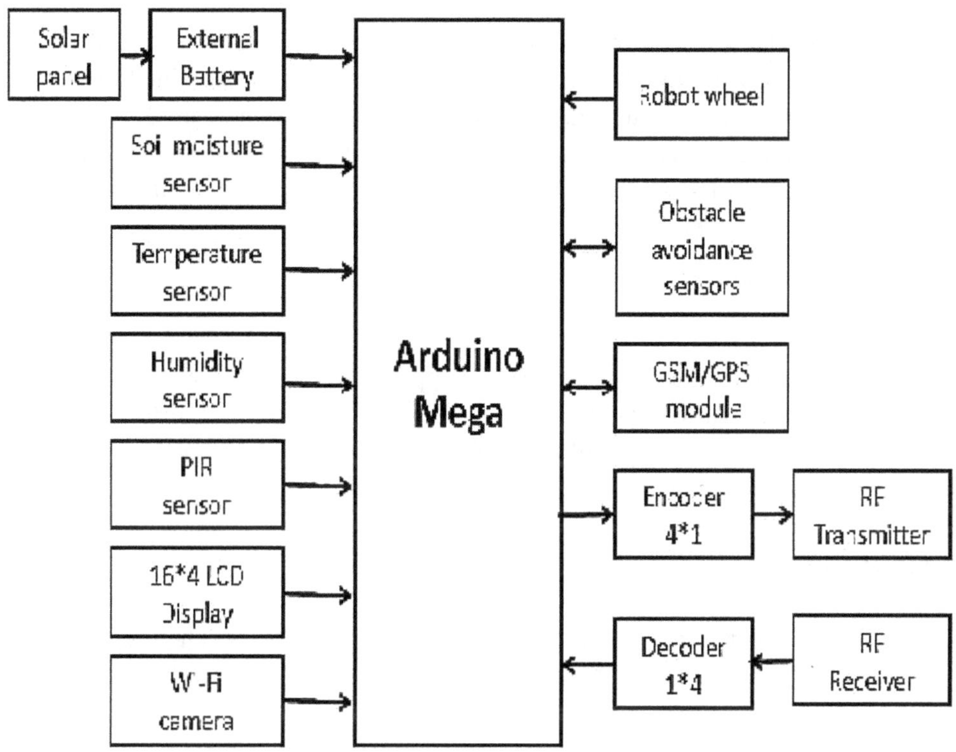

Figure 2. Block diagram of system

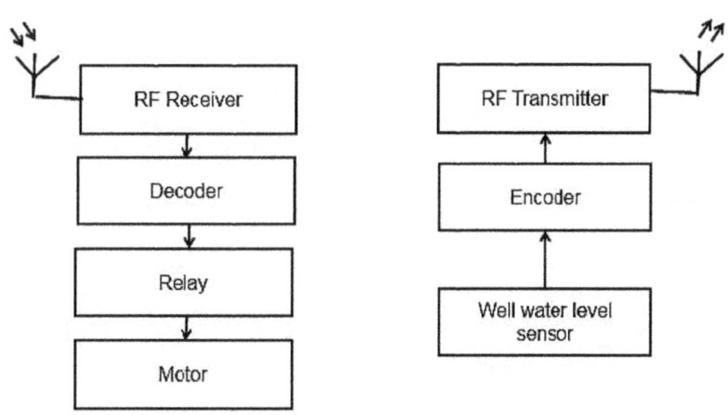

Figure 2 shows that the sensors that give information to Arduino. It forwards the information to the other parts of system as per the program that is inserted in it .The other parts of system include LCD, relays, DC motors and GSM.

J. Power Supply

The present chapter introduces the operation of power supply circuits built using filters, rectifiers and voltage regulators. Starting with an AC voltage a steady DC voltage is obtained by rectifying the AC voltage, then filtering to a DC level and finally regulating it to obtain a desired fixed DC voltage. The regulation is usually obtained from an IC voltage regulator unit, which takes a DC voltage and provides somewhat lower DC voltage which remains the same even if the input DC voltage varies or the output load connected to a DC voltage changes (Santhi et al., 2019).

K. Transformer

The potential transformer will step down the power supply voltage (0-230V) to (0-6V) level. Then the secondary of the potential transformer will be connected to the precision rectifier, which is constructed with the help of Op-amp. The advantages of using precision rectifier are it will give peak voltage output as DC; rest of the circuits will give only RMS output.

L. Bridge Rectifier

When four diodes are connected in a circuit it is called as bridge rectifier. The input to the circuit is applied to the diagonally opposite corners of the network and the output is taken from the remaining two corners (Udayakumar et. al., 2016).

After one half-cycle, the polarity across the secondary of the transformer reverses, forward biasing the diodes D2 and D5 and reverse biasing the diodes D1 and D3. Current flow will be from point A through D5, then through RL, through D2, through the secondary of T1, and finally back to point A. This path is indicated by broken arrows. Waveform (3) and (5) can be observed across D2 and D5 (Agrawal, & Singhal, 2015). The current through RL is always in the same direction. While flowing through RL this current develops a voltage corresponding to the wave form (5). Since current flows through the load (RL) during both half cycles of the applied voltage. This bridge rectifier is full wave rectifier.

M. IC Voltage Regulators

Voltage Regulator is one of the most important and commonly used electrical components. Voltage regulators are responsible for maintaining a steady voltage across an Electronic system. Voltage fluctuations may result in undesirable effects on an electronic system, so maintaining a steady constant voltage is necessary according to the voltage requirement of a system. Even a slight increase in voltage may result in failure of the entire system by causing damage to other components too. For avoiding damage in such situations voltage regulators are used for regulated power supply (Al-Ali et. al., 2015).

N. LCD Display

Liquid crystal display is used in similar applications where LEDs are used. The applications include display of numeric and alphanumeric characters in dot matrix and segmental display. When a potential is applied across the cell, charge carries following through the liquid disrupt the molecular alignment and produce turbulence. When liquid is not activated scattering in all directions, the cell appears too bright. This phenomenon is called dynamic scattering (Nallani, & Hency, 2015).

WORKING PRINCIPLE

The basic working principle of this system is easy to understand. The system is divided into smaller circuitries. First one is solar circuit. It provides DC power to the components whenever needed. Second circuit is the sensor network. Moisture sensors are submersed into soil and connected back with the main system. The sensors give values of moisture content of the soil and these values can be read on the LCD. Another circuit is the GSM module. This is also connected with Arduino and is responsible for sending information to the user about all the operation taking (Nallani, & Hency, 2015). In the code, there are basically two threshold values that are defined by the user (i.e.) upper and lower. The actual value of water content in soil is read by the moisture sensors which are submersed in soil. The code compares this value with the two user defined threshold values (Uddin et. al., 2012).

If actual value happens to be lesser than the lower threshold value, the code will generate a signal that will turn the motor on. The process will be autonomous and the dried part of soil gets moisturized. The values of moisture level are constantly compared with the threshold values in code and if actual moisture value crosses the upper threshold value then the code will send the signal for the motor to be turned off. The process starts with sensors reading the value and displaying them on LCD. As the value of moisture falls below the lower threshold point, the motor starts and if the content climbs the upper threshold value the motors switches off. In either case, the user will get an SMS of the undergoing processes. The status of the motors will also be displayed on the LCD (Rajpal et. al., 2011).

Figure 3. Connections between basic components

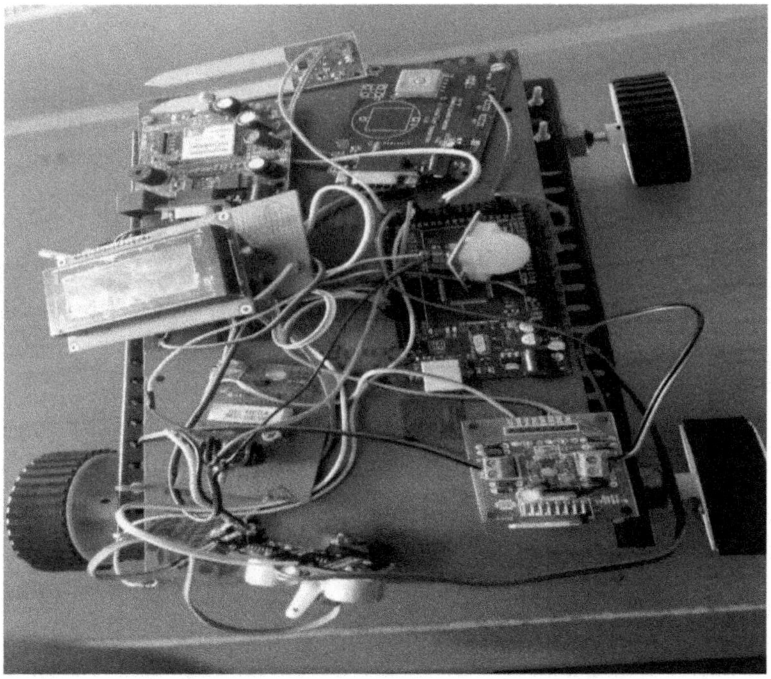

Figure 4. The constructed prototype

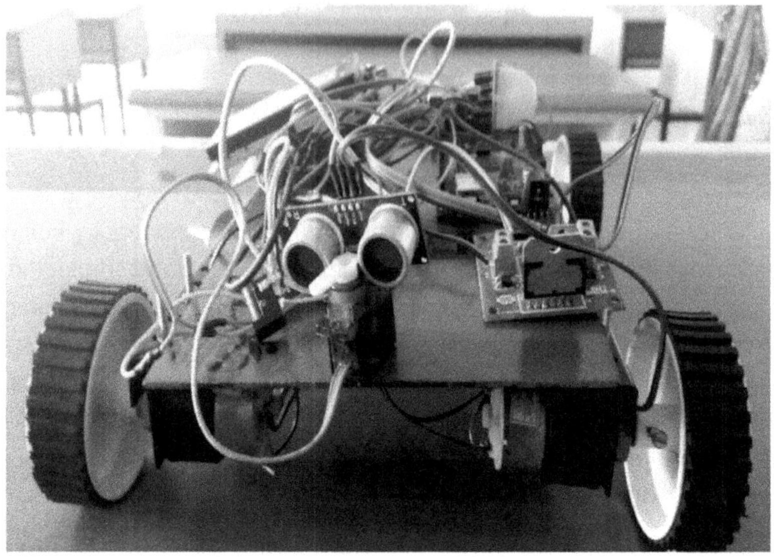

Figure 5. Simulation of Circuit for DC Motor and Sensors

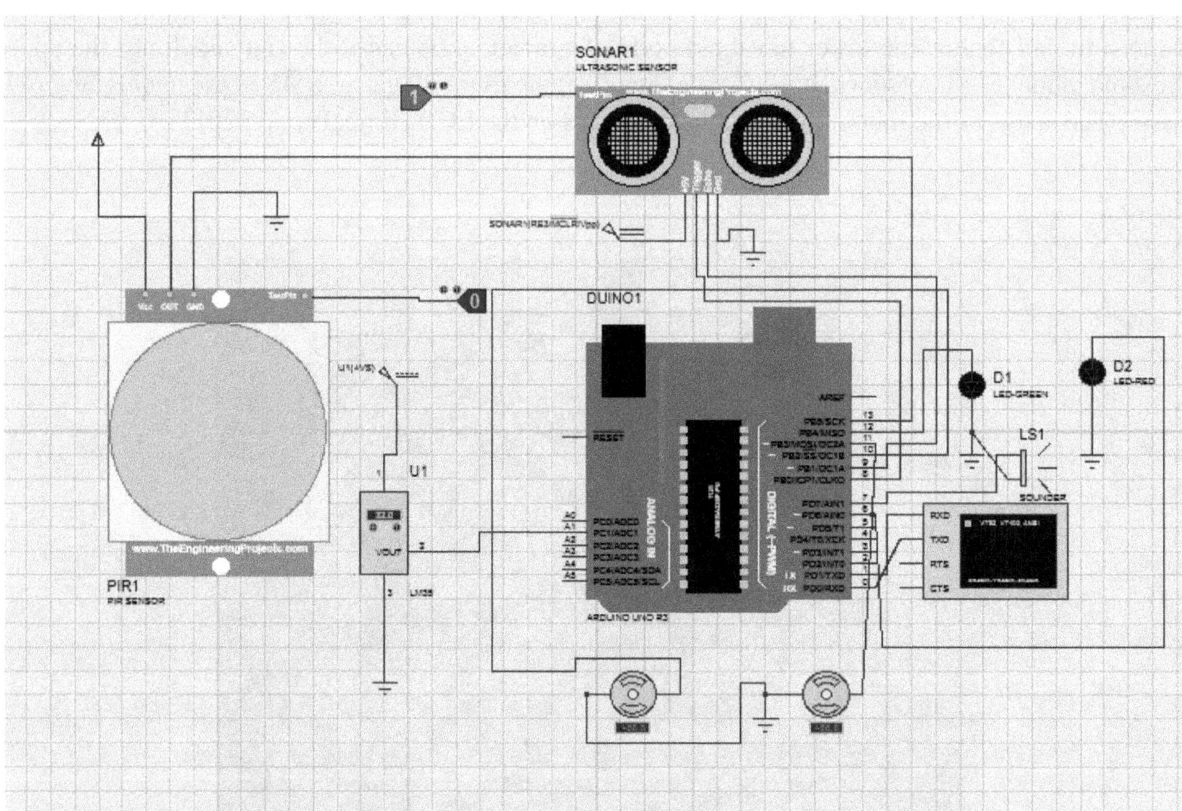

Figure 3 & 4 show the pictures of original prototype. Fig. 4 gives the whole model in which two plants are also shown. Both plants have moisture sensors immersed in them. A separate motor is also connected to each plant which turns on when water is demanded by soil of plant. In the Fig 3, a more elaborate form of is seen in which the basic components can be visualised. The components are connected to each other and shown in the basic scheme of this paper. Fig. 5 shows the simulation of circuit for DC motors and sensors.

CONCLUSION

The use of automatic irrigation method would allow us to save the excess water which may be wasted during manual methods and also monitor crop field consistently by using wireless camera. Further it improvises the process of irrigation and makes it a reliable one. The provision of water to the fields is done in a more effective way using this technique. Moreover, electricity issue can also be resolved by using solar energy. Thus, this method has an upper hand over all other methods of irrigation because of its consistency and usability. This improvisation in food production technology and security greatly enhances the opportunities to increase the economic growth in India. With the use of minimum resources, the proposed system can save a lot of water and electricity hence economically favourable.

REFERENCES

Aarthi, R., Shaik, A., & Khadir, A. (2015). An efficient method of irrigation using sensors. *International Journal of advanced research in computer and communication engineering, 4*(7).

Agrawal, N., & Singhal, S. (2015). Smart drip irrigation system using raspberry pi and Arduino.

Al-Ali, A. R., Qasaimeh, M., Mamoun, A. M., Radder, S., & Zualkernan, A. (2015). ZigBee-based irrigation system for home gardens. *Communications*.

Alsayid, B., Jallad, J., Dradi, M., & Al-Qasem, O. *(2013). Automatic irrigation system with pv solar tracking. Int. J Latest Trends.*

Bircher, S., Skou, N., Jensen, K. H., Walker, J. P., & Rasmussen, L. (2012). A soil moisture and temperature network for SMOS validation in Western Denmark. In 2015 International Conference Computing, Communication & Automation (ICCCA), Denmark. *Hydrology and Earth System Sciences, 16*(5), 1445–1463. doi:10.5194/hess-16-1445-2012

Nallani, S., & Berlin Hency, V. (2015). *Low power cost effective automatic irrigation system. Optimization and Sciences* (pp. 224–228). EEECOS.

Rajpal, A., Jain, S., Khare, N., & Shukla, A. K. (2011). Microcontroller-based automatic irrigation system with moisture sensors. In *Proceedings of the International Conference on Science and Engineering* (pp. 94-96).

Rani, M. U., & Kamalesh, S. (2014). Web based service to monitor automatic irrigation system for the agriculture field using sensors. In *2014 International Conference on Advances in Electrical Engineering (ICAEE).*

Santhi, S., Udayakumar, E., & Gowthaman, T. (2019). SOS Emergency Ad Hoc Wireless Network. In *Computational Intelligence and Sustainable Systems* (pp. 227–234). Cham, Switzerland: Springer.

Shahin, A. P., & Hate, S. G. (2016). *Automatic irrigation system using wireless sensor network. 2015 International Conference* Signal Processing, and their Applications (ICCSPA).

Sharvin, R. (2016). Automated Irrigation System Using X-Bee and LabView. In *3rd International Conference on Electrical, Electronics, Engineering Trends, Communication, Optimization and Sciences (EEECOS).*

Udayakumar, E., ... (2016). Automatic Battery Replacement of Robot, *7th International Conference on Advances in Natural and Engineering (ICECE).* IEEE. Retrieved from https://www.arduino.cc/en/Main/arduinoBoardMega. *Indian Journal of Science and Technology, 8*(23).

Udayakumar, E., Ramesh, C., Tamilselvan, S., Yogeshwaran, K., & Kanagaraj, T. (2016). Foot Pressure Measurement by using ATMEGA 164 Microcontroller. *Advances in Natural and Applied Sciences, 10*(13), 224-229.

Uddin, J., Reza, S. T., Newaz, Q., Islam, T., & Kim, J. M. . (2012). Automated irrigation system using solar power. *Electrical & Computer Applied Sciences, AENSI Publications, 9*(7), 2015, pp. 33-38.

Chapter 14
Smart Water Level Monitoring System for Farmers

Nalina Suresh
University of Namibia, Namibia

Valerianus Hashiyana
University of Namibia, Namibia

Victor Panduleni Kulula
University of Namibia, Namibia

Shreekanth Thotappa
JSS Science and Technology University, India

ABSTRACT

With advancement in technology and ever-changing weather conditions, accurate and affordable water level measurement systems has become necessary for farmers. This therefore brings about the need for a system incorporating the use of IoT technology that will monitor water levels at a cost-effective price with accurate and dependable results. The prototype will monitor water levels on a regular basis and the data captured will be stored in a database to help farmers improve the way they manage their water resource. Farmers will be able to monitor the water levels from any location at any given time. This chapter focuses on a Smart Water Level Monitoring System for Farmers and provides a smart way to manage water resources on farms in the most cost-effective and convenient manner for farmers.

INTRODUCTION

Water is the most important resources for human survival and is essential for agriculture, industry and domestic consumption. According to Nepomilueva (2017), the shortage of water affects around 2.8 billion people around the world at least one month out of every year. Water shortage can be defined as a lack of sufficient water, or not having access to safe water supplies (Paulson, 2015). Namibia is a country

DOI: 10.4018/978-1-5225-9246-4.ch014

where more than 1.5 million people rely on subsistence farming and the country is experiencing its worst drought in recent history, which has led to the government declaration of a national drought emergency (Crisp, 2017). Thus, it is essential for Namibians to conserve water.

Water shortage is affecting the productivity of the agricultural sector in arid and semi-arid regions at large. Considering the case of Namibia, this sector consumes about 75% of the water in the country, with the commercial agriculture being the largest sub-sector and communal farmers being the least consumptive (Dirkx et al., 2008). It is difficult for staple crops such as wheat, mahangu and white maize to mention few, be grown as they require large volumes of water for growth, resulting in household food security being compromised. This is because Namibia depends on cereals as our staple food. According to Pérez-Hoyos et al. (2017), Vision 2030 of the country emphasizes the need to increase agricultural productivity in order to achieve food security.

Rainfall harvesting is a way of combating water shortages but, Pérez-Hoyos et al. (2017) argued that only 2% of Namibia's land receives sufficient rainfall to grow crops, most rivers flow only occasionally, hence the need for irrigation and other innovative technologies such as water level monitoring systems that can help with the efficient use of water.

Storing water in water tank reduces the need for the Namibian government to construct new dams. People in the rural areas especially in the farming sector store water in tanks. Most water tanks can be seen at schools and hospitals in these areas. The water is used for cooking bathing and farming. Although people in these areas have water tanks, they do not get notified on the current status of the water level in the tanks and these results in water shortage for long period of time. The methods of monitoring water levels in water storage tanks in rural areas are simple and cheap but not accurate nor dependable.

This research is based on the current challenges that the agricultural sector faces in terms of shortages of water in their water tanks. Thus, this research study focused on providing an innovative technology that will help farmers manage their water resource. This will reduce or prevent water shortages on their farms. Farmers are more likely to be motivated and active with their agricultural activities if they have water access, as they can see that these activities have a chance of succeeding.

In this chapter, we look at how a Smart Water Level Monitoring System (SWLMS) can be designed that will monitor water levels on a regular basis and the data captured will be stored in a database to help farmers improve the way they manage their water resource. Farmers will be able to monitor the water levels from any location at any given time. The Prototype will measure water levels in real time and give feedback to the farmers, warning them on when water levels are at critical points through a short message services (SMS).

LITERATURE REVIEW

There are various advanced technologies for monitoring water. The researcher believes that traditional methods are time-consuming and labor-intensive and is one of the reasons that result in water shortage. Relevant stakeholders do not get alerted early enough on the current water level resulting in water shortages. Previous studies have covered a variety of solutions on how to monitor water levels effectively. This chapter will focus on three Remote Sensor Networks that have been developed to monitor water, outlining the significance of water monitoring systems.

Tank Water Level Monitoring System Using GSM Network

This system is created by Johari et al. (2015) as an alternative way to prevent unexpected shortage of water supply. The researchers observe that the student hostels received water from tank at the roof top of the building and that there where was no early warning system. Furthermore, Johari et al. (2015) conducted a research to monitor and alert the person-in-charge through Short Message Service (SMS). This study is one of many that explore the use of mobile phones as a way to reduce water shortages. The limitation of their research is that their system only uses the SMS platform to alert the person in charge. Furthermore, the system does not have a web-based platform that will enable the end user to monitor the current water status at any given time instead of just waiting for the water level to be at a critical point to know the status.

Automatic Water Level Control System

Eltaieb, & Min (2015) from china contribute to the research on ways technology can reduce water shortages. The motivation for research was to use computing techniques in creating a barrier to water wastage. Their proposed system is cost effective, energy saving and also helps the environment water cycle. The system used a microcontroller, Arduino to automate the process of water pumping in a tank with the ability to detect the level of water in a tank, turn the pump on or off and display the status on an LCD screen. If the level inside the tank is low, the pump will automatically switch ON and this protects the motor from dry running. When the tank is low a beep sound is generated. However, this system notifies the person in charge through a beep sound that the water level is a critical point. This alert can only be heard when the person being alerted is near the tank, whereas the SWLMS will alert you wherever you are via SMS and in addition the user can view current water status on his smartphone or laptop.

Wireless Sensor Networks for Water Quality Monitoring and Control Within Lake Victoria Basin: Prototype Development

The research carried out by Gupta et al. (2016) on Wireless Sensor Networks for Water Quality Monitoring. Their research was driven on the need for effective and efficient monitoring, evaluation, and control of water quality in the Lake Victor basin. The researchers observed that the current traditional methods people use for collecting water samples, testing and analyses in water laboratories are costly and lack capability for real-time data capture, analyses and that the information got from the testing are not sent to the relevant stakeholders on time for them to make informed decisions. This prototype monitors water temperature, dissolved oxygen, pH, and electrical conductivity in real-time and informs relevant stakeholders through web-based portal and mobile phone platforms (Gupta et al., 2016). Whereas, the proposed prototype aims at solving the same issues, but as the Water level Monitoring System. Both the systems aim at informing relevant stakeholders about the current water status in very timely way so that the stakeholders can make decisions based on the information they received, but the former system has limitation of not having security features. The stakeholders can access the results with any cereals leaving there, the information vulnerable to hackers (Deepiga, & Sivasankari, 2015).

The common objectives of the above literatures done by different researchers from all corners of the globe is to alert stakeholders and people in charge of the current water status. The researchers believe that alerting people in a more prompt and effective way will reduce the water shortages. They also considered the accuracy and how cost effective the systems are.

The researcher's SWLMS will overcome and improve the limitations on the traditional techniques that are used to monitor water level. In addition, it has a security feature for Authentication. The benefits of this security feature are that it validates farmer's right to access the information. The water level monitoring system captures data in the database that can be accessed by the farmer at any given time or location. In the future, the system will be developed further in terms of having a mobile application (App) for all mobile platforms (Aly, 2016).

METHODOLOGY

Design Science is used in the creation of technological artifacts that impact people and organization (Bichler, 2006). Design Science Research Methodology (DSRM) focuses on problem solving and takes a simple view of the people contexts in which the designed artifact must function. The prototype has been developed using a DSRM. DSRM is important in a discipline such as Information Technology as well as the creation of successful prototypes (Peffers et al., 2007).

Design Science Research Methodology is motivated by the desire to improve an environment by introducing new and innovative artifacts. DSRM is both a process and a product (Peffers et al., 2007). Figure 1 illustrates the design science research methodology process model.

Figure 2 summarizes the Design Science Research Methodology (DSRM) as a form of three closely related cycles of activities.

Figure 1. Design Science Research Methodology (DSRM) process model.

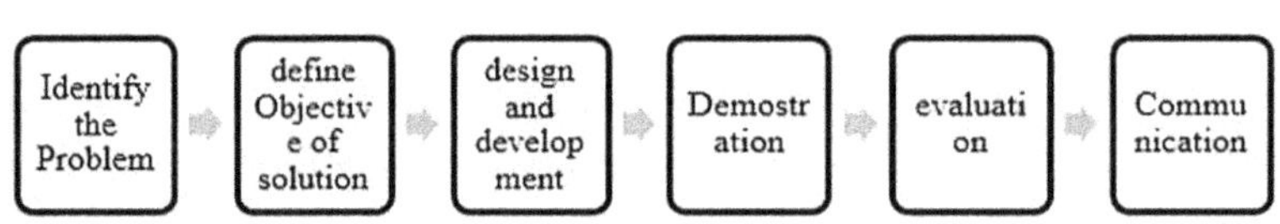

Figure 2. Three Cycle View of Design Science Research process

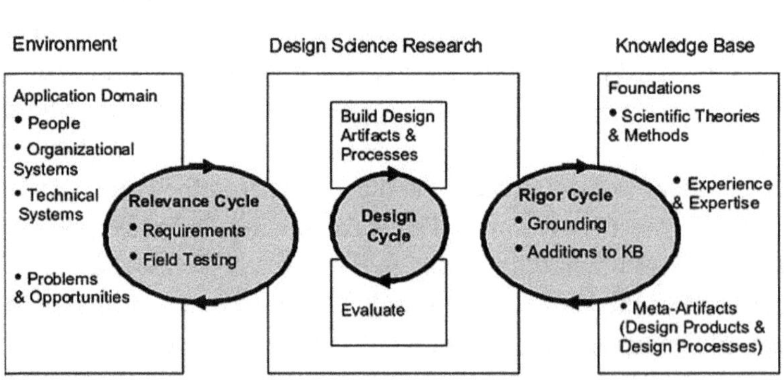

Design Science Research Process

The researchers Peffers et al. (2007) used the design science research process (DSRP) model that has three objectives: The research should be consistent with prior literature; It should provide a nominal process model for conducting Design Science research; and it should provide a mental model for presenting and appreciating Design Science research in Information System (Hevner, 2007). The process includes six steps namely, viz; Identify problem and motivation, defining the objectives of solutions, design and development, demonstration, evaluation and communication.

Identify Problem and Motivation

In driest regions, there is a need for appropriate water monitoring system. Furthermore, the researcher observed that the uncontrolled ways of monitoring water on farms result in water shortages. The water shortages are generally caused by farmers not being informed early enough about their water status. Famers use traditional methods to monitor the water level, which are not only inaccurate but are unreliable too. Thus, there is a need to develop a monitoring system that will reduce water shortages on farms. The water monitoring system will not only alert the user when the water level is at the critical point but has the capability of real time monitoring.

Defining the Objectives of Solutions

The objective of the study is that the researcher believes that the SWLMS will reduce water shortages and increase productivity on farms. The SWLMS will alert the person in charge early enough, on the water levels, in order to make well informed decisions and will provide a platform for the famer to monitor the water status of the tank from any location. This implies that the farmer does not have to be on the farm to monitor his water levels. The main objectives of the study were to develop a prototype that will monitor water levels on regular basis.

Design and Development

The third step is the development of the artifact. After successfully developing the artifact it will be evaluated according to the functional specifications (Peffers et al., 2007). In this step, the researcher began designing of the prototype. It started with collecting project requirements, whereby farmers use the system, resulting in requirement documentation (Peffers et al., 2007). This documentation will be used to design a prototype, which if successful will be improved upon, to develop detailed System architecture and determine the system functionality.

The prototype is able to monitor water in real time and notify farmers when the water level is low. The farmers use their credentials to access a web portal, where they can monitor the water tanks in real-time.

Demonstration

The Researcher then used this opportunity to demonstrate feasibility, efficacy, usability and accuracy of the prototype. The Prototype was tested in the presence of farmers, using a water bucket to represent a tank. The demonstration was a simulation of a real-world example where the researcher asked one of the famers to be in a different room away from the water tank; reason being farmers have other activities and therefore, will not always be close to their water tanks all the time. This step demonstrated the use of the prototype to solve the problems that the researcher identified. The researcher then tested the accuracy of the ultrasonic sensor. This test ensured that the sensor does not give inaccurate results. The researcher commenced with testing the NEXMO SMS API to be able to notify the farmer when the water level reaches a critical point, which is 20%. The prototype was tested and worked according to the research objectives. The selected participants were satisfied with the results, as they got SMS alerts on their phones in real-time. They received alerts regardless of the type of phone they had.

Evaluation

This phase of the DSRM process is finished when the IT artifact satisfies the requirements and solves the water shortage problem (Peffers et al., 2012). The researcher observed and monitored how efficient the prototype supports the solution to the problem. This meant analyzing the objectives of a solution to actual observed results from the use of the prototype in the demonstration.

The system was tested in the field and worked accordingly. Field testing of the prototype confirmed the system functionalities and its practical application in an actual environment. It is shown in appendix B. The prototype was tested by farmers at any convenient site. The measured water status data is recorded on to an excel sheet and can be viewed via the web-based portal. Farmers got an SMS alert when the water level reached a critical point.

After the evaluation process was complete, the researcher satisfied all the proposed objectives by developing a prototype that monitored water levels on a regular basis, asses the effectiveness of alerting the user on their current water status. In addition, the prototype provided a platform for the famers to monitor the water status from any location, at any given time and analyze stored data. Therefore, there was no need to iterate on the process model.

Communication

The researcher communicated the problem and the importance of the prototype to farmers and other researchers.

SYSTEM DEVELOPMENT

Design phase

System Architecture

The WLMS will have simple system architecture as shown in Figure 3. It is composed of the water tank where the ultrasonic sensor and GSM shield will be placed, webserver that will store the captured data, internet and cellphone that will receive the captured data.

The prototype is composed of two devices: Arduino Uno, Ultrasonic sensor as shown in Figure 4 & 5 The ultrasonic sensor sends ultrasonic waves to the water surface which reflects the waves back to the sensing head; the time it takes to receive the waves provides it with the distance.

Figure 3. System architecture

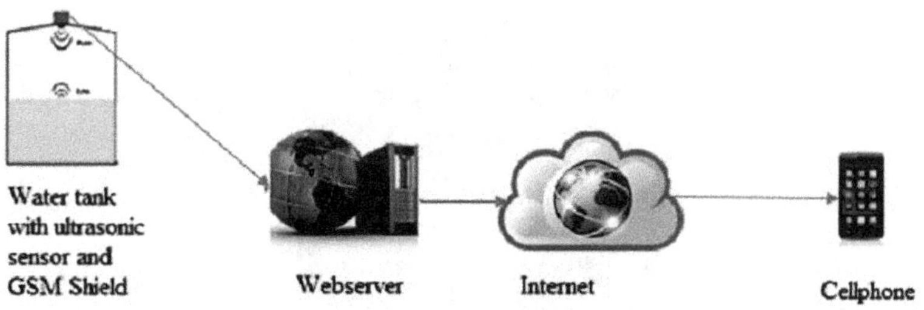

Figure 4. Arduino Uno

Figure 5. Ultasonic sensor

Figure 6. Web-portal tabs to display water status

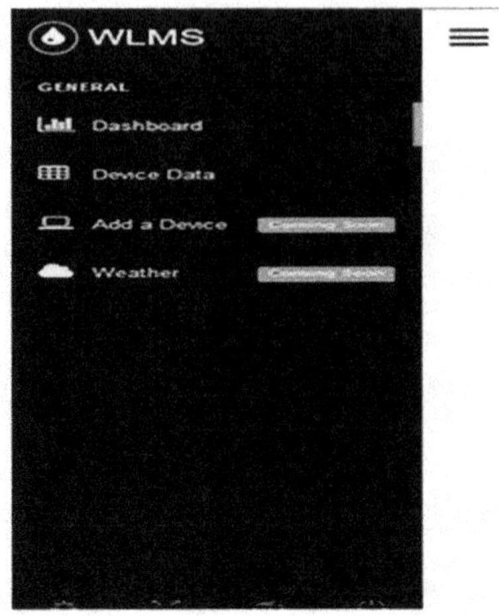

The WLMS prototype uses the NEXMO SMS API to send out alerts to the farmer on the current water status. The ultrasonic sensor is responsible for calculating the water distance, this distance is then recorded on an excel sheet via Parallax Data Acquisition tool (PLX-DAQ). PLX-DAQ is a software add-on for Microsoft Excel that makes it simple to create log excel sheets. The prototype was developed using Node.js. Node.js server then gets the file from the desktop and there after recorded data can be accessible on the website. If the recorded data is below 20% the node.js server request Nexmo to send out an alert to the farmer via the SMS.

Farmers are able to view the water status of their water tanks over the internet, where they are required to login using their credentials. These credentials are created and stored on firebase. The information on the web portal is simple to understand as it only displays relevant information about the water status. The web-portal has two tabs as shown in Figure 6. The home page shows you the water status in a graphical view where the user has two options either bar graph or line graph. The second tab on the home page (device data), displays the water status in a holistic view.

Data Flow Diagram

This section shows the visual representation of the sequence of steps and decisions needed to perform process for the SWLMS. The Figure 7 (A, B & C) shows the data flow diagram of the microcontroller functions. It illustrates the process of acquiring data from the sensor that is placed on top of the water tank. Figure 8 illustrates how the captured data in the excel sheet is read and how alerts are send out to relevant stakeholders using the Nexmo API. The system administrator is responsible for registering all WLMS users. Figure 9 depicts the flow diagram of the end user's interaction with the prototype. The farmers are prompted to login with their credentials, which are provided by the system administrator. After successfully login, Farmers can the view their water status. Figure 7 (A, B & C), depicts the flow diagram from user login all the way through to the web-portal, where the farmer can view the current water status and water status history.

Figure 8 shows a Use Case diagram that identified and clarified the organized system requirements and user's interaction. The description of this activity diagram is explained below.

Figure 7. (A) Data capturing flow diagram; (B) SMS alert flow diagram; and (C) User interaction diagram

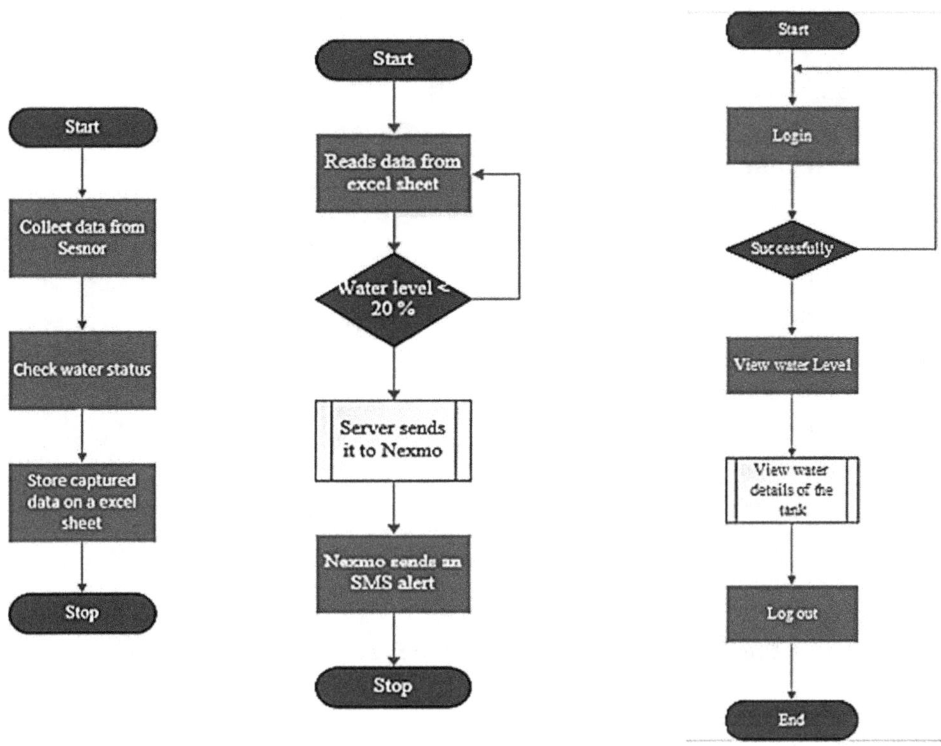

Figure 8. Use Case diagram

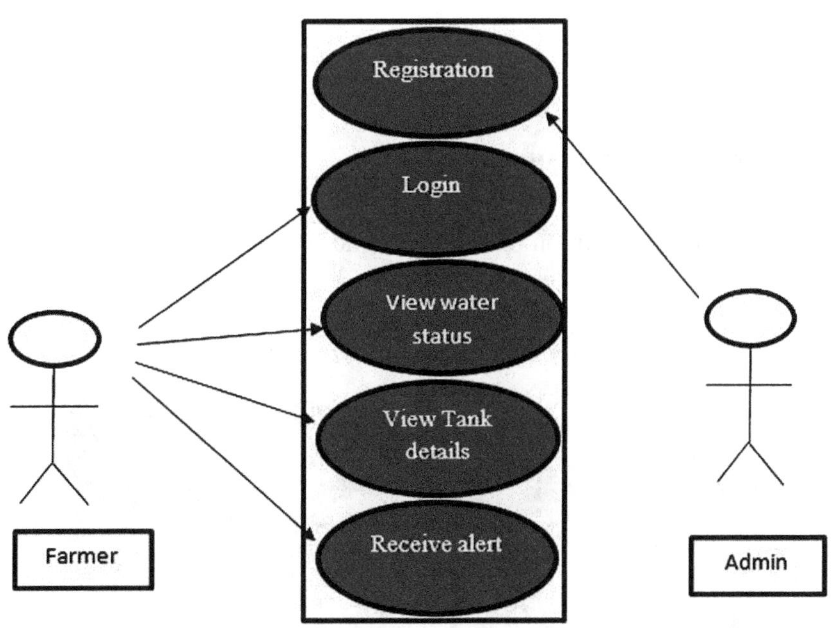

Description of Use Case for the WLMS Prototype

```
i. Registration
   Actor     Farmer
   Goal      To access the web portal
   Description     The farmer is required to login to view the water status
ii.Login
   Actor     Farmer
   Goal      To access the web portal
   Description     The farmer is required to login to view the water status
iii.  View water status
   Actor     Farmer
   Goal      Display water status
   Description     The farmer is able to view his current water level
iv.View details of the tank
   Actors    Farmer
   Goal      Access the excel
   Description     The farmer is able to access the excel sheet to view water
level data
v. Receive notifications
   Actors    Farmer
   Goal      Alert the farmer
   Description     The farmer is alerted through short message services on his
water status
```

PROTOTYPE TESTING AND FINDINGS

Figure 9 shows the successful installation of the ultrasonic sensor on top of water container.

Figure 9. Ultrasonic sensor installed on top of the prototype tank

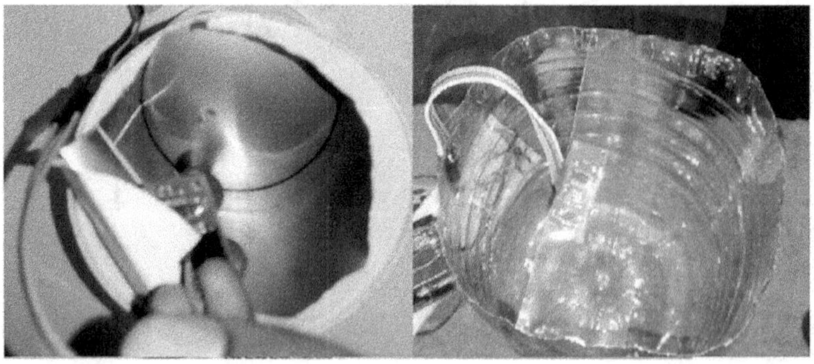

Figure 10. Login Screen for SWLMS

Figure 10 is the login form where the user will proceed to home page where he will be able to view the water status. The web-portal has two tabs as shown in Figure 6. The home page shows you the water status in a graphical view where the user has two options either bar graph or line graph as depicted on Figure 11 & 12. The second tab on the home page (device data), displays the water status in a holistic view. You can select which option you one on the top left corner.

The homepage displays the water status of the water tank. Water status is displayed in a line graph or bar graph.

User has to be registered on Nexmo and Firebase. Firebase is used to create an account for the farmer. The credentials the system administrator creates are then used to login into the website. Firebase is

Figure 11. Bar graph view of the water status

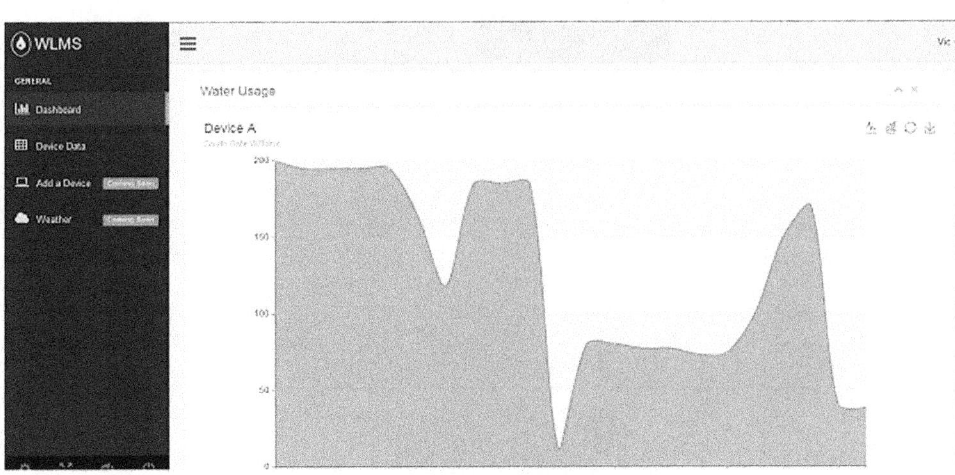

Figure 12. Line Graph view of the water status

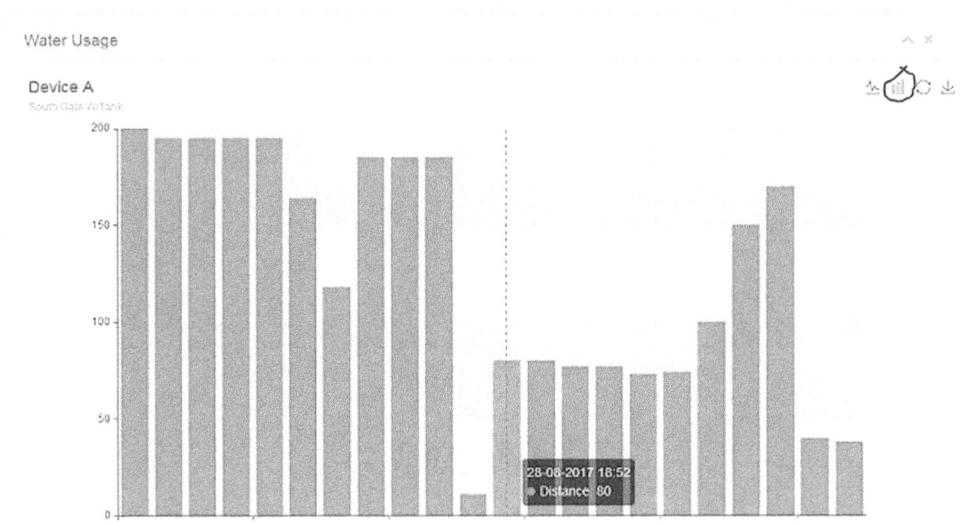

Figure 13. System administrator creates login into the Firebase Website

Figure 14. A holistic view of the water status

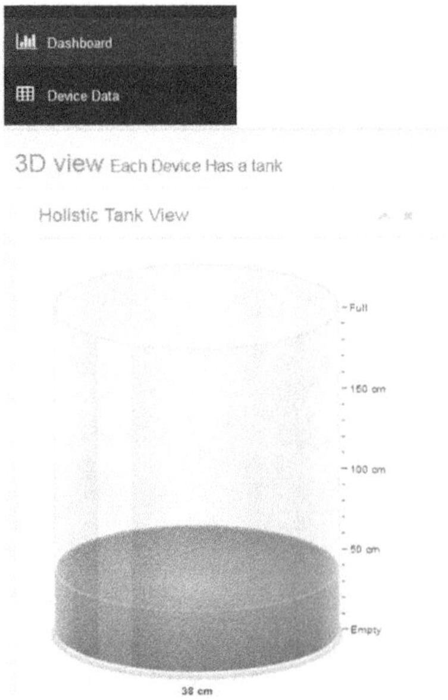

Figure 15. Shows the critical point of the water tank and its SMS alert

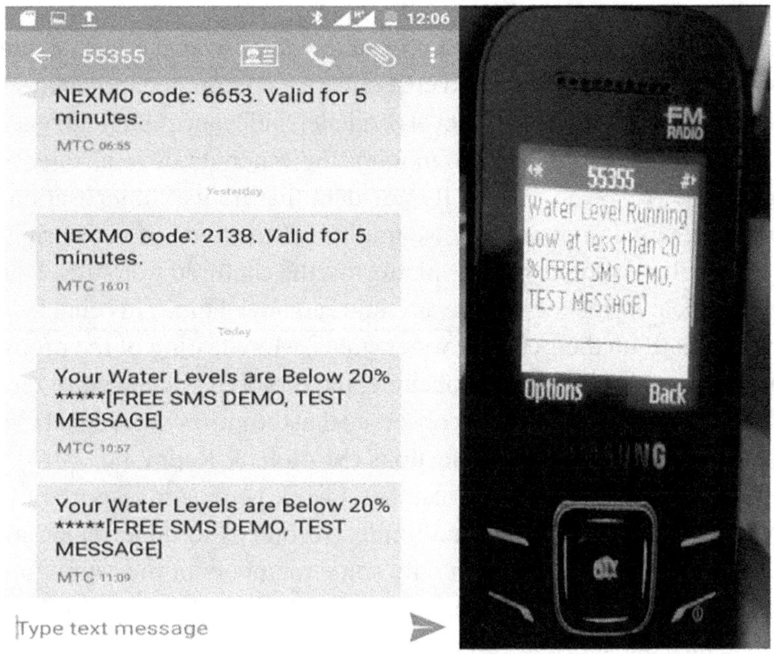

Figure 16. Illustrates the captured data and its view deployed on an excel sheet

A	B
Time	Distance_in_c
13-11-2017 12:53:29 AM	0
13-11-2017 12:54:29 AM	6
13-11-2017 12:55:29 AM	18
13-11-2017 12:56:29 AM	18
13-11-2017 12:57:29 AM	18

only used for creating users as shown Figure 13. Nexmo SMS API is used send out alerts. So users are required to sign up and register their numbers. Nexmo will provide an api key and api secret key that is used in the node.js server.

The user can view the water status in a holistic view by selecting the device data tab as depicted in the Figure 14.

The water status of the Figure 14 holistic view is at 40cm of 200cm which is 20%. This is the critical point of the water tank. When this happens, the user will get a SMS alert as shown Figure 15.

The captured data and can be view on a excel sheet as shown in Figure 16. This data can be used to analyze past water levels. The excel sheet records the distance from the sensor to the water head.

DISCUSSION, LIMITATIONS, FUTURE WORK AND CONCLUSION

During the study, the researcher developed and piloted the demostrated a low cost real time water level monitoring system prototype. The system used cost effective equipment and open source software with the objective of the system to monitor water levels and alert relevant stakeholders.

After testing the prototype, it is possible to monitor the water status in a water tank, alert users and view past data in real-time (Ragavan, 2016). It is evident that no user intervention is needed in water monitoring anymore. Farmers do not have to use traditional methods of measuring water levels. With the use of web-based portal and mobile phones platforms, the captured water level data is accessible and displayed in simple graphical formats available anytime and anywhere. This enables farmers prevent the risk of being caught off guard on their current water status. More testing of the prototype has to be done to identify areas that will require fixing. To conclude, the prototype managed to reach all its objectives but with room for improvement. The researcher observed that the prototype taught farmers basic computer literacy and the wonders of modern ICT applications (Morton, & Redmond, 2015).

Although there is much to be done, the researcher's work generated important findings in the field of remote sensor networks. The researcher had limited resources to undertake a more comprehensive study. There were challenges of working with only some members in the region under study and does not cover all water resources on farms that lead to exclusion of the other members of the community, challenges such as language barriers and computer literacy (Isala, 2016). The researcher could not manage to get a GSM Shield and alternatively opted for the Nexmo SMS API as the tool to send out alerts to farmers. The SWLMS is using the demo version of the Nexmo API and is limited to the number of SMS alerts it can send out, for testing purposes the researcher had to use different phone numbers and email address every time the Nexmo limit is reached. The SMS API requires an email address in order to sign up. The researcher had to make countless dummy email accounts, during the testing phase. In addition, farmers are required to have internet access in order to view the water status on the web portal. Nexmo API requires internet to send out SMS alerts.

SWLMS could not be implemented using Unstructured Supplementary Service Data (USSD) allowing farmers to view their water status via USSD. Smartphones are not accessible to majority of farmers, thus making it difficult for them to monitor levels at any given time. The data captured by the Ultrasonic sensor could be stored in a database and analyzed off real time. Further, The SWLMS can be improved by addressing the security features; Authentication mechanisms that validates the farmer's right to access the information; data integrity. Furthermore, this work can be developed as an applications (App) and made available on the app store and hosted on Google play store or Google Cloud. In addition, the unforeseen situations such as water tank leakage and theft can be controlled. Field testing will be done using a solar panel to confirm whether the SWLMS will be feasible even during harsh weather conditions.

Water is one of the most important natural resource that humans need for their survival, but unfortunately a huge amount of water is being wasted by uncontrolled use ("Windhoek water savings still crucial," 2018). SWLMS substantially benefit for efficient management of water from farmers. The main objective of the SWLMS prototype was to monitor water levels on regular basis, asses the effectiveness of alerting the farmers on the water status, to provide a platform for the farmer to monitor their water level from any location at any given time and to analyze stored data in the database. Despite the limitations the prototype has, it had accurate results and proved to be an efficient and safe way to monitor water. Farmers no longer have to use traditional methods of monitoring water in the tanks as they can view the water status from any location using their smartphones. In this project the researcher's explained the design and testing of the prototype. The authors emphasized on how cost effective and efficient the prototype is. Although the prototype was initially for all water sources on farms, it can only be implemented on water tanks. It was observed that water shortages on farms are mainly caused by uncontrolled ways of monitoring water, therefore constant monitoring of water levels in the tank is highly recommend (Faustine et al., 2014). The study concludes that objectives are achieved by creating a successful artifact using a DSRM approach (Weber, 2012). The prototype was able to monitor the tank water status and send out SMS notification when water level was at critical level and holistic view when the water level is at 40cm of 200cm which is 20%.

REFERENCES

Aly, W. H. F. (2013). Wireless Sensor Networks for Water Management that supports Differentiated Services. *International Journal of Scientific and Engineering Research.*

Bichler, M. (2006). Design science in information systems research. *Wirtschaftsinformatik, 48*(2), 133–135. doi:10.100711576-006-0028-8

Crisp, B. R. (2017). *The Routledge handbook of religion, spirituality and social work.* UK: Taylor & Francis. doi:10.4324/9781315679853

Deepiga, M. T., & Sivasankari, A. (2015). Smart water monitoring system using wireless sensor network at Home/Office. *International Research Journal of Engineering and Technology, 2*(4), 1305–1314.

Dirkx, E., Hager, C., Tadross, M., Bethune, S., & Curtis, B. (2008). Climate change vulnerability and adaptation assessment Namibia. Final Report. March.

Eltaieb, A. A. M., & Min, Z. J. (2015). Automatic Water Level Control System. *International Journal of Science and Research, 4*(12), 1505–1509.

Faustine, A., Mvuma, A. N., Mongi, H. J., Gabriel, M. C., Tenge, A. J., & Kucel, S. B. (2014). Wireless sensor networks for water quality monitoring and control within Lake Victoria basin: Prototype development. *Wireless Sensor Network, 6*(12), 281–290. doi:10.4236/wsn.2014.612027

Gupta, N., Kumar, S., Kumar, M., Mathur, D., & Jilani, E. (2016). *Wireless Water Level Controller Using Zigbee, V*(Iv), 79–81.

Hevner, A. R. (2007). A three cycle view of design science research. *Scandinavian Journal of Information Systems, 19*(2), 4.

Isala, S. (2016). *The impact of green schemes on the livelihood of communities in the Kavango region, Namibia* (Doctoral dissertation, JKUAT).

Johari, A., Wahab, M. H. A., Latif, N. S. A., Ayob, M. E., Ayob, M. I., Ayob, M. A., & Mohd, M. N. H. (2011). Tank water level monitoring system using GSM network. *International Journal of Computer Science and Information Technologies, 2*(3), 1114–1115.

Morton, S., & Redmond, P. (2015). The odd one out: Revisiting and investigating the gender imbalance in ICT study choices. *Journal of Applied Computing and Information Technology, 19*(1), 2–3.

Nepomilueva, D. (2017). Water scarcity indexes: water availability to satisfy human needs.

Peffers, K., Rothenberger, M., Tuunanen, T., & Vaezi, R. (2012, May). Design science research evaluation. In *International Conference on Design Science Research in Information Systems* (pp. 398-410). Berlin, Germany: Springer.

Peffers, K., Tuunanen, T., Rothenberger, M. A., & Chatterjee, S. (2007). A design science research methodology for information systems research. *Journal of Management Information Systems*, *24*(3), 45–77. doi:10.2753/MIS0742-1222240302

Pérez-Hoyos, A., Rembold, F., Kerdiles, H., & Gallego, J. (2017). Comparison of Global Land Cover Datasets for Cropland Monitoring. *Remote Sensing*, *9*(11), 1118. doi:10.3390/rs9111118

Ragavan, E., Hariharan, C., Aravindraj, N., & Manivannan, S. S. (2016). Real time water quality monitoring system. *International Journal of Pharmacy and Technology*, *8*(4), 26199–26205. doi:10.15680/ijircce.2015.0306016

Water crisis continues while dam levels rise - The Namibian. (n.d.). Retrieved from https://www.namibian.com.na/161330/archive-read/Water-crisis-continues-while-dam-levels-rise

Weber, S. (2012). Comparing Key Characteristics of Design Science Research as an Approach and Paradigm. In PACIS (p. 180).

What Is Water Scarcity? | Fluence. (n.d.). Retrieved from https://www.fluencecorp.com/what-is-water-scarcity/

Windhoek water savings still crucial - Local News - Namibian Sun. (n.d.). Retrieved from https://www.namibiansun.com/news/windhoek-water-savings-still-crucial2018-01-12

World Bank Group. (2014). *World development indicators 2014*. World Bank Publications.

Chapter 15
Smart Agricultural Practice for India

Dinesh Goyal
Poornima Institute of Engineering and Technology, India

Ravindra Singh Rajput
G. B. Pant University of Agriculture and Technology, India

ABSTRACT

The rapid growth of food needs due to the increasing worldwide population is raising the requirement for smart agriculture. Smart agriculture employs superior technologies such as decision support system, expert system, IoT, GPS, machine learning, robotics, and application of connected devices. Smart agriculture supports an automated farming system that includes a collection of data related to the farming area, and then analyzes the data so that the farmer can make the right decisions in order to grow high-quality products. In smart agriculture, farming-related data are collected using some unusual instruments like sensors, cameras, microcontrollers, and actuators. Then the collected data of farming areas are transferred via the internet to the farmer for decision making. This chapter is aimed to discuss critical topics for the implementation of smart agriculture in India.

THE INDIAN AGRICULTURE

Agriculture is the science, art, or manner of cultivating the soil, yielding crops, and keeping livestock and in diversifying scales, the manufacture, and marketing of the resulting products. Agriculture has been connected with the production of essential food crops, at now agriculture more than traditional farming and includes fisheries, dairy, forestry, fruit cultivation, vegetable production, flowers, and medicinal plant growing, poultry, beekeeping, mushroom, etc. The present time agriculture also included processing, marketing, and distribution of field crops and livestock products. Thus, agriculture could be related operation to as the promotion, production, processing, and distribution of farming products. Agriculture plays an essential role in the life cycle of an economy of the nation. Agriculture is the spine of the economic operation of a given nation. In an extension of providing foods and raw materials, agriculture also contributes to employment opportunities for a large number of the population. The notable expectations from agriculture are:

DOI: 10.4018/978-1-5225-9246-4.ch015

- Food security
- Source of raw material for industries
- Employment possibilities
- Contribution to national revenue and significance to the international trade

According to Zion Market Research, the global smart agriculture market was USD 5,098 million in the year 2016 and is expected to reach around USD 15,344 million by the end of 2025 increasing at a CAGR of more than 13.09% between 2017 and 2025.

The climate of India is a typical monsoon, and the average annual rainfall of the whole country is approximately 1170 mm but place-to-place it significantly varies. There are 20 agro-ecological zones in India based on characteristics of the soil, physiographic features, bio-climatic features, and the plant growing period length. All characteristics closely correlated with rainfall, so that rainfall acting a critical role in the climate as well as agriculture of India.

Table 1. Drought in India [1]

Year	The Reason of Drought and Its Impact
1837	The short monsoons in the area between Agra, Allahabad, and Delhi.
1866	The monsoon failed in Orissa due to it one million people lost their lives, which was one-third population of Orissa state that time.
1896	Due to the shortage of rains in the south part of Uttar Pradesh, lost of around 7.5 to 10 million people happened.
1899	A drought happened in the western and central part of India due to fewer monsoons, after those prices of foods commodities overgrow.
1943	Four million people in eastern India lost their lives for hunger in the drought.
1966	Due to the shortage of monsoon in Bihar and Orissa, and around 50 million people were affected.
1969	The monsoon failed in Uttar Pradesh, Haryana, Rajasthan, Gujarat, Madhya Pradesh, Tamil Nadu, Andhra Pradesh, and Karnataka, around 15 million people were affected.
1970	The monsoon failed in Rajasthan and Bihar, and around 17.2 million were people affected.
1972	Due to the shortage of rains Himachal Pradesh, Uttar Pradesh and Rajasthan for that around 50 million people were affected.
1979	The short monsoons in and Uttar Pradesh, Punjab, Himachal Pradesh and eastern part of Rajasthan, around 200 million were affected.
1982	The monsoon failed in Punjab, Himachal Pradesh and Rajasthan for that approximately 100 million people were affected.
1983	Due to the shortage of rains in Bihar, Rajasthan, Tamil Nadu, West Bengal, Kerala, Karnataka, and Orissa, and around 100 million people were affected.
1987	Due to the shortage of rains northwestern and eastern parts of India 300 million.
1992	Due to the shortage of rains Rajasthan, Orissa, Gujarat, Bihar, and Madhya Pradesh, a large number of people were affected.
2000	Due to the shortage of rains Andhra Pradesh, Orissa, Rajasthan, Gujarat, and Madhya Pradesh, More than 100 million were affected.
2013	the drought of Maharashtra happened primarily owing to low rainfall from June to September a large number of people were affected.
2015	Due to the shortage of rains drought happened in Maharashtra and it affected around 90 million farmers, which is strict to the population of a europian country like Sweden.

Extreme weather events in India in the past 10 years. Retrieved from https://www.downtoearth.org.in/news/extreme-weather-events-in-india-in-the-past-10-years-46450

A drought is an environmental or natural disaster in a particular region which creating prolonged shortages in the water supply and whether atmospheric. Just like a human, the life of animals and plants also depend upon the water, When a drought happened the water, and food supply can be limited or discontinued, and their habitat can be damaged. Sometimes the damage for a short time, and their habitat and food supply come back to the normal situation after the drought. According to the world's worst record in food disaster happened in 1943 during British-ruled India known as the Bengal Famine; an estimated 4 million people in the eastern part of India lost their lives due to hunger.

GREEN REVOLUTION IN INDIA

Green revolution started in India in the 1960s under the leadership of Prime Minister Smt. Indira Gandhi (1917-1984). It leads to an increase in food grain production, mainly wheat. The high yielding varieties of crops were engineered by teams of Nobel price holder scientist Dr. Norman Borlaug (1914 – 2009) and Indian scientist Dr. M S Swaminathan (born 1925). After the green revolution, some significant achievements are shown in table 2.

The Green Revolution in India was a duration when Indian agriculture system was transformed into an industrial-like system of agriculture with the adoption of modern methods and techniques such as the use of high yielding varieties of seeds or seedlings, improved farm machinery, modern irrigation facilities, effective chemical pesticides, and fertilizers. That was part of the more massive Green revolution is initiated by Dr. Norman Borlaug, to increase agricultural productivity in developing countries. The primary outcomes of the Green Revolution for the scenario of India are:

- Introduce the new and high yielding variety of seeds and seedlings.
- Implement the use of active chemical fertilizers, pesticides, and weedicides in order to reduce agricultural losses.
- Use of a large number of chemical fertilizers in order to increase agricultural production amount.
- Use of efficient agricultural machinery like an irrigation pump, tractor, cultivators, seed drills, threshers, and harvester for saving time and control wastage.
- Selection of disease or insect resistance varieties so that production will enhance.
- To use of modern techniques for enhancement of livestock farming.

Table 2. Agricultural commodity production in India [2]

Agricultural Commodity	1960-61	1980-81	2013-14
Rice (m. tonnes)	35.00	54.00	106.50
Wheat (m. tonnes)	11.00	36.00	95.90
Total Cereals (m. tonnes)	69.00	119.00	24.50
Total Pulses (m. tonnes)	13.00	11.00	19.30
Sugarcane (m. tonnes)	110.00	134.00	350.00
Cotton (m. bales)	6.00	7.00	36.70
Jute & Mesta(m. bales)	4.00	8.00	11.60
Oilseeds (m. tonnes)	7.00	9.00	32.90

Agricultural Research Data Book. 2017, retrieved from http://iasri.res.in/agridata/17data/HOME_17.HTML

According to FAOSTAT data (published on FAO website 25.5.2017 and Agriculture Research Data 2017), after the green revolution Indian achieved second position in the production of Rice (21.20% share in world) next to China, and second position in the production of Wheat (13.15% share in world) next to China, fourth position in production of Maize (2.18% share in world) next to USA, China and Brazil. First position in the production of pulses (22.22% share in the world), and a second position in Groundnuts (14% share in the world) next to China.

Finally, it is concluded that India achieved a position to control over hunger of their people and become an exporter country of agricultural commodities. According to Data (Agricultural Statistics at a glance 2015, Ministry of Economics and Statistics, Ministry of Agriculture and Farmer Welfare) and Agriculture Research Data 2017 in the year 2014-15, India exported has been exported agricultural commodities amount of Rs 18964454.70 million, which was 12.55% of national income that time.

In the Annual Report 2017-18, Department of Animal Husbandry, Dairying, and Fisheries, Ministry of Agriculture and Farmers Welfare, Government of India. During the year 1916-17 with the 165.40 million ton milk production India became fist position in the world, with 88.14 billion ton eggs production India became the third position in the world, and with the 114.09 lakh ton, fish production India became the second position in the world.

SIDE EFFECTS OF THE GREEN REVOLUTION IN INDIA

Green Revolution is one of a unique event in Independent India's agricultural history. It has saved us from the disasters of hunger. However, green revolution farming methods also provided some unwanted side effects; some of those are serious.

Use of an Enormous Amount of Pesticides

During the green revolution many pesticides used, some of them are very toxic to humans, animal as well as plants. Used pesticides and chemicals are currently available in the drinking water. These materials are harmful and can cause serious health problems. In 2008, A researchers of Punjabi University discovered DNA damaged in 30 percent of Indian farmers who treated plants with herbicides and pesticides. Use of chemicals also destroyed the environment.

Loss of Genetic Diversity

In the conventional farming practice, farmers plant a lot of varieties of crops, but farmers using green revolution farming methods plant some selected crop varieties in favor of those that produce high yields with quality. This kind of cultivation causes an undesirable loss in crop genetic diversity. At present, about 75 percent of their rice fields contain only ten varieties of plants. It is a significant drop compared to the thirty thousand rice varieties that were planted 50 years in the past.

Inter-Crop Imbalances

The outcome of the green revolution is primarily focused on food-grains crops. Although all food-grains including rice, wheat, jowar, maize, and bajra have raised from the green revolution, the rice and wheat benefited the most. It has some ignored areas that are coarse cereals, oilseeds, and pulses.

Unemployment

Apart from in Punjab, and to some area in Haryana, farm mechanization under Green Revolution has produced extensive unemployment among agricultural laborers in the rural areas.

Use of Chemical Fertilizers

In a case study of Punjab, M.K. Sekhon and Manjeet Kaur of Punjab Agricultural University Ludhiana has warned against the extreme use of groundwater, chemical fertilizers, and pesticides. That will lead to large-scale depletion of groundwater and will adversely influence the health of the soil.

KEY PROBLEMS IN INDIAN AGRICULTURE

There are 20 agro-ecological zones in India. Classification of zones is based on the soil characteristics, physiographic features, bio-climatic features, and the plant growing period length. The characteristics of agro-ecological closely correlated with rainfall and, that rainfall acting a critical role in the agriculture for India.

The following points will become known problems in Indian agriculture.

Instability of Weather

Indian agriculture is a 'chance of monsoon.' It is at the sympathy of the weather god. The instability of weather makes an uncertainty in agriculture production. Further less production will be affected by agricultural markets and increasing the prices of food, food products, and edible oils.

Natural Disaster

Earthquakes, windstorm, and floods are some caused to significant losses. In the history of India, every two to three years faced a new natural disaster situation. Some major disaster that happened in the last few years are Mumbai (July 26, 2005), West Bengal (2009), Ladakh (August 6, 2010), Uttarakhand (June 2013) and, Jammu and Kashmir (September 2014). Uttarakhand disaster, official, estimates the death at 5,700. It is believed that over 10,000 people died in an acute weather event.

Figure 1. Agro-ecological zones in India
https://www.mapsofindia.com/my-india/india/drought-in-india-before-and-after-independence

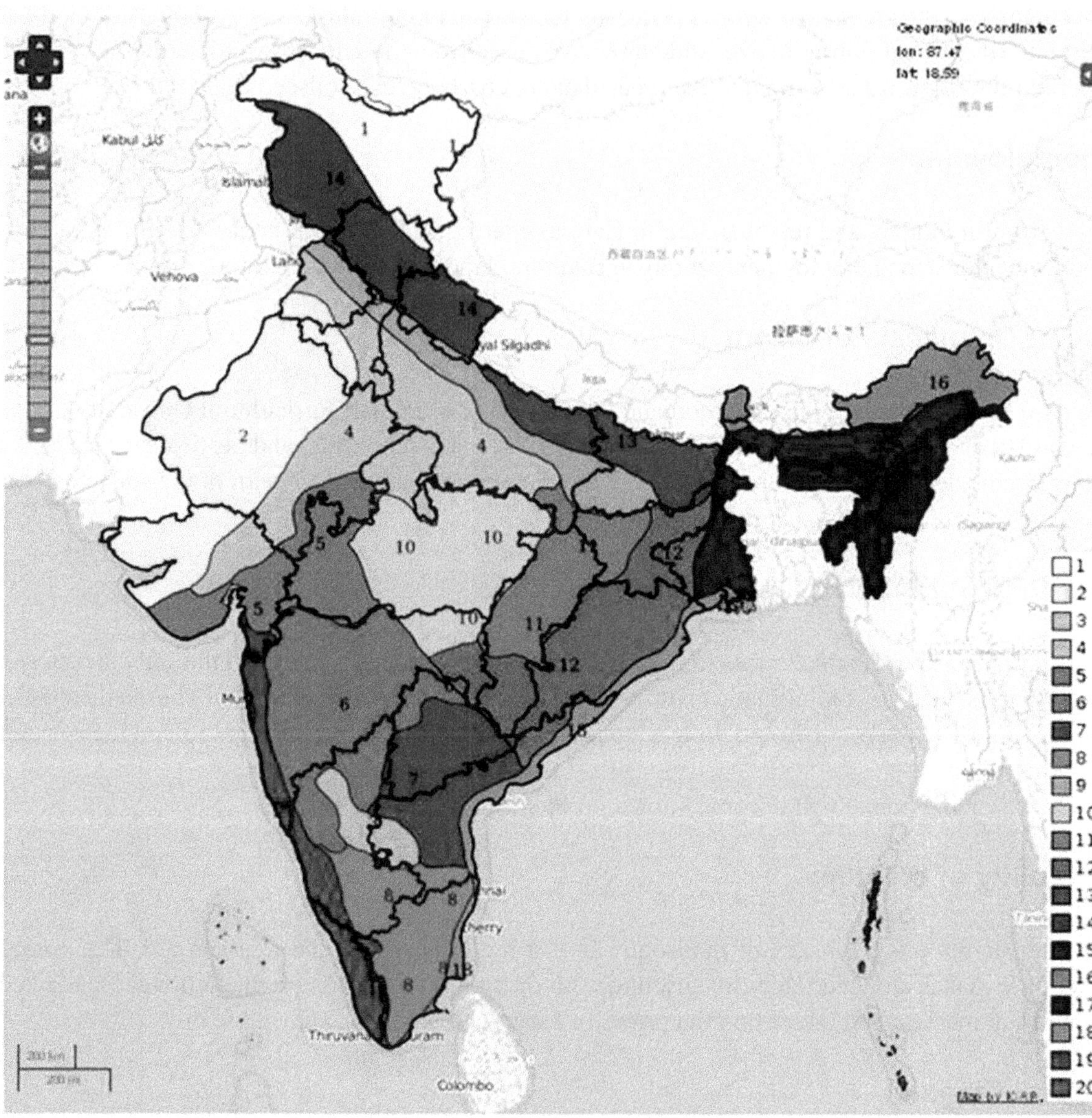

Selection of Cropping Pattern

Generally, each ecological zones of India has the capacity of conducting two or three crops in a year. Environmental requirements of each crop are different even variety to variety. Due to unstable weather and mismatch selection of cropping pattern may case of losses.

Type of Land Ownership of Farmers

Land-holding in India of farmers is small and also scattered so it may be difficult for the implementation of policies. Use of advanced farm machinery also tricky if the size of agricultural land is small and not in regular shape.

Social-Economic Status of Agricultural Laborers

Social-economic status of Indian laborers is not very good. According to a study, the majority of laborers belongs to below poverty line. The average annual saving of labor is around Rs. 25,000. The main problems of laborers are seasonal unemployment and lack of alternative of work. (Deepti Atulker 2009)

The Systems and Techniques of Farming

Last several years, various farming techniques adopted by Indian agricultural system, some of them known as traditional, modern, inorganic, organic, Precision, etc. but the aim of all system and techniques get better crop yield. As discussed earlier, India divided into 20 agro-ecological zones so that farming techniques might involve different zone to zone.

Inadequate use of Manures and Fertilizers

Plants required nutrition to grow, and every plant has a limit to take nutrients. When fertilizer or manures supply to plant, plants absorb only required nutrient amount. The excess nutrients remain in the soil, and it may be harmful to the soil. Without a precise study or soil test, it is not possible to define the supply amount of manures and fertilizers for a particular crop.

The Use of Poor Class Seeds

Good seed always a priority of farmers. Selection of a variety of seeds and seedlings is always a big problem for Indian farmers because the variation of climate condition and agro-ecological required a specific variety of seeds.

Inadequate Water Supply

In India, all the agriculture area is not under irrigation. Irrigation is the most critical operation for farming of crop. India situated in a tropical area, where rainfall is unreliable and uncertain. Farmers entirely not depend upon rainfall. Canals, groundwater, ponds are some other sources, but indecently they also depend upon rainfall

Inadequate Use of Efficient Farm Equipment

Sufficient types of equipment are available for ploughing, sowing, irrigating, pruning, weeding, harvesting and transporting for crops if farmers do not use equipment it a case of wastage of human labor and increase production cost. It is mainly the case where agricultural land small.

Agricultural Marketing and Prices

Agricultural marketing is not a pleasant condition in rural India. In the absence of a rural market, the farmers have to depend upon local traders and intermediaries for the clearance of their agricultural produce and not gain good price.

Inadequate Storage Facilities in India

Storage facilities in the rural under is developing the condition. Cold store required for preserving fruit and vegetable crops after harvesting. At present, there are many agencies engaged in storage activities like Food Corporation of India, Central Warehousing Corporation.

Inadequate Transport

There is a lack of cheap and capable means of transportation in the country. However, the various plan is also working on it. Even at present, there are thousands of villages not connected with major roads or with the agricultural market.

EARLIER EFFORTS FOR IMPROVEMENT OF INDIAN AGRICULTURE USING ICT

Some chronological efforts of implementation of Information Technology for enhancement of agriculture in India are shown in the Table 3.

Based on table-3, it is concluded key areas where ICT use required, are crop production, crop protection, crop improvement, animal farming, storage and transportation of farm produce, post-harvest and value additions of farm products and advisory of agricultural marketing.

Crop Production

Crop production is the area of agriculture and farming that deal production process of agricultural and horticultural crops. Planning and preparation for farming land, selection of cropping pattern, use of manures, fertilizers, and water, use of farm equipment covers in this area. Smart techniques are required for the preparation of land for crops, in-depth investigation of soil for determination of manures and fertilizer, planting and sowing, controlled irrigation, monitoring of day-to-day farm activity, and harvesting crop without losses. Advantages of the use of smart technology in this area are saving time and money, improve productivity, and reduce the use of chemical like fertilizers.

Crop Protection

Crop protection is the area of agriculture and farming that deals with protection of crop with harmful insects, nematodes, animals and birds, management of crop diseases, controlling of weeds. Smart technology is required for protection of crop to minimize the use of pesticides, insecticides and weed control chemical. Robotics helps control the weed. Advantages of smart technology in this sector are producing chemical free products.

Table 3. Application of Information Technology for enhancement agriculture

Researchers	Significant Contribution
Pinaki Charaborty et al. (2008)	An expert system for crop protection was developed that focused on pesticide advisory, pest and disease management for Indian agriculture.
Pinaki Charaborty et al. (2008)	Expert system for the management of malformation disease of mango with prescribing treatment packages was introduced.
Haritha Akkineni et al. (2010)	Developed an expert for farmers which provides advisory services for decision support on crop-related issues. In the proposed system the climatic information, soils types, seeds types, pesticides used and yield information are some essential constraints.
P. Mercy Nesa Rani et al. (2011)	Introduce an expert system for various Indian crops like rice, wheat, tomato, rapeseed and mustard, and mango. The purpose expert system was diagnosed with various pests and management.
Sudeep Marwaha (2012)	An online expert system AGRIdaksh developed that provide information on diseases, insects, weeds, nematodes, and physiological disorders to farmers.
S.J Yelapure and R V Kulkarni (2012)	Discussed need of an expert system and found to the primary purpose of expert systems are information support and decision making.
S Mohan, E Praveen Kumar, B Paulchamy (2013)	Explained a vision for crop production automation in future and disused various automatic system like Cheamen (seedbed preparation), Griepentrog (seed mapping), Shibusawal (seed placement), Godwin (reseeding), Bak and Jokobsen (crop scouting), Pedersen (weed mapping), Norremak and Griepentrog (Robotic weeding), Lund, Logaard and Graglia (micro spraying)
Rachana P. Koli et al. (2013)	Introduced an android application for farmers to checking most suitable crops for a particular place.
Vidya Kumbhar, T P Singh (2013)	Found six types of DSS system is required in India for land and water management, soil and water management, crop production, soil moisture estimation, simulation for application of fertilizers, integrated pest management.
Yukikazu Murakami, 2014	Developed a web-based system for cultivation and cost management for agriculture, the aim of this system optimizing farming.
Nilesh R Patel et al. (2014)	A proposed microcontroller-based monitoring system, system monitor different environmental parameters like soil moisture, relative humidity, and atmospheric temperature with wireless transmitting system.
Divya Joy et al. (2014)	Discussed various expert system like AGRIDASK, EXOWHEM, Expert systems for Crop productions, expert systems for crop productions of barley, pineapple crops, etc. Also categorized experts system into four categories Rule-based, Knowledge-based, Fuzzy based, Artificial Neural Network and Ontology-based.
M C S Geetha (2015)	Integrated data mining work in the Agriculture, and summarized using the Artificial neural network, K-means, Fuzzy sets, Decision tree and support vector machine in the agriculture decision making.
M K Gayatri et al. (2015)	The proposed use of IoT in Indian Agriculture.
Anand Nayyar et al. (2016)	Introduced a Smart IoT based agriculture stick for assisting farmers to get live data of temperature and soil moisture for environment monitoring
H. L. Kushwaha et al. (2016)	Conducted a comparative study for robotics approach in precision agriculture and concluded autonomous robot has the potential to work on precision agriculture. The agricultural robot also has the potential to take off a load of labor shortage and to increase productivity.
K. Ravisankar (2017)	proposed Big Data analytics for decision making in agriculture
Ramesh Babu Palepu et al. (2017)	presented the role of data mining from the perspective of soil analysis to increase agriculture productivity
Sachin Chawhanl et al. (2017)	Developed a hybrid model using data mining techniques to improve agriculture productivity. Soil, weather, and water were some parameters in the model.
Rupinder Singh et al. (2017)	Presented a review of different data mining techniques that were helpful to estimate the impact of different meteorological parameters on various crop yields.
Dhakne Aniketl et al. (2017)	Proposed an E-agriculture system to direct communication between farmer traders to avoid intermediate agents and help the farmer to gain a fair price for their products.
Jharna Majumdar et al. (2017)	Using data mining, regression and optimization techniques on soil and climatic data to forecast production price.
N H Rao (2018)	Explored the application of big data analytics in climate-smart agriculture and also described how big data analytics is useful to increase productivity with climate change effect.
Manabi Katoch (2017)	The success story of five Indian farmers using smart agriculture.
R.S. Rajput et al. (2018)	Developed a Sugarcane productivity model using Multiple Regression with Genetic algorithms.

Crop Improvement

Crop improvement is the area of selection of seeds and seedling, preparation of high-quality seeds and seedling. Study of genetics level also belongs to this area. Smart technology helpful for the in-depth investigation of crop plant characters up to DNA sequence label and identified find disease or insect resistance varieties for a particular agro-ecological zone. Advantages of smart technology in this sphere are producing quality and also reduce the use of the chemical.

Animal Farming

Milk, meat, eggs, fishes, honey, gum, lakh, silk are some products those received from the animal. Smart technology solution required for controlling and monitoring animals, feed supply, cleaning animal and its living area with minimized human involvement, value addition like meat, milk products will be manufactured using smart technology. Advantages of smart technology in this area hassle-free animal farming and minimize losses.

Storage and Transportation of Farm Produce

Storage in the surplus quantity of agriculture commodity, products received from livestock, storage methods, maintains storage and transportation loss. Smart technology solutions required for the modernization of food warehouse to reduce storage losses, ordination transportation of food especially milk, vegetables, fruits, and meat. The main advantage of smart technology in this domain is to maintain quality during storage and transportation.

Post-Harvest and Value Additions of Farm Products

Effective harvesting methods for mention quality and controlling loss, persevering farm products, value addition of farm products like sauce, jam, cheese, yogurt. Smart technology is required for the organization of production of value-added products like sauce, jam, cheese, yogurt and also packaging. Smart technology also required for post-harvesting work like picking of fruits and vegetables, harvesting crops, grading products and packaging. Advantages of this area maintain quality and reduce post-harvest losses.

Advisory of Agricultural Marketing and Price

Advisory for best selling time and market to farmers. In this area smart technology required for planning of farming in the context of finance, advisory of best selling time, selling place so farmers can get maximum profit, advisory regarding insurances, etc. Advantages of smart technology involved in this area farmers gain maximum profit.

Figure 2. Modules for implementation of ICT in the agriculture
http://www.economicsdiscussion.net/essays/green-revolution-essays/essay-on-green-revolution-in-india/17559

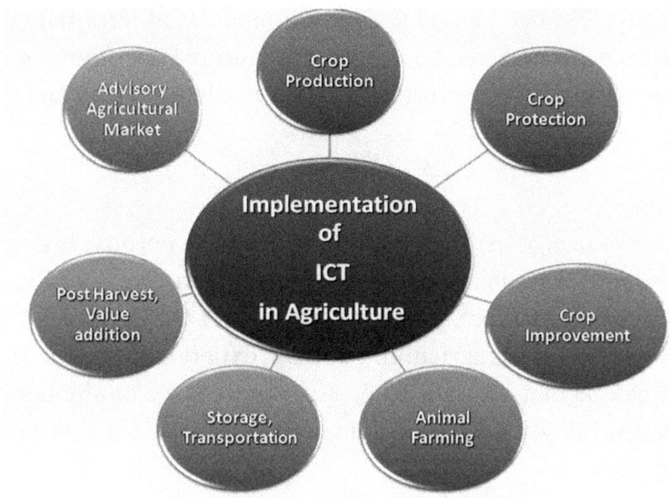

AVAILABLE TECHNIQUE FOR SMART AGRICULTURE

Smart agriculture can be described as a method of conducting agricultural activities using high-level technologies. The smart agriculture employs emerging technologies such as Decision Support System, Expert System, Information System for farmers, Internet of Things (IoT), Global Positioning System (GPS), Big Data, Artificial Intelligence, Robotics and Application of connected devices. Smart agriculture is an innovative approach to conducting farming exercises by reducing human efforts and producing quality products by making maximum utilization of the available resources. The smart agriculture used a wide range of applications, e.g., crop planning, yield monitoring, crop inspection, field mapping, irrigation management, milk harvesting, and many others. Some essential points those are untouched during the green revolution.

- To increase production in quantity using optimized inputs
- Improvement in quality of the agriculture produce, quality means good in taste, well in size and shape, without using harmful chemicals
- Product free from insects and diseases.
- Minimum environment impact
- Minimum waste during excess production, transportation or storage
- The minimum accidental risk for farm machinery or farming operation
- Cost optimization, not to burden on farmers or not on the consumer.

Decision Support System

A decision support system (DSS) is a computer programme based information system used to decide on an organization or a business. DSS can be used in agriculture for different places where what-if type analysis required like decision-making, selection, timely problem-solving and efficiency improvement.

Expert System

An expert system is software using Artificial Intelligence, whose purpose is to use facts and rules, taken from the domain knowledge experts in a particular field, to help make decisions and solve problems. In the case of implementation in agriculture, an agriculture expert system is an artificial intelligence based system that converts the knowledge of agriculture domain expert in a specific subject into software programs. These programs can be merged with other programs (based on the knowledge of other domain experts) and finally developed software used for questions (queries) submitted through a computer without using domain expert.

OnFarm provides an expert system solution that has Agronomic Models, Pest and Disease Models provide SMS and Voice Alerts and capture data using IoT.

Global Positioning System

Global Positioning system enables real-time data collection of position. At present, GPS-based applications in precision farming are being used for farm investigation, planning, field mapping, soil sampling and investigation, tractor guidance, crop inspection, variable rate applications, and yield mapping. GPS is also helpful to work during low visibility circumstances such as fog, rain, dust and night.

Geographic Information Systems

As awake the Geographic Information Systems is a computer-based tool that analyzes stores, manipulates and visualizes geographic information, usually in a map. The planning, development, and implementation of site-specific farming have been made possible by integrating the Global Positioning System and Geographic Information Systems.

Remote Sensing

Remote sensing is the method of acquiring information about an object or phenomenon without making any physical contact with it. That often requires the use of aerial sensor technologies in order to capture and analyze objects on the Earth, usually on the surface. Currently, Remote Sensing is helpful to estimate water resources, monitoring the land situation and surface temperature.

Internet of Things

Internet of Things (IoT) is a system of interrelated connecting computing devices with the ability to transfer digital data over a network without requiring any human-to-computer interaction. The primary purpose of IoT devices is to generate real-time data that can be analyzed and use to create desired business outcomes. Real-time monitoring and automation of agricultural equipment will be possible through the application of IoT.

Big Data

Big data is data that concerned with large volume, different types, growing nature, and generating with multiple resources. Currently, Healthcare, Administration in Public sector, Retail marketing, Manufacturing, Personal location data, Fact-based decision making, Improvement in customer experience, Improvement in sales, Innovation of new products, Reducing risk in various sectors, Quality improvement of product and services are some significance fields of big data applications (Yojna Arora, Dinesh Goyal, 2015 & 2016). Big data technology has capabilities to fetch result from a large data set. Integration of big data with a decision support system, expert system, IoT, GIS will be more beneficial compared to useful for better results in Agriculture.

Artificial Intelligence, Machine Learning, and Deep Learning

An artificial intelligence is a part of computational science that highlights the formation of intelligent machines that work like a human. A subset of Artificial intelligence is known as Machine learning. Machine learning is the utilization of artificial intelligence techniques that provides computer systems the ability to learn automatically without using explicitly programmed. Machine learning concentrates on the development of computer programs that can access data and use it to determine for themselves. The primary aim is to enable computers to learn automatically without human interference. Deep Learning is a subfield of machine learning concerned with algorithms inspired by the composition and function of the brain called artificial neural networks. A subset of machine learning is known as deep learning, and deep learning has networks competent in learning unsupervised from data that is unstructured or unlabeled. Deep learning is also known as Deep Neural Learning or Deep Neural Network. Agriculture advisory system can be developed using Machine learning or deep learning. Some application like identification weed plants within a crop, deficiency of nutrients, grading can be developed using deep learning.

Management Information System

A Management Information System (MIS) is a computerized software with the database management system, procedure, and reports generating tool in such a way that it generates regular reports on operations for every level of management personnel in an organization. It is typically also possible to obtain special reports from the system quickly. MIS plays a significant role in the organization; it creates an impact on the organization's functions, performance, and productivity. A Farm Management Information System is also useful for the efficient management of agricultural operations.

Robotics: Robotics is a way to perform tasks using machines in place of human beings. Robots are widely used in industries and manufacture sector for repetitive tasks. Some developed countries are using robotics solutions in the crop seeding, plantation, irrigation, crop monitoring, thinning, pruning, picking, and harvesting. Company *PrecisionHawk* provides farmers robotic solution for crop monitoring. Company *BoniRob* provides a detailed monitoring robotic solution for weeding and fertilizing operation. The benefit of the use of robotic in crop production is the minimal use of chemicals like herbicides. *Robot-Assisted-Precision-Irrigation* reduces water by targeting specific plants. A robot named *Rowbot* developed for the supply of fertilizer like urea. It is directly supplied area to the plant root so to consumption of urea reduce. *The robocrop* robot uses computer vision to detect plants and weed. Computer vision technology is used by Micro-spraying robots to discover weed plants and then spray a targeted drop of herbicide. *Lely* provides robotics based smart solution for dairy farming for Milking, Feeding, Housing, and caring Animal health and Management systems. Neelima Mishra et al. (2018) explained some methodology about robots that are capable for working in hotels and restaurants. Same robots after some changes will be applied for agricultural farms.

Unmanned Aerial Vehicles

UAVs, drones for agriculture, Today agriculture becomes one of the major industries to incorporate UAVs. Drones are being used in agriculture to heighten a range of agricultural practices. The ground-based and aerial-based drones are being used in agriculture for irrigation, crop monitoring, crop health estimation, spraying, planting, soil, and field analysis. The significant benefits of using drones cover crop health imaging, GIS mapping, it is ease of use and saves time, and the potential to increase quantity yields with quality. *Precision Hawk* is a trade that uses drones for collecting valuable data via a series of sensors that are used for mapping, and surveying of farming land. These drones perform in-flight monitoring, imaging, and observations. The farmers enter the details of the field to be surveyed and select an altitude or ground resolution, the drone data, helpful for estimating plant health indices, plant counting, yield prediction, plant height measurement, field water posing mapping, drainage mapping, weed mapping, and so on.

Connected Devices

Some sensing technologies are used in the precision agriculture, those providing data that assists farmers to investigate, monitor and optimize cropping operation. Sensors, e.g., NPK sensor, detects the presence of crucial nutrient elements in the soil like Nitrogen, Phosphorus, and Potassium, The soil moisture

Figure 3. Available Techniques for smart Agriculture

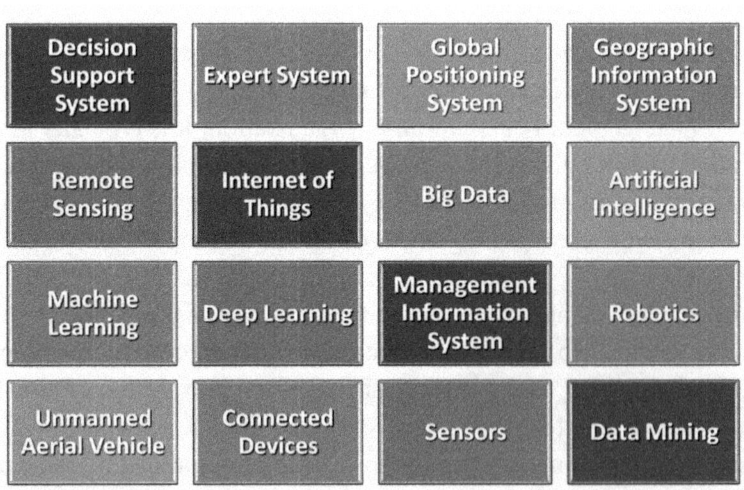

sensor, is used to measure the water content of the land. A temperature sensor is a sensing device that provides for temperature measurement through an electrical signal.

FRAMEWORK FOR IMPLEMENTATION OF SMART AGRICULTURE IN INDIA

In the scenario of Indian Agriculture, implementation of smart agriculture has to divide into the following five sub-projects:

- Smart Greenhouse
- Smart Farm
- Smart Fisheries, Poultry and Dairy
- Smart Horticulture
- Smart Food Warehousing, Transportation and Marketing

Smart Greenhouse

Smart Greenhouse is a self-regulating, control environment through micro-controller for optimal plant growth. The climatic setting inside the greenhouse, such as temperature, humidity, luminosity, soil moisture, CO_2 level is monitored continuously. Automatic actions can trigger any small variations in these climatic conditions. The automatic actions appraise change and take corrective action thus maintaining optimal conditions for plant growth.

Tools and Techniques Required

Greenhouse, Sprinkler based irrigation system, Fans, Air conditioner, Electricity power distribution system with proper backup, Sensors (humidity, temperature, moisture, NPK etc), Microprocessor (data gathering from sensors, and actuator for executing actions), Expert system for proposed crops, Communication network and Servers.

Methodology

In the proposed system, the greenhouse will be divided into similar clusters and in each cluster equipped with a set of sensors. All sensors will be connected through a microprocessor-based system. This system is known as a data gathering node. The data gathering note will be equipped with internet connectivity. Data gathering node will continuously be collecting time stamp data from sensors and transfer to the central server. The central server may be located in the data center or even in the cloud. The central server will be equipped with a data storage system and an expert system. An expert system will be analyzed time stamp data after analysis experts system will generate decisions. The corresponding command related to decisions will be transferred to the second microprocessor node. Second microprocessor node which is capable of control instruments like a fan, water sprinkler, and air conditioner, will operate accordingly.

Advantages of smart greenhouse practice are that the plants are free from outside effects like harmful insects or diseases. Minimized losses occur due to unfavorable climatical conditions. Smart greenhouse produces quality based production because plants grow within the most favorable situation. Off-season vegetables may also be cultivated to gain profit. Limitations of the smart greenhouse are high setup and running costs. Uninterrupted power and internet services are the basic requirements of this project. This practice is not suitable for grain and horticulture crops. It is only applicable for vegetables and flowers production which have short cultivation duration and small plants. Crops that have required pollination through an external source like honeybee or insect are also not suitable for this project. Farmers can estimate production and control over production losses.

Some Available Technologies

Atul (2016) proposed a project for Smart Greenhouse, which smart greenhouse is an ordinary greenhouse but equipped with sensors and actuators. The sensors and actuators are connected to the microcontroller. The microcontrollers send data and receive commands from a control center hosted in the AWS cloud. Users can control smart greenhouse operations through a dashboard type software application on the mobile, laptop of the tablet. Users can also provide voice commands to the greenhouse.

The proposed model used two types connecting devices sensor node and actuator node, and both nodes are Intel Edison microcontroller-based board with a NodeJS application running. The NodeJS application capture data from the associated sensors every second and readings by AWS IOT service. The control center hosted in this example in AWS cloud, control center has three application controller web applications, streaming web server and monitoring lambda functions. Control center send data to the actuator node after the judgment of the rule base. Actuator nodes are responsible for performing other devices like e.g., automatic sprinkler, electric fan, ventilation system, light system, etc.

Figure 4. Smart Greenhouse using AWS
https://www.hackster.io/synergy-flynn-9ffb33/smart-greenhouse-the-future-of-agriculture-5d0e68

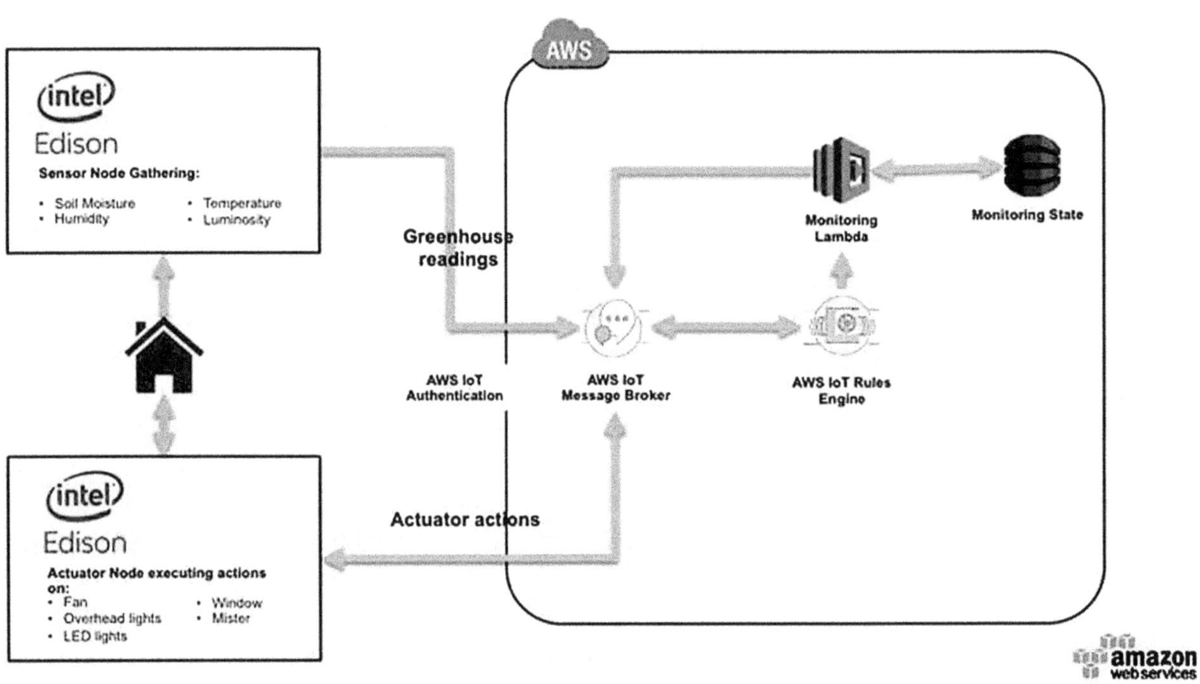

Smart Farm

Smart Farm represents the application of modern Information and Communication Technologies (ICT) into the Farm of an individual farmer.

Tools and Techniques Required

Agriculture Farm, Sprinkler or drip based irrigation system, Electricity power distribution system with proper backup, Sensors (humidity, temperature, moisture, NPK etc), Microprocessor (data gathering from sensors, and actuator for executing actions), Expert system for proposed crops, Communication network, Servers, MIS for Farm,. Robotics based or automatic machinery.

Methodology

The next green revolution is a world based upon the combined application of ICT solutions such as precision equipment, the Internet of Things (IoT), Sensors, GPS, Big Data, Unmanned Aerial Vehicles, robotics, etc. Smart farm operation can be divided into three subparts as:

- **Management Information Systems of Farm:** A managed systems for collecting, processing, storing, and disseminating field data those are useful for agricultural operations and functions.
- **Precision Agriculture:** Optimized management techniques to improve economic returns reduce inputs and environmental impact. It includes Expert System, Decision Support Systems for farm operation management with the inclusion of modern techniques.
- **Agricultural automation and robotics:** The manner is applying some techniques like robotics, automatic control, and artificial intelligence at all levels in the agricultural operation for improvement of product quality and quantity. Use of UAV in the optimized spraying quantity of water, insecticide or fertilizer. Use UAV in planting, seeding, and Inspection of crops. Deep learning-based identification of weeds and removes it from the farm without using the chemical. Deep learning-based identification of water and nutrient requirement. Use robotic based harvesting system for minimized harvesting losses.

Advantages of smart farm practice are that the quick action mechanism available for fighting from outside effects likes harmful insects or diseases. Compare to traditional farming fewer production losses occurs. The smart farm produces better quality compared to traditional. Smart farm mechanism is capable of working day and night. Limitations of the smart Farm are high setup and running costs. Uninterrupted power and internet services are the basic requirements of this project. Farmers can estimate production and control over production losses.

Some Available Technologies

Meghan Brown (2018) proposed a smart farm as the farm is equipped with sensors, IoT, robotic and automatic systems. Driverless tractors will be used to prepare fields with accurate labeling and drones for imaging fields, planting, seeding and spraying from the air. Robotic system will be used for weeding, crop maintenance, and harvesting. Sensors based Automatic system will be used for irrigation and

Figure 5. Components of Smart Farm

drainage. Advantages of this project are minimizing losses during maintenance and harvesting, improve quality to minimize the use of chemicals. Limitations of this project are uninterrupted power and internet service required; high setup cost, and running costs.

Smart Fisheries, Smart Poultry and Smart Dairy

Tools and Techniques Required

Animal shade, Fish pound, Electricity power distribution system with proper backup, Sensors (humidity, temperature, moisture, pH etc), Microprocessor (data gathering from sensors, and actuator for executing actions), Expert system for Animal husbandry or fisheries, Communication network, Servers, MIS for Animal Farm,. Robotics and automatic control types of machinery.

Methodology

The sensors, actuators, microcontroller equipped pond will be helpful for aquaculture. Monitoring and control of aquatic environment will be enhanced quality as well quantity of fisheries products, the sensors, actuator, and microcontroller based environmental control system also useful for poultry and animal shade. GPS and IOT based system can be used for monitoring the health and movement of animals. Robotics can be used for cleaning of animals and animal's shades.

Advantages of Smart Fisheries, Smart Poultry and Smart dairy practice are that the quick action mechanism available for environmental effects. The smart system gives better attention to animals compared with traditional. Smart farm mechanism is capable of working day and night. Limitations of the smart system are high setup and running costs. Uninterrupted power and internet services are the basic requirements of this project. Farmers can estimate profit and control over production losses.

Some Available Technologies

Rupali B Mahale et al. (2016), proposed a method of chicken farming by intelligent chicken farming using an embedded system for control temperature, humidity, air quality, light intensity, ventilation window. This smart system can effectively control the farm from any location and reduces cost time and human resources. One of the advantages of this system poultry farm is safe from outside infections. Lata S. Handigolkar et al. (2016) proposed the IoT system for environmental control comprising Sensors, Arduino UNO, Raspberry Pi, and Actuators. Rupali B. Mahale and S. S. Sonavane (2016) proposed an embedded system for poultry farming for controlling environmental parameters.

Large animal farm proprietors can be utilized wireless IoT applications to collect data regarding the location and health monitoring of their animals. This information helps them in identifying animals those are not well or sick so that they can be separated from the herd to prevent the spread of disease. JMB North America is an organization that offers cow monitoring solutions for cattle producers. One of its solutions helps the cattle owners observe cows that are pregnant and about to give birth.

Smart Horticulture

Tools and Techniques Required

Orchards of horticultural crops, Drip based irrigation system, Electricity power distribution system with proper backup, Sensors (humidity, temperature, moisture, NPK etc), Microprocessor (data gathering from sensors, and actuator for executing actions), Expert system for proposed crops, Communication network, Servers, MIS for Orchards,. Robotics and automatic control of farm machinery.

Methodology

Harvesting of fruits, pruning of fruits tree, the spray of insecticides, etc. is difficult for big trees. The robotic system is useful for harvesting fruit without loss and also capable of grading and packaging. The robotic system is also useful for pruning of trees. Unmanned Aerial Vehicles are useful for spraying and preserving with birds and bats. IoT based system can be implemented for drip irrigation of fruit trees.

Some Available Technologies

Mohd Saiful Azimi Mahmud et al. (2018) proposed an IoT based system for mushroom production. Mushroom cultivation required temperature, humidity, and CO_2 in a significant amount. The IoT and automated control system are useful for quality mushroom production.

Smart Food Warehousing, Transportation, and Marketing

Tools and Techniques Required

Food Warehouse, Packaging boxes, transportation containers with freezers, Electricity power distribution system with proper backup, Sensors (humidity, temperature, moisture), Microprocessor (data gathering from sensors, and actuator for executing actions), Expert system for food storages, Communication network, Servers, MIS for Food warehouse, Robotics and automatic control machineries for food warehouse and packaging.

Methodology

Farmers are facing huge losses during storage and transportation. Shreyas et al. (2017) proposed a project uses raspberry pi which acts as a microcontroller as well as server and sensors like temperature, moisture, smoke, and a light sensor. All these sensors can be controlled merely using mobile through a web application for minimization of losses robotics also use for warehouse operation. IoT based environment-controlled vehicle will be used for transportation. Agriculture marketing advisory system will be developed with the help of big data that help sell farm produce market.

Some Available Technologies

Mirjana Maksimović et al. (2015) illustrated the significance of IoT instrument, defined concepts and, proposed of low-cost solution based on IoT for real-time food traceability and monitoring in food during the transportation.

REFERENCES

Akkineni, H., & Junapudi, V. (2010). Agriculture wrapped with social networks, data mining, and mobile computing to boost up crop productivity. *International Journal of Biological Sciences and Engineering. 1*(1), pp. 43:48.

Aniket, D., Mayur, D., Mayur, P., & Aakash, R. (2017). E-Agriculture Information Monitoring System using Data Mining. *International Journal of Advanced Research in Computer and Communication Engineering*, *6*(5).

Arora, Y., & Goyal, D. (2015). Big Data Technologies: Brief Overview. *International Journal of Computer Applications, 131*(9), 1-6.

Arora, Y., & Goyal, D. (2016). Big Data: A Review of analytics methods & Techniques. *Proceedings of 2nd International Conferences on Contemporary Computing and Informatics, 225-230*. IEEE. 10.1109/IC3I.2016.7917965

Atul (2016). Smart Greenhouse: The future of agriculture. Retrieved from https://www.hackster.io/synergy-flynn-9ffb33/smart-greenhouse-the-future-of-agriculture-5d0e68

Atulker, D. (2009). Soci-Economic status of Agricultural Landless labors in Indore Block of Indore District of Madhya Pradesh. (MS Thesis Jawaharlal Nehru Agriculture University). Retrieved from http://krishikosh.egranth.ac.in/bitstream/1/5810019780/1/T-82174.pdf

Balasubramanian, A. (2013). Technical Report on Agro-Ecological zones of India. Retrieved from https://www.researchgate.net/publication/314206350

Brown, M. (2018). *Smart Farming—Automated and Connected Agriculture,* engineering.com. Retrieved from https://www.engineering.com/DesignerEdge/DesignerEdgeArticles/ArticleID/16653/Smart-FarmingAutomated-and-Connected-Agriculture.aspx

Chakraborty, P., & Chakrabarty, D. K. (2008). *An example of agricultural expert systems being used in India. Georgian Electronic Scientific Journal of Computer Science and Telecommunications, 1,* 15.

Chakraborty, P., & Chakrabarty, D. K. (2008). A brief survey of computerized expert systems for crop protection being used in India. *Progress in Natural Science, 18,* 469-473.

Chawhan, S., Ghodichor, P., & Paunikar, K. (2017). Survey on Data Mining for Increasing Agriculture Productivity. *International Journal for Research in Applied Science and Engineering Technology*, *5*(3).

Divya, J., & Sreekumar, K. (2014). A Survey on Expert System in Agriculture. *International Journal of Computer Science and Information Technology 5*(6). 7860-7864.

Food and Agriculture Organization of the United Nations. (2015). Retrieved from http://www.fao.org/nr/water/aquastat/countries_regions/IND

Gayatri, M. K., Jayasakthi, J., & G.S., & Mala, G. A. (2015). Providing Smart Agriculture Solutions to Farmers for better yielding using IOT. *IEEE International Conference on Technology Innovations in ICT for Agriculture and Rural Development.* (pp 40-43). IEEE.

Geetha, M. C. S. (2015). A survey on Data mining Techniques in Agriculture. *International Journal of Innovative Research in Computer and Communication Engineering*, *3*(2), 887-892.

Global Position System. Retrieved from https://www.gps.gov/applications/agriculture/

Green Revolution. Retrieved from http://www.economicsdiscussion.net/essays/green-revolution-essays/essay-on-green-revolution-in-india/17559

Katoch, M. (2017). *5 Farmers Who Prove Smart Agriculture Can Make you Rich.* Retrieved from https://www.thebetterindia.com/125477/kisan-diwas-successful-farmers-lucrative-business

Koli, R. P., & Jadhav, V. D. (2015). Agriculture Decision Support System as Android Application. *International Journal of Science and Research*, *4*(4), 903-906.

Kumbhar, V., & Singh, T. P. (2013). A comprehensive study of application of decision support system in agriculture in Indian Context. *International Journal of Computer Application. 63*(14), 6-11.

Kushwaha, H. L., Sinha, J. P., Khura, T. K., Kushwaha, D. K., Ekka, U., Purushottam, M., & Singh, N. (2016, December). Status and Scope of Robotics in Agriculture. In *International Conference on Emerging Technologies in Agricultural and Food Engineering.* Agricultural and Food Engineering Department, IIT Kharagpur.

Lata, S. (2016). IoT Based Smart Poultry Farming using Commodity Hardware and Software. *Bonfring International Journal of Software Engineering and Soft Computing*, *6*(Special Issue), 171–175. doi:10.9756/BIJSESC.8269

Lee, K. (2018). *Harmful Effects of the Green Revolution.* Retrieved from https://sciencing.com/harmful-effects-green-revolution-8587115.html

Lee, H. J., & Kim, M. (2018). The Internet of Things in a Smart Connected World. doi:. doi:10.5772/intechopen.76128

Mahale, R. B., & Sonavane, S. S. (2016). Smart Poultry Farm: An Integrated Solution Using WSN and GPRS Based Network, *International Journal of Advanced Research in Computer Engineering & Technology, 5*(6).

Mahmud, M. S. A., Buyamin, S., Mokji, M. M., & Abidin, M. S. Z. (2018). Internet of Things based Smart Environmental Monitoring for Mushroom Cultivation. *Indonesian Journal of Electrical Engineering and Computer Science, 10*(3), 847-852.

Majumdar, J., Naraseeyappa, S., & Ankalaki, S. (2017). Analysis of agriculture data using data mining techniques: application of big data. *Journal of Big Data, pp. 4*(1), 20.

Maksimović, M., Vujović, V., & Omanović-Miklić anin, E. (2015). Application of internet of things in food packaging and transportation. Int. J. Sustainable Agricultural Management and Informatics, 1(4), 333-350.

Marwaha, S. (2012). AGRIDAKSH- A Tool for developing the online expert system. *Agro-Informatics and Precision Agriculture (AIPA 2012)*. 17-23.

Mishra, A. S., & Deep, V. (2014). Expert System in Agriculture: An Overview. *International Journal of Science Technology & Engineering., 1*(5), 45–49.

Mishra, N., Goyal, D., & Sharma, A. D. (2018). Issues in Existing Robotic Service in Restraurants and Hotels. *Proceeding of Third Int.l Conf. on Internet of Things and Connected Technologies (ICIoTCT)*, MNIT Jaipur, India. doi:10.2139srn.3166508

Mohan, S., Kumar, E. P., & Paulchamy, B. (2013). *Certain Investigation of Precision Agriculture Robot using Lab view.* IEEE International conference on current trends in Engineering and Technology, India. pp. 319-322. 10.1109/ICCTET.2013.6675975

Mondal, P. (2018). *Five Major Demerits or Problems of Green Revolution in India.* Retrieved from http://www.yourarticlelibrary.com/green-revolution/5-major-demerits-or-problems-of-green-revolution-in-india/20954

Murakami, Y. (2014). iFarm: Development of Web-based System of Cultivation and Cost Management for Agriculture. In *2014 Eighth International Conference on Complex, Intelligent and Software Intensive Systems, (pp. 624-627)*. 10.1109/CISIS.2014.89

Nayyar, A., & Puri, V. (2016, November) Smart Farming: IoT Based Smart Sensors Agriculture Stick for Live Temperature and Moisture Monitoring using Arduino, *Conference Paper on Cloud Computing & Solar Technology*. doi:10.1201/9781315364094-121

Palepu, R. B., & Muley, R. R. (2017). An Analysis of Agricultural Soils by using Data Mining Techniques. *International Journal of Engineering Science and Computing*. October, 15167-15177.

Patel, N. R., Choudhari, P. G., Kale, P. D., Patel, N. R., Raut, G. N., & Bherani, A. (2014). Smart Design of Microcontroller based monitoring system for Agriculture. IEEE International Conference on Circuit, Power and Computer Technology. pp 1710-1713.

Rajput, R. S., Pant, A., & Kumar, S. (2018). Development of Forecasting Model for Sugarcane Productivity using Multiple Linear Regression with Genetic Algorithm. *Periodic Research. 7*(2), 124-128.

Rani, P. M. N., Rajesh, T., & Saravanan, R. (2011). Expert Systems in Agriculture: A Review. *Journal of Computer Science and Applications*, *3*, 59-71.

Rao, N. H. (2018). Big Data and Climate Smart Agriculture - Status and Implications for Agricultural Research and Innovation in India. *Proceedings of the Indian National Science Academy*. 10.16943/ptinsa/2018/49342

Ravisankar, K., Sidhardha, K., & Prabhadevi, B. (2017). Analysis of Agricultural Data Using Big Data Analytics. *Journal of Chemical and Pharmaceutical Sciences*, *10*(3), 1132-1135.

Shreyas, B. et al. (2017). Real-time monitoring in agricultural warehouse using IOT. *International Research Journal of Engineering and Technology*. *4*(4).

Singh, R., & Singh, G. (2017). *Review: Role of Data Mining in Agriculture Yield Analysis*. Paper presented at International Conference on Soft Computing Applications in Wireless Communication - SCAWC 2017.

Smart Greenhouse Remote Monitoring Systems. (n.d.). Retrieved from https://www.postscapes.com/smart-greenhouses

Mahale, R. B., & Sonavane, S. S. (2016). Smart Poultry Farm Monitoring Using IOT and Wireless Sensor Networks. *International Journal of Advanced Research in Computer Science*, *7*(3), 187–190.

What is Smart Farming? (n.d.). Retrieved from https://www.smart-akis.com/index.php/network/what-is-smart-farming/

Yelapure, S. J., & Kulkarni, R. V. (2012). Literature Review on Expert System in Agriculture. *International Journal of Computer Science and Information Technology*. *3*(5), 5086-5089.

Zion Market Research. (n.d.). Smart Agriculture Market by Agriculture Type. Retrieved from https://www.zionmarketresearch.com/report/smart-agriculture-market

Chapter 16
Intelligent Sockets for Home Automation and Security:
An Approach Through IoT and Image Processing

Bharath V. N.
JSS Science and Technology University, India

Adyanth H.
JSS Science and Technology University, India

Shreekanth T.
JSS Science and Technology University, India

Nalina Suresh
University of Namibia, India

Ananya M. B.
JSS Science and Technology Universit, India

ABSTRACT

The intelligent sockets are an advancement in approach to better the features and convenience offered by the existing switchboards. All updates to the board are done via a separately kept server for the web interface which connects to the home network. The features provided to the user can be bettered progressively via software updates. Features like timers which work in both automatic and manual mode, security aspect via surveillance and facial recognition, overload and usage logging with the help of the current sensor is provided. The data is also verified with the actual meter for accuracy and as a check for tampering. The data so gathered can also be used for prediction using machine learning. System first classifies various types of analog meters. Right now, the lbph classifier is trained to detect analog meter with needle and Analog meter with text readings.

DOI: 10.4018/978-1-5225-9246-4.ch016

INTRODUCTION

Automation has crept into our everyday life and is helping us in myriad of ways. Very aggressive development integrating smartness in everyday appliances like washing machines, microwave ovens, air conditioners, refrigerators among others tell us this story. The smartness integrated in these can be temperature sensors and timers for periodic usage, saving power when not in use, among other helpful features. The authors are trying to implement an idea of such a smart device which is not limited to the device in question, but as a general socket switchboard used every day. Central power saving can be implemented, while retaining the locally controllable feature along with addition of remote control and other mentioned features.

Safety mechanisms can be built into each switchboard with current meters and power consumption can be logged onto the server. This will be used to keep track and optimise the power consumption. As switch boards are present everywhere, access control is provided, security mechanisms like motion detection when away, auto turn off after daylight, built in fire alarms, complete power down remotely when required and such can be easily implemented and integrated. Further expansions can be made to incorporate machine learning to smartly turn the lights on and off from known patterns to conserve energy.

Raspberry Pi is used as the central controlling unit and as server. NodeMCU is provided at each switch board for local control and to communicate with the server. Relay modules are used for mechanical control of the switches. Current sensor is used to measure the current consumption and power consumption, which is finally logged onto the server. RFID can be used for the child mode protection for driving high power devices like grinders. Protection against overload and high current is provided.

Since multiple cameras can be incorporated into the raspberry pi, tampering of the meter can also be detected along with a meter reading from the meter itself using image processing techniques. This can be used as a cross verification of the values already being logged in all the sockets as a check for the data collected using the current sensor being used as well as a security measure to prevent tapping of connections. To make the system more aesthetic touch screen can be provided at each switchboard along with timers for user interface.

EXISTING TECHNOLOGY AND ITS REVIEW

Literature Survey

- Santoro et al. proposed an innovative socket that helps in demand and response analysis and plan the generation of power. They evaluated total harmonic distortion and sent the calculated data to the monitoring station. There is no active logging system and security system (Santoro, Calderaro, Galdi, & Piccolo, 2016).
- Pawar et al. proposed an made an effort to measure power consumption with the use of Zigbee. The setup consisted of op-amp circuits to calculate continuous voltage and current sensor to calculate the instantaneous power. The cost of implementation was higher due to the use of Zigbee and decrease in the range (Pawar, & Vittal, 2017).
- Xu et al. proposed a system using low power microcontroller MSP430 to achieve wireless home intelligent socket. The system works by connecting and controlling via SMS and a mobile app.

Some of the disadvantages are lack of security system, making it viable to dangers (Xu, & He, 2017).

- Xiao et al. proposed a system to link independent electric appliances independently to gather information and control those subsystems at a low cost and power consumption. It was not extensible to large scale systems, hence could not be scaled (Xiao, Liu, & Hu, 2016).
- Mr. Atonu Ghosh put forward a system using Raspberry Pi that controls the tasks at home such as switching appliances on & off over internet using a personal computer or a mobile or a tablet through the browser. Some of the drawbacks are unavailability of low power mode for child access, Raspberry Pi is used in every socket which increases the cost (Ghosh, 2016).
- Yan-Rong Tong et al. proposed a system which controls appliances connected to smart sockets. The disadvantages are power consumption was not measured, no protection against high voltage/ current fluctuations (Tong, & Li, 2017).
- Mon-Chau Shie et al. proposed a system where sockets were equipped with ZigBee wireless transceivers and the devices connected could be controlled. Some of the disadvantages are no intrusion system, no timers for the sockets (Shie, Lin, Su, Chen, & Hetahaean, 2014).

METHODOLOGY AND IMPLEMENTATION

Operation Flowchart

See Figure 1.

System Description

Intelligent sockets are not a replacement to the existing switchboard, rather a modification to the existing hardware. Not only does it connect the switches to the internet, it also provides remote accessibility, helping in efficient usage of power in the household.

It provides security as it stores the live feed from the camera on the server. It also may prevent fire hazards and overload due to short circuit in the sockets. The control of switches from the server can be through voice. For instance, to switch on a fan, all it takes is for the user to say "turn fan on". We have also implemented timer on the switches so that a device can be switched off after a specified time interval. This is of two methods:

- **Auto Timer:** Set once and forget. The intelligent hardware turns on at the given time and turns off automatically, repeating day after day.
- **Manual Timer:** This is particularly helpful for routine tasks like charging a phone. Overcharging is not recommended, thus set the timer for 2 hours, and the phone stops charging after the set interval expires.

Wi-Fi

Wi-Fi is a technology for wireless local area networking with devices based on the IEEE 802.11 standards. Wi-Fi is a trademark of the Wi-Fi Alliance, which restricts the use of the term Wi-Fi Certified to

Figure 1. Process Flow

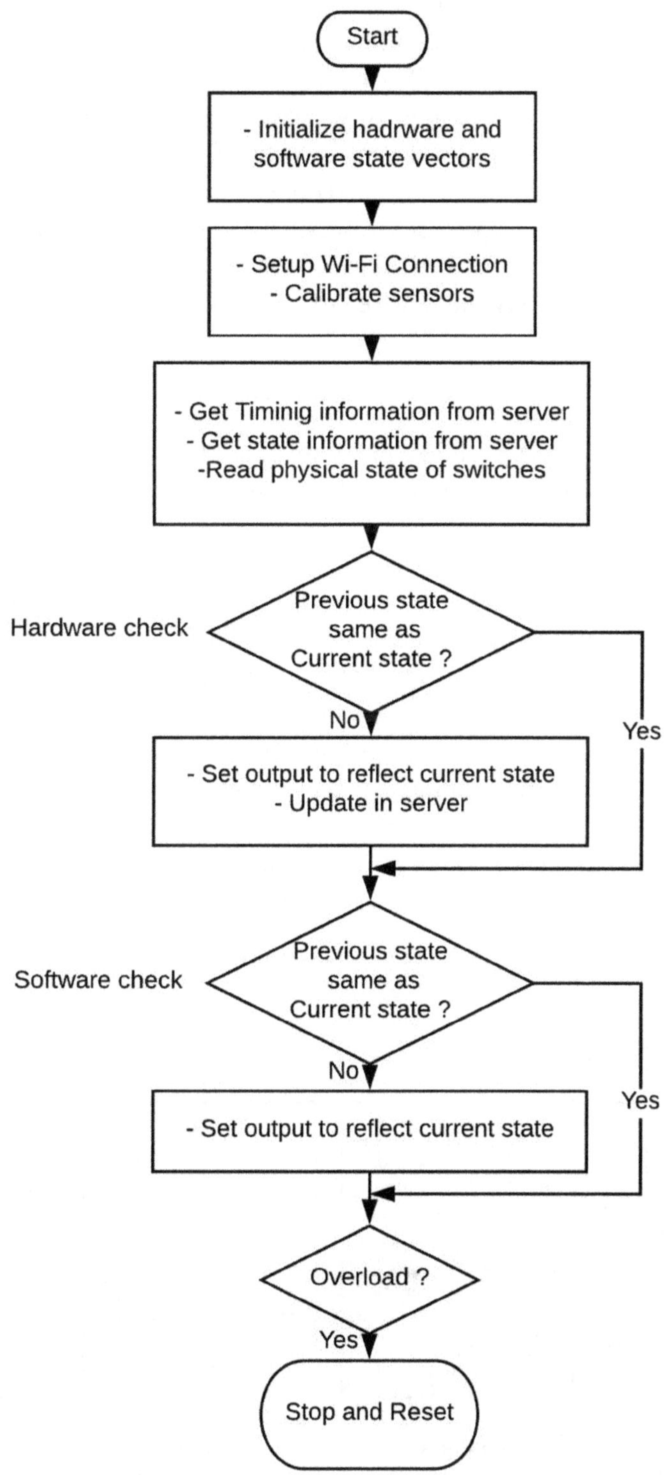

products that successfully complete interoperability certification testing. An access point or hotspot has a range of about 20 meters (66 feet) indoors and a greater range outdoors. Hotspot coverage can be as small as a single room with walls that block radio waves, or as large as many square kilometres achieved by using multiple overlapping access points. Wi-Fi most commonly uses the 2.4 gigahertz (12 cm) UHF and 5.8 gigahertz (5 cm) SHF ISM radio bands. Anyone within range with a wireless modem can attempt to access the network. Because of this, Wi-Fi is more vulnerable to attack of eavesdropping than wired networks. This was somewhat compensated by Wired Equivalent Privacy or WEP. Wi-Fi Protected Access is a family of technologies created to protect information moving across Wi-Fi networks and includes solutions for personal and enterprise networks. Security features of Wi-Fi Protected Access constantly evolve to include stronger protections and new security practices as the security landscape changes.

ESP8266-12E

The ESP8266, whose block diagram is shown in Figure 2, is a low-cost Wi-Fi microchip with full TCP/IP stack and microcontroller capability produced by Shanghai based Chinese manufacturer, Espressif Systems.

The chip first came to the attention of western makers in August 2014 with the ESP01 module, made by a third-party manufacturer, Ai-Thinker. This small module allows microcontrollers to connect to a Wi-Fi network and make simple TCP/IP connections using Hayes-style commands. However, at the time there was almost no English-language documentation on the chip and the commands it accepted. The very low price and the fact that there were very few external components on the module which suggested that it could eventually be very inexpensive in volume, attracted many hackers to explore the module, chip, and the software on it, as well as to translate the Chinese documentation.

The ESP8285 is an ESP8266 with 1 MiB of built-in flash, allowing for single-chip devices capable of connecting to Wi-Fi. The successor to these microcontroller chips is the ESP32.

Figure 2. ESP8266 Block Diagram

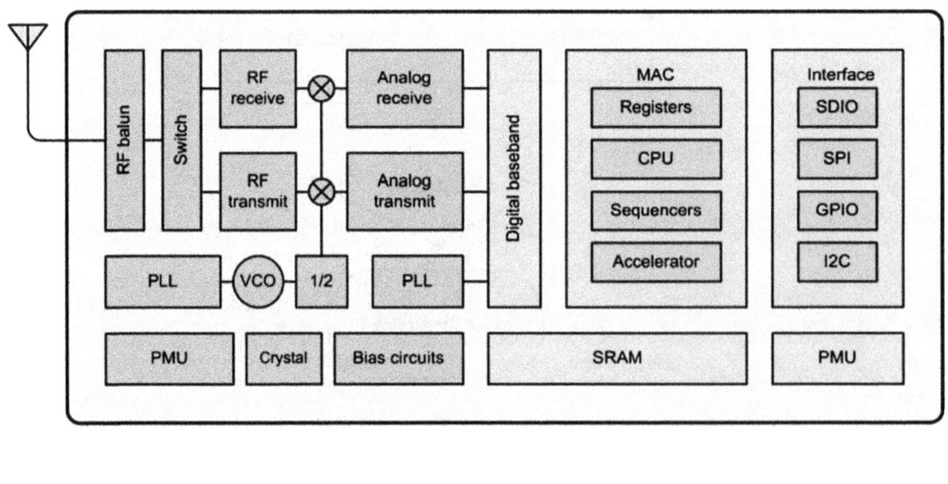

Hall Effect

The Hall effect is the production of a voltage difference (the Hall voltage) across an electrical conductor, transverse to an electric current in the conductor and to an applied magnetic field perpendicular to the current. It was discovered by Edwin Hall in 1879.

The Hall coefficient is defined as the ratio of the induced electric field to the product of the current density and the applied magnetic field. It is a characteristic of the material from which the conductor is made, since its value depends on the type, number, and properties of the charge carriers that constitute the current. The Hall effect is due to the nature of the current in a conductor. When a magnetic field is present, these charges of the current flow experience a force, called the Lorentz force. When such a magnetic field is absent, the charges follow approximately straight, 'line of sight' paths between collisions with impurities, phonons, etc. However, when a magnetic field with a perpendicular component is applied, their paths between collisions are curved, thus moving charges accumulate on one face of the material. This leaves equal and opposite charges exposed on the other face, where there is a scarcity of mobile charges. The result is an asymmetric distribution of charge density across the Hall element, arising from a force that is perpendicular to both the 'line of sight' path and the applied magnetic field. The separation of charge establishes an electric field that opposes the migration of further charge, so a steady electric potential is established for as long as the charge is flowing.

Secure Socket Layer - SSL

SSL (Secure Sockets Layer) is a standard security protocol for establishing encrypted links between a web server and a browser in an online communication. The usage of SSL technology ensures that all data transmitted between the web server and browser remains encrypted. An SSL certificate is necessary to create SSL connection. One would need to give all details about the identity of the website and the company as and when chosen to activate SSL on the web server. Following this, two cryptographic keys are created - a Private Key and a Public Key. SSL or TLS (Transport Layer Security) certificates are data files that bind a cryptographic key to the details of an organization. When SSL/TLS certificate is installed on a web server, it enables a secure connection between the web server and the browser that connects to it. The website's URL is prefixed with "https" instead of "http" and a padlock is shown on the address bar.

HTML5

HTML5 is a markup language used for structuring and presenting content on the World Wide Web. It is the fifth and current major version of the HTML standard.

It was published in October 2014 by the World Wide Web Consortium (W3C) to improve the language with support for the latest multimedia, while keeping it both easily readable by humans and consistently understood by computers and devices such as web browsers, parsers, etc. HTML5 is intended to subsume not only HTML 4, but also XHTML 1 and DOM Level 2 HTML.

HTML5 includes detailed processing models to encourage more interoperable implementations; it extends, improves and rationalizes the markup available for documents, and introduces markup and application programming interfaces (APIs) for complex web applications. For the same reasons, HTML5 is also a candidate for cross-platform mobile applications, because it includes features designed with low-powered devices in mind.

Many new syntactic features are included. To natively include and handle multimedia and graphical content, the new <video>, <audio> and <canvas> elements were added, and support for scalable vector graphics (SVG) content and MathML for mathematical formulas. To enrich the semantic content of documents, new page structure elements such as <main>, <section>, <article>, <header>, <footer>, <aside>, <nav> and <figure>, are added. New attributes are introduced, some elements and attributes have been removed, and others such as <a>, <cite> and <menu> have been changed, redefined or standardized.

WebSpeech API

The new JavaScript Web Speech API is a HTML5 API which makes it easy to add speech recognition to your web pages. This API allows fine control and flexibility over the speech recognition capabilities in Chrome version 25 and later. This specification defines a JavaScript API to enable web developers to incorporate speech recognition and synthesis into their web pages. It enables developers to use scripting to generate text-to speech output and to use speech recognition as an input for forms, continuous dictation and control. One such live text to speech translation demo is shown in Figure 3 The JavaScript API allows web pages to control activation and timing and to handle results and alternatives. At the first launch of speech recognition, Chrome needs to ask the user for permission to use the microphone, in which case onstart only fires when and if the user allows permission. Pages hosted on HTTPS do not need to ask repeatedly for permission, whereas HTTP hosted pages do.

SYSTEM REQUIREMENTS

The following are being used in our project. These are either essential or make the end experience nicer by helping the user.

Figure 3. WebSpeech API Demo

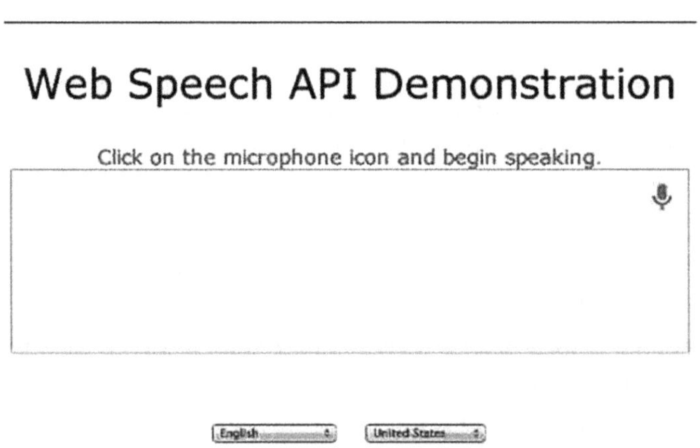

Hardware Requirements

The following are the hardware components being used.

Raspberry Pi

Used for the LAMP server as a database and control point.

The Raspberry Pi is a powerful credit-card sized micro-computer that can do anything a desktop computer can do. The board was designed to help both kids and adults learn how to program and get started in the world of computers, all at an affordable price, making the Raspberry Pi a fantastic tool for coding in languages such as Python and Scratch. The single-board computer comes with built-in ports allowing it to interact with the outside world – it can accept accessories such as a keyboard, mouse, monitor or camera, as well as connect to the Internet, and more. There are many versions available. We have used a Raspberry Pi Model B+ as shown in Figure 4 but the same can be done a more economical Raspberry Pi Zero.

The Raspberry Pi lets you browse the web; take videos and photographs; play music and high-definition videos; or if you just need to work on a word-processor, it can do that too. Being Internet-connected gives the Raspberry Pi the ability to let makers create Raspberry Pi projects such as weather stations; voice assistants; game consoles – the possibilities are endless.

We have used this multi-functional computer to host a LAMP server. This includes phpmyadmin for easy control of databases and html, php, css and js files to provide User Interface (UI) and User eXperience (UX) to the user. Some of the PHP files are important for the modules to communicate with the server itself.

Figure 4. Raspberry Pi 1 B+

NoIR Pi Camera Module

The Raspberry Pi camera module is an integrated board easily attachable to give a static photo or a video input to the Raspberry Pi. The camera consists of a small (25mm × 20mm × 9mm) circuit board, which connects to the Raspberry Pi's Camera Serial Interface (CSI) bus connector via a flexible ribbon cable, shown in Figure 5. The camera's image sensor has a native resolution of five megapixels and has a fixed focus lens. The software for the camera supports full resolution still images up to 2592x1944 and video resolutions of 1080p30, 720p60 and 480p60/90.

The one used was a NoIR camera, which means that it has no IR blocking filter and can be used in the dark with IR lights since it is being used for surveillance.

NodeMCU

Development board with ESP8266-12E for each switch board for up to 6 switches

The NodeMCU, shown in Figure 6, is an open-source firmware and development kit that helps us prototype IOT products within a few Lua script lines. But it is also fully compatible with Arduino IDE, thus exposing it to millions of libraries already available for the Arduino. This makes selection of components easier.

Coding for this is possible in both Arduino C++ style and NodeJS style asynchronous, API event driven Lua language.

Relay Module

To control the actual load or devices connected to the sockets.

A relay, shown in Figure 7, is an electromagnetically operated switch. The electromagnetic operation provides electrical isolation between the control and the power circuit. The relay is, in turn, controlled and driven by an optocoupler isolator to prevent damage to microcontrollers due to high reverse voltage of the coils during turn off.

Figure 5. Raspberry Pi NoIR Camera Module V2

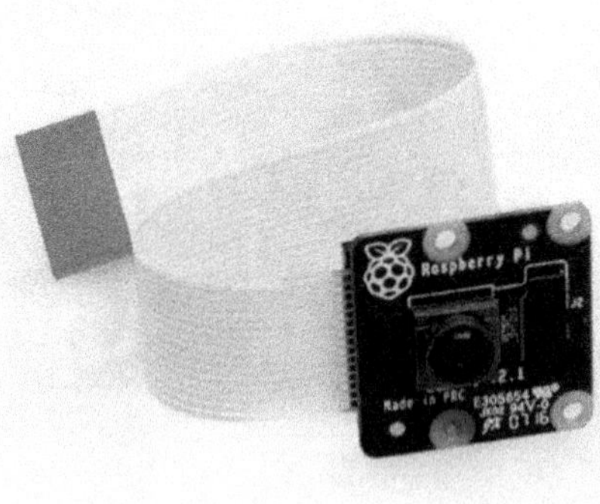

Figure 6. NodeMCU Amica board

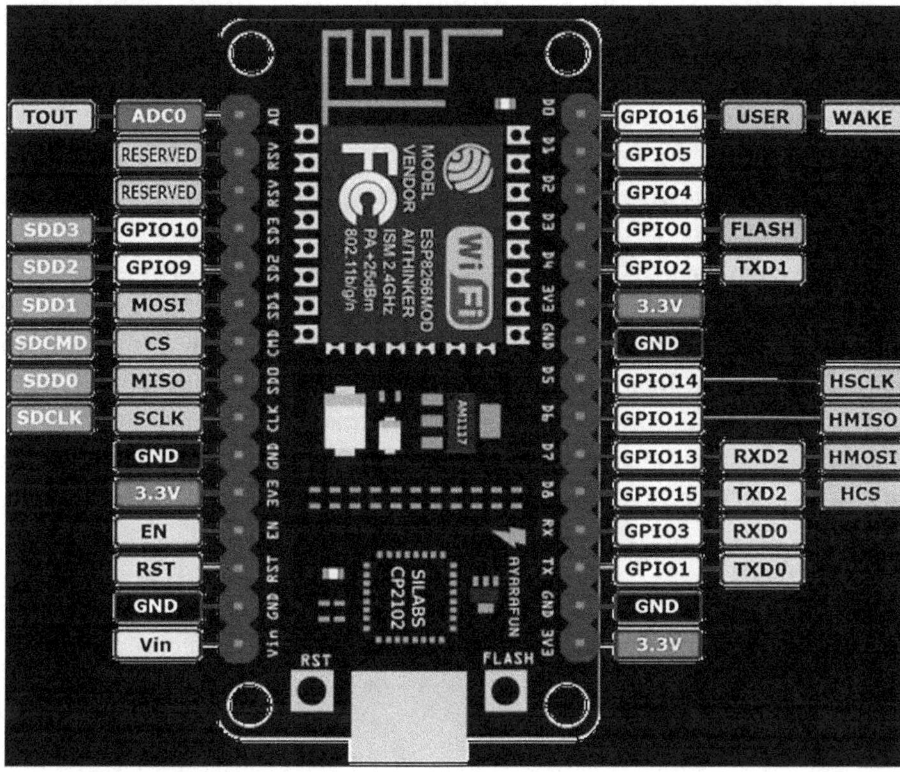

Figure 7. 4 channel Relay board

The relay used here are easily driven by 3.3V logic of the NodeMCU board. The relay is changed from Normally Closed (NC) to Normally Open (NO) by pulling the input pin of that channel LOW. Each relay can be used to control one device.

Hall Effect Current Sensor

To measure and log currents flowing into the board and to analyse power consumption and check for overload.

Current sensors are either open or closed loop. Open-loop current sensors measure AC and DC currents and provide electrical isolation between the circuit being measured and the output of the sensor. Being less expensive, these are generally preferred in battery powered circuits given their low operating power requirements. Closed-loop sensors also measure AC and DC currents and provide electrical isolation and are often the sensors of choice when high accuracy is essential.

The ACS712-30A shown in Figure 8 is a Hall Effect Sensor providing isolation for voltages up to $2.1kV_{rms}$ on the pins and up to 350V on the lead pins.

Software Requirements

The following softwares are required for the smooth operation of the device.

Raspbian: Debian for Raspberry Pi

Raspbian is an open source operating system based on Debian optimized for the Raspberry Pi hardware. However, Raspbian provides more than a pure OS: it comes with over 35,000 packages, pre-compiled software bundled in a nice format for easy installation on Raspberry Pi.

Raspbian uses PIXEL, Pi Improved Xwindows Environment, Lightweight as its main desktop environment as of the latest update, shown in Figure 9. It is composed of a modified LXDE desktop environment and the Openbox stacking window manager with a new theme and few other changes. The distribution is shipped with a copy of computer algebra program Mathematica and a version of Minecraft called Minecraft Pi as well as a lightweight version of Chromium as of the latest version.

Figure 8. ACS712 30A Current Sensor

Figure 9. Raspibian with PIXEL Desktop Environment

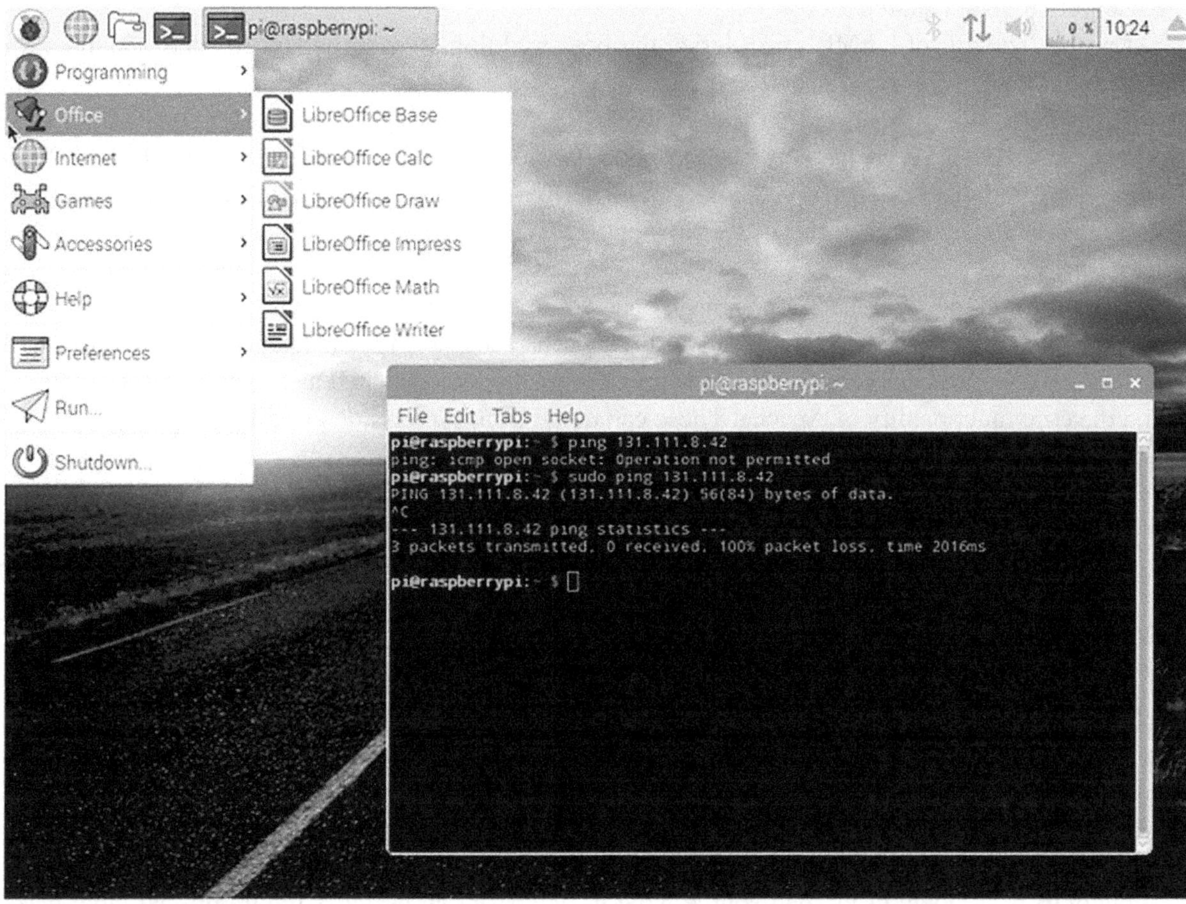

LAMP Server with JS

Used for backend operation of the boards and front-end access and control to the user.

LAMP is an archetypal model of web service stacks, named as an acronym of the names of its original four open-source components: the Linux operating system, the Apache HTTP Server, the MySQL relational database management system (RDBMS), and the PHP programming language. The LAMP components are largely interchangeable and not limited to the original selection. As a solution stack, LAMP is suitable for building dynamic web sites and web applications.

Since its creation, the LAMP model has been adapted to other components, though typically consisting of free and open-source software. For example, an equivalent installation on the Microsoft Windows family of operating systems is known as WAMP and an equivalent installation on macOS is known as MAMP.

1. Linux: Linux is a Unix-like computer operating system assembled under the model of free and open source software development and distribution. Most Linux distributions, as collections of software based around the Linux kernel and often around a package management system, provide complete LAMP setups through their packages. According to W3Techs in October 2013, 58.5%

of web server market share was shared between Debian and Ubuntu, while RHEL, Fedora and CentOS together shared 37.3%

2. Apache: The role of LAMP's web server has been traditionally supplied by Apache, and has since included other web servers such as Nginx.

The Apache HTTP Server has been the most popular web server on the public Internet. In June 2013, Netcraft estimated that Apache served 54.2% of all active websites and 53.3% of the top servers across all domains. In June 2014, Apache was estimated to serve 52.27% of all active websites, followed by nginx with 14.36%.

Apache is developed and maintained by an open community of developers under the auspices of the Apache Software Foundation. Released under the Apache License, Apache is open-source software. A wide variety of features are supported, and many of them are implemented as compiled modules which extend the core functionality of Apache. These can range from server-side programming language support to authentication schemes.

3. MySQL: MySQL's original role as the LAMP's relational database management system (RDBMS) has since been alternately provisioned by other RDBMSs such as MariaDB or PostgreSQL, or even NoSQL databases such as MongoDB.

MySQL is a multi-threaded, multi-user, SQL database management system (DBMS), acquired by Sun Microsystems in 2008, which was then acquired by Oracle Corporation in 2010. Since its early years, the MySQL team has made its source code available under the terms of the GNU General Public License, as well as under a variety of proprietary agreements.

MariaDB is a community-developed fork of MySQL, led by its original developers. PostgreSQL is also an ACID-compliant relational database, unrelated to MySQL.

MongoDB is a widely used open-source NoSQL database that eschews the traditional table-based relational database structure in favour of JSON-like documents with dynamic schema (calling the format BSON), making the integration of data in certain types of applications easier and faster.

4. PHP: PHP's role as the LAMP's application programming language has also been performed by other languages such as Perl, Python and even NodeJS.

PHP is a server-side scripting language designed for web development but also used as a general-purpose programming language. PHP code is interpreted by a web server via a PHP processor module, which generates the resulting web page. PHP commands can optionally be embedded directly into an HTML source document rather than calling an external file to process data. It has also evolved to include a command-line interface capability and can be used in standalone graphical applications.

PHP is free software released under the terms of PHP License, which is incompatible with the GNU General Public License (GPL) due to the restrictions PHP License places on the usage of the term PHP.

Python is a widely used general-purpose high-level programming language. Python supports multiple programming paradigms, including object-oriented, imperative, functional and procedural paradigms. It features a dynamic type system, automatic memory management, a standard library, and strict use of white-space. Like other dynamic languages, Python is often used as a scripting language, but is also used in a wide range of non-scripting contexts.

Node.js is an open-source, cross-platform JavaScript run-time environment that executes JavaScript code server-side. Historically, JavaScript was used primarily for client-side scripting, in which scripts written in JavaScript are embedded in a webpage's HTML and run client-side by a JavaScript engine in the user's web browser.

Node.js lets developers use JavaScript for server-side scripting—running scripts server-side to produce dynamic web page content before the page is sent to the user's web browser. Consequently, Node.js represents a "JavaScript everywhere" paradigm, unifying web application development around a single programming language, rather than different languages for server side and client-side scripts.

Though .js is the conventional filename extension for JavaScript code, the name "Node.js" does not refer to a particular file in this context and is merely the name of the product. Node.js has an event-driven architecture capable of asynchronous I/O. These design choices aim to optimize throughput and scalability in web applications with many input/output operations, as well as for real-time Web applications (e.g., real-time communication programs and browser games).

The Node.js distributed development project, governed by the Node.js Foundation, is facilitated by the Linux Foundation's Collaborative Projects program.

Arduino IDE

To compile the C++ code and burn the hex file to the NodeMCU.

The Arduino Integrated Development Environment - or Arduino Software (IDE) contains a text editor for writing code, a message area, a text console, a toolbar with buttons for common functions and a series of menus. It connects to the Arduino and Genuino hardware to upload programs and communicate with them.

Programs written using Arduino Software (IDE) are called sketches. These sketches are written in the text editor and are saved with the file extension.ino. The editor has features for cutting/pasting and for searching/replacing text. The message area gives feedback while saving and exporting and also displays errors. The console displays text output by the Arduino Software (IDE), including complete error messages and other information. The bottom right-hand corner of the window displays the configured board and serial port. The toolbar buttons allow you to verify and upload programs, create, open, and save sketches, and open the serial monitor.

The Arduino Software (IDE) uses the concept of a sketchbook: a standard place to store your programs (or sketches). The sketches in your sketchbook can be opened from the File - Sketchbook menu or from the Open button on the toolbar. The first time you run the Arduino software, it will automatically create a directory for your sketchbook. You can view or change the location of the sketchbook location from with the Preferences dialog.

Support for third-party hardware can be added to the hardware directory of your sketchbook directory. Platforms installed there may include board definitions (which appear in the board menu), core libraries, bootloaders, and programmer definitions. To install, create the hardware directory, then unzip the third-party platform into its own sub-directory. (The "Arduino" can't be used as the sub-directory name or it will override the built-in Arduino platform.) To uninstall, simply delete its directory.

The board selection has two effects: it sets the parameters (e.g. CPU speed and baud rate) used when compiling and uploading sketches; and sets and the file and fuse settings used by the burn bootloader command. Some of the board definitions differ only in the latter, so even if you've been uploading successfully with a particular selection, you'll want to check it before burning the bootloader.

The Arduino IDE includes support for the NodeMCU ESP8266-12E module via the Arduino esp8266 core provided open source on GitHub [5]. So, this was used to upload code to the NodeMCU module.

The text editor Sublime Text 3 was used to write all the required codes for the server's HTML, PHP, CSS and JavaScript code, the Arduino core for NodeMCU and the libraries for the wireless connection to the server.

RPi-Cam-Web-Interface

Used for remote monitoring to provide live feed along with peace of mind to the user.

RPi Cam Web Interface is a web interface for the Raspberry Pi Camera module. It can be used for a wide variety of applications including surveillance, DVR recording and time lapse photography. It is highly configurable and can be extended with the use of macro scripts. It can be opened on any browser, even on a smartphone.

It contains the following features:

- View, stop and restart a live-preview with low latency and high framerate. Full sensor area available as shown in Figure 10.
- Control camera settings like brightness, contrast, etc. live.
- Record full-hd videos and save them on the sd-card packed into mp4 container while the live-preview continues.
- Do timed or continuous video recording with splitting into fixed length segments.
- Take single or multiple (timelapse) full-res pictures and save them on the sd-card (live-preview holds on for a short moment).
- Preview, download and delete the saved videos and pictures, zip-download for multiple files.
- Trigger captures by motion detection using internal or external detection processes.
- Shutdown/Reboot your Pi from the web interface.

LIBRARIES

The following libraries being used provide functionality to the NodeMCU development board.

Libraries Written

The WiFi connection to the database was done with a custom library to retrieve and dump values into the databases.

Libraries Used

The following libraries make it easier to interface external hardware and process information.

Figure 10. RPi-Cam-Web Interface

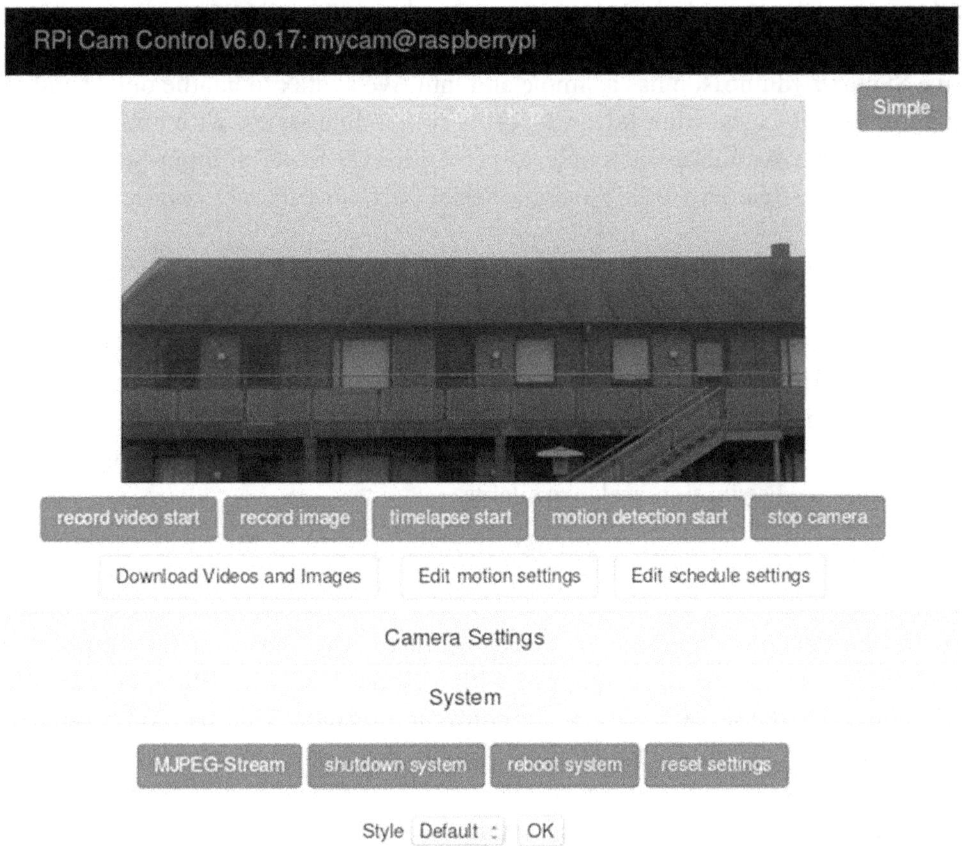

ACS712 Current Sensor Library

This is an Arduino library to interact with the ACS712 Hall effect-based linear current sensor. This library includes DC and RMS AC current measuring along with support for ACS712-05B, ACS712-10A, ACS712-30A sensors.

ESP8266WiFi

WiFi interface library for ESP8266 This library provides built in support for using the WiFi functionality of the ESP8266 modules directly from the Arduino IDE. This comes with features to communicate over WiFi using TCP and UDP, set up HTTP, mDNS, SSDP and DNS servers.

ArduinoJSON

JSON - The JavaScript Object Notation - Parser Library This library is a go-to for all JSON parsing applications. It includes the following features:

- Serialize and Deserialize ArduinoJson supports both JSON serialization and deserialization.
- Fixed Memory Allocation ArduinoJson uses a fixed memory allocation, allowing to work on devices with very little RAM. It also works on both stack and heap memory.
- Intuitive Syntax ArduinoJson has a simple and intuitive syntax to handle object and arrays.
- Implicit or Explicit Casts ArduinoJson supports two coding styles, with implicit or explicit casts.
- Parse from Stream ArduinoJson is able to parse directly from an input Stream or std::istream. This includes the streaming ability from a Serial port, an Ethernet connection or a WiFiClient connection.
- Compatible with Flash Strings ArduinoJson works directly with strings stored in program memory (PROGMEM). We can also use Flash String as input.
- Print to Stream ArduinoJson is able to print directly to a Print or std::ostream. This includes the streaming ability from a Serial port, an Ethernet connection or a WiFiClient connection.
- Header Only ArduinoJson is a header-only library, meaning that all the code is the headers. This greatly simplifies the compilation as we don't have to worry about compiling and linking the library. A single file distribution is also available.

Design Process

As shown in the block diagram in Figure 11, we have built a LAMP server on the Raspberry Pi. Further, each node, which is a switchboard, is equipped with a NodeMCU with ESP8266-12E Wi-Fi module which will the connect to the server to receive the configuration data. This also controls the relay module to turn the devices connected on and off along with monitoring overload conditions using Hall current sensors as shown. All logging information is saved in the server, i.e. the Raspberry Pi itself. The web interface which dynamically changes its page width to be compatible with the mobile platform is also hosted here. The whole system will take care of both physical switch presses as well as remote control instructions along with timer service.

The major components required and used are Raspberry Pi and web cam. The camera captures the images of the Analog measuring instruments at regular intervals and sends it to RaspberryPi2.0forima geprocessinganddetection. The Raspberry Pi then sends the detected values along with other required parameters to the cloud server or database. We have used SQL DB here to simulate cloud server features and functions.

Figure 11. Block diagram of system

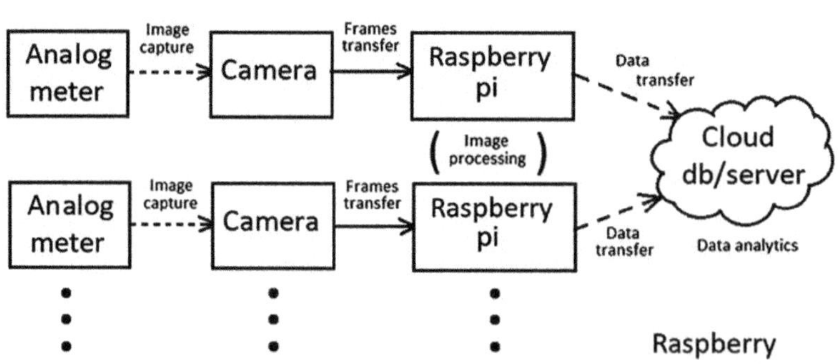

LAMP - Server Side

This server is the main backend and frontend to our project. This has a MySQL database containing 3 tables.

1. State
2. Log
3. Time

The state table holds the state of the sockets connected as 0 for ON and 1 for OFF. Modifying this results in the change of state of the relay. Also, the change in state of hardware switches are reflected in this table. Switching remotely over the network also uses the same table for updation. This holds both the state of the socket, and helps to change the state together with changing the state.

The log table is for the current sensor to dump its data about the real-time current being used by the sockets. This is used to calculate the power consumed by the board.

The time table is used to set a timer to individual sockets to auto turn on and off. The times are associated with an ID to distinguish between the sockets and the times are provided with an hour interval for 24 hours.

NodeMCU - Client Side

The NodeMCU client resides in each of the switchboxes implemented. It is the heart of the device. This follows a particular flow of operation. The flowchart in Figure 4.2 gives a better visual idea of the flow of the algorithm given below.

- Maintain a previous and current state vector for both hardware and software switches.
- Initialize the ports and the Wi-Fi connection.
- Do forever,
    - The current reading is checked to be under the threshold, along with sending it to the database every minute.
    - The timing information is retrieved every 5 minutes and set as required.
    - Get the status of all the switches from the database.
    - Do for each hardware switch,
        - Read the state of the switch.
        - If there is a change from the previous state, indicating an actual switch press, update the previous state and set the socket state accordingly.
        - Send the new state to the server and change the software vectors.
        - If there is any change in the software state, indicating a virtual switch press, update the previous software state and set the socket state accordingly.

The working can be broken into two parts.

1. Initialize - The previous and current state of hardware and software maintaining vectors are initialized. The Wi-Fi connection is established to the predefined Access Point.

2. Loop - This is the place where the processor resides most of the time. The continuous state of checking and updating the values of the database is done. This is a two-way process.

Along with this, there is also retrieval of current timestamp and time intervals from the database, check for validity, and turn on or off according to the requirement. The overcurrent detection also takes place here. Once detected, the device turns off all sockets and waits for its manual reset indefinitely.

Classification Using CNN

The classification is a problem of identifying which of a set of categories the new data/observation belongs to. Classification is done by collecting the data called as training data and devising a pattern in the training data to further predict the category the test data belongs to. Here we have to classify image as image with needle reading or image with digit reading. In supervised learning, a classifier model is trained with a set of training features and their corresponding labels. This model generalizes on the training data and predicts the test labels on the test data. The process of categorizing the test data on a previously learned model in a supervised classification problem is called as Classification. The classification problem in this scenario is a supervised classification. The different types of classifiers used for supervised classification are K-Nearest Neighbours Classifier, Support Vector Machines (SVM), Neural networks, etc.

Convolutional Neural Networks take advantage of the fact that the input consists of images and they constrain the architecture in a more sensible way. In particular, unlike a regular Neural Network, the layers of a ConvNet have neurons arranged in 3 dimensions: width, height, depth. (Note that the word depth here refers to the third dimension of an activation volume, not to the depth of a full Neural Network, which can refer to the total number of layers in a network.) For example, the input images in CIFAR-10 are an input volume of activations, and the volume has dimensions 32x32x3 (width, height, depth respectively). As we will soon see, the neurons in a layer will only be connected to a small region of the layer before it, instead of all of the neurons in a fully-connected manner. Moreover, the final output layer would for CIFAR-10 have dimensions 1x1x10, because by the end of the ConvNet architecture we will reduce the full image into a single vector of class scores, arranged along the depth dimension.

Image Pre-Processing Techniques for Detecting Needle

Next step is to detect the needle in the analog meter. This is done using Gaussian Blur, thresholding methods and Hough transform. Gaussian Blur is used to remove low frequency noise in the image. It is a widely used effect in graphics software, typically to reduce image noise and reduce detail. The visual effect of this blurring technique is a smooth blur resembling that of viewing the image through a translucent screen, distinctly different from the bokeh effect produced by an out-of-focus lens or the shadow of an object under usual illumination. Gaussian smoothing is also used as a pre-processing stage in computer vision algorithms in order to enhance image structures at different scales—see scale space representation and scale space implementation.

After segmentation binary thresholding is used for setting a proper threshold to detect the needle. The Binary Thresholding function creates a raster output that divides your raster into two distinct classes. The algorithm behind the Binary Thresholding function, the Otsu method, was designed to distinguish between background and foreground in imagery by creating two classes with minimal intraclass variance

(Otsu 1979). When working with a raster dataset that has a unimodal distribution, Binary Thresholding divides the data into two distinct classes. It creates a high-value class, displayed with white pixels, and a low-value class, displayed with black pixels.

After setting the binary threshold, Hough Transform is used to detect the needle. The Hough transform is a technique which can be used to isolate features of a particular shape within an image. Because it requires that the desired features be specified in some parametric form, the classical Hough transform is most commonly used for the detection of regular curves such as lines, circles, ellipses, etc. A generalized Hough transform can be employed in applications where a simple analytic description of a feature(s) is not possible.

Digit Segmentation and Recognition

Next to detect meters with digit readings, we have to segment the digits first. Next, we have to recognize each of the digits. Segmentation is done using Otsu's thresholding, Adaptive thresholding, finding area of contours and Hough circles. In Otsu's method we exhaustively search for the threshold that minimizes the intra-class variance (the variance within the class), defined as a weighted sum of variances of the two classes:

$$\sigma_\omega^2 = \omega_0(t)\sigma_0^2(t) + \omega_1(t)\sigma_1^2(t)$$

Adaptive thresholding typically takes a gray-scale or colour image as input and, in the simplest implementation, outputs a binary image representing the segmentation. For each pixel in the image, a threshold has to be calculated. If the pixel value is below the threshold it is set to the background value, otherwise it assumes the foreground value.

There are two main approaches to finding the threshold:

1. The Chow and Kaneko approach
2. Local thresholding

The assumption behind both methods is that smaller image regions are more likely to have approximately uniform illumination, thus being more suitable for thresholding. Chow and Kaneko divide an image into an array of overlapping sub-images and then find the optimum threshold for each sub image by investigating its histogram. The threshold for each single pixel is found by interpolating the results of the sub-images. The drawback of this method is that it is computational expensive and, therefore, is not appropriate for real-time applications. An alternative approach to finding the local threshold is to statistically examine the intensity values of the local neighbourhood of each pixel. The statistic which is most appropriate depends largely on the input image. Simple and fast functions include the mean of the local intensity distribution or average of maximum or minimum pixel values. Example of Threshold value where max is maximum pixel value and min is minimum pixel value.

T = (max+min)/2 The dot in the analog meters can be detected using the concept of Hough circles. In a two-dimensional space, a circle can be described by: (x−a)2+(y−b)2 = r2 where (a, b) is the center of the circle, and r is the radius. If a 2D point (x, y) is fixed, then the parameters can be found according to (1). The parameter space would be three dimensional, (a, b, r). And all the parameters that satisfy (x,

y) would lie on the surface of an inverted right-angled cone whose apex is at (x, y, 0). In the 3D space, the circle parameters can be identified by the intersection of many conic surfaces that are defined by points on the 2D circle. This process can be divided into two stages. The first stage is fixing radius then find the optimal center of circles in a 2D parameter space. The second stage is to find the optimal radius in a one-dimensional parameter space. Then the segmented digits are recognized using a trained CNN model. The dataset used for training is the Chars74 K dataset. This training is similar to the training employed in classifying the type of analog meter.

CNN for Digit Recognition

Once image segments are detected, recognizing the digits in each of them was the next step. This is done using a machine learning model trained using Convolutional Neural Network (CNN). The dataset used for training is CHARS74k dataset. In this dataset, symbols used in both English and Kannada are available.

In the English language, Latin script (excluding accents) and Hindu-Arabic numerals are used. For simplicity we call this the "English" characters set. The dataset consists of:

- 64 classes (0-9, A-Z, a-z)
- 7705 characters obtained from natural images
- 3410 hand drawn characters using a tablet PC
- 62992 synthesized characters from computer fonts This gives a total of over 74K images (which explains the name of the dataset). We have used 10 classes of numbers in the dataset for training. LeNet architecture is again used here for training.

TESTING, RESULTS AND COMPARISONS

Testing

The testing of the product was done by starting with the operation of the NodeMCU itself. The relay module was not being driven by the 3.3V supply of the NodeMCU. So, the 5V supply was used, along with 3.3V logic for control. The analog pin of the NodeMCU had an issue with the Arduino core library as mentioned in the GitHub Issue Tracker [6] which basically results in fluctuating ADC value. This prevented us from using the analog current meter for accurate current measurement and logging. Instead, we used the existing mechanism for overload protection exceeding the limitations.

Results

The following images show the final results achieved by our project.

The Hardware

The Figure 12 shows the hardware of intelligent sockets. We currently have added four sockets to be controlled. The board has NodeMCU, relay module and current sensor built into it. The working can be easily seen as the mosquito repellent connected to socket 1 is turned ON even as its corresponding switch is OFF.

Figure 12. Switchboard Hardware

The Web Interface

The web interface is the main component of our project as it allows users to control it from any part of the world.

Authentication

The Figure 13 shows the authentication asked by the webpage as it would be a security issue to allow anyone to login and use the switches. If authentication fails, no control is available.

Figure 13. User Authentication

Figure 14. Home page

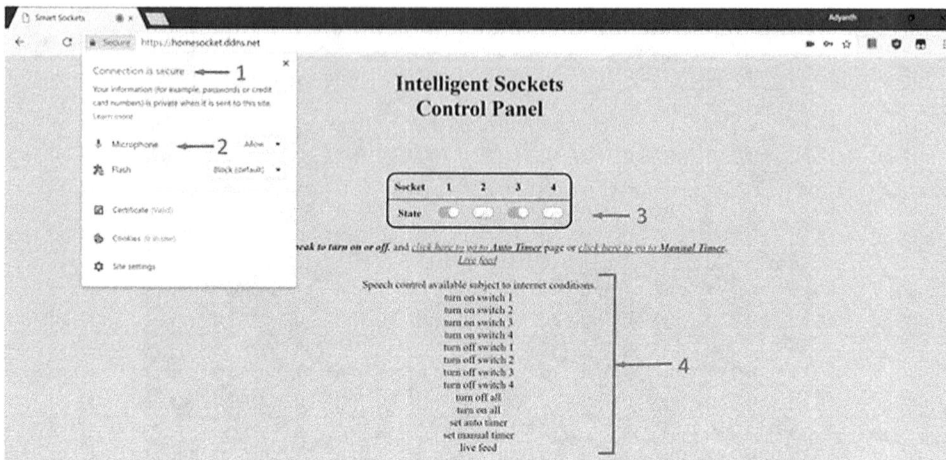

Home Page

Once authentication is successful, the main control home page of the server as in Figure 14 is displayed. This scales according to the display size as seen in Figure 15. Some of the points that can be noticed are:

1. A HTTPS connection: Ensures data cannot be hacked while being used to control the devices.
2. Microphone access: It is given to the page for easy, client-side voice control.
3. Virtual switches: These reflect the state of the hardware sockets, and also used for control by clicking on it or by voice.
4. Voice commands: The predefined voice commands can be used to turn the switches ON and OFF. Additionally, it can also be used for navigating to other features such as timers and camera feed.

Timer Control

There are two types of timers available for the user as shown below.

1. Auto Timer: This mode allows the user to automatically turn the device ON and OFF on a daily schedule. The user can, for example, select a time slot of 05:00 to 08:00 in the morning and 06:00 to 07:00 in the evening for socket 1 as shown in Figure 16.
2. Manual Timer: This mode allows the user to manually set a timer from current time, for example, for the next one minute as shown in Figure 17. Once the timer expires, the switch automatically turns OFF without user intervention.

Live Camera Feed

The camera is on the server which is available to the user for streaming. This is also used for comparing against known faces for intruder detection. This interface as shown in Figure 18, which provides the user a peace of mind while not at home. The interface also allows image and video recordings, motion detection and more.

Figure 15. Home page as seen in a smartphone

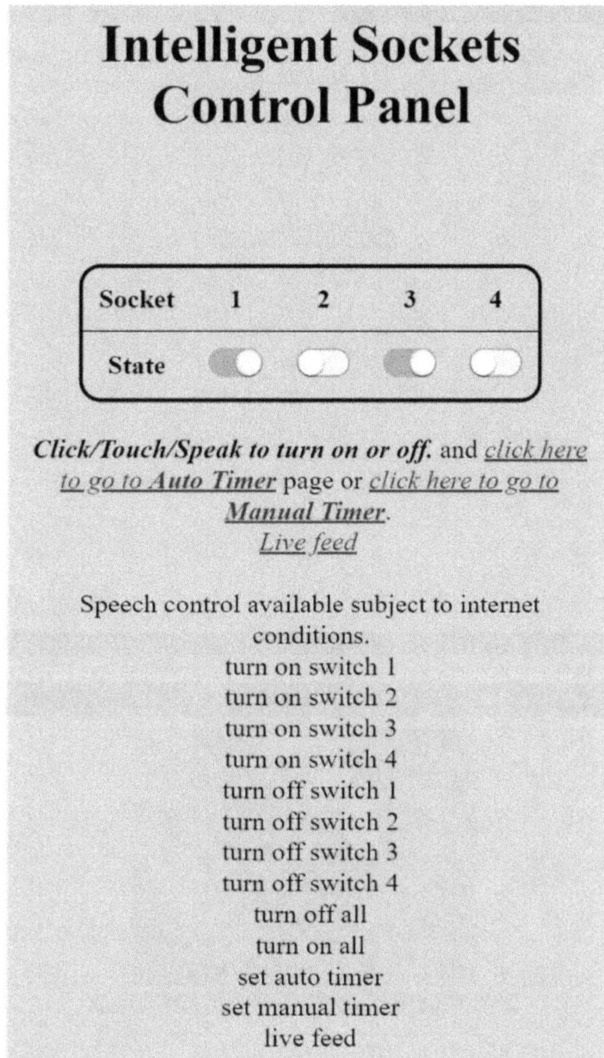

Figure 16. Auto Timer

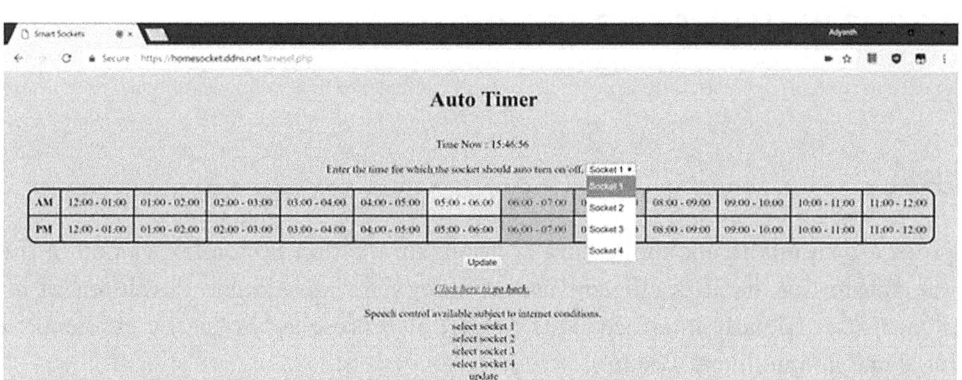

Figure 17. Manual Timer

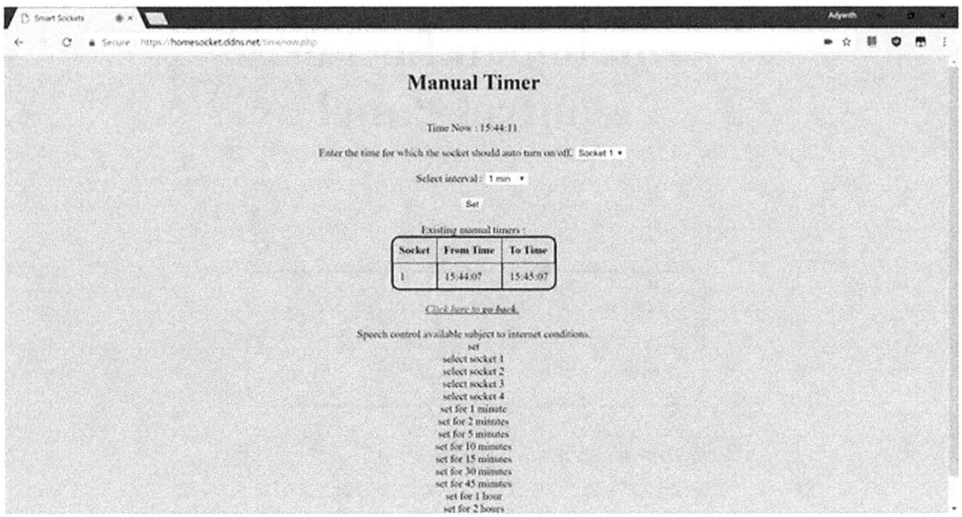

Figure 18. Live Feed Interface

COMPARISONS WITH EXISTING TECHNOLOGY

See Table 1.

Advantages

- This project is an intelligent application of home automation principles, as one of the base lines of home automation itself is efficient use of energy for sustainable development along with a provision of user-friendly interface. This project also presents an energy efficient-user friendly, integrated and an intelligent system.

Table 1.

Features	[1]	[2]	[3]	[4]	[5]	[6]	[7]	Proposed
Local remote control	✓	✓	×	✓	✓	✓	✓	✓
Global remote control	×	×	×	✓	✓	✓	✓	✓
Short circuit protection	×	×	×	✓	×	×	✓	✓
Power measurement	×	✓	✓	✓	×	×	✓	✓
Meter tamper check	×	×	×	×	×	×	×	✓
Surveillance	×	×	×	×	×	✓	×	✓
Motion detection	×	×	×	×	✓	×	×	✓
Face detection and authentication	×	×	×	×	×	×	×	✓
Auto and manual mode timers	×	×	×	✓	×	✓	✓	✓

- The timer modes enable the users to have a hassle-free remote control over the switches. The Automatic Timer mode enables the switching action of appliances in a regular specific time for a specified time interval. The Manual timer mode allows the user to switch on the supply to a device for any desired interval of time.
- One of the major upper hand this project has is that it makes use of the prevailing hardware which make it easier for the users to transit from the present to an efficient future.
- The adaptation of the face-id using digital image processing principles also provides an efficient access control for high wattage devices which also server as child lock for the operation of these devices.
- This project overcomes the major bottleneck of internet-based system which is security. The live feed from the pi-cam installed at the entryways makes the system secure. It is very cost effective, as it does not require any much of new hardware.

Drawbacks

- The remote control has to be actuated from the signals from internet. The delays in the processing lead to a significant delay in the response.
- There is no redundancy for the internet dependant control i.e. if the internet fails for any reason the remote control will be void, only the traditional switching mechanism exits.
- The security if not implemented properly for the Wi-Fi network, the system may be hacked since it is available over the local network to anyone connected. So, anyone can control the switching system of the house.

CONCLUSION

The following features were achieved at the end:

- Easy access of devices over the network/internet.
- Timing control.
- Safety from overload and short circuit protection.
- Intrusion system.
- Face detection and authentication.
- Computer Vision meter reading

The authors have used a LAMP server to control the devices remotely. A web interface provides cross platform support alongside manual control and predefined voice commands. Two implementations for timers are provided so that the devices may be on for the specified time duration or on a timer. Overload protection has been achieved with the usage of hall sensor which switches off the whole circuitry in case of an overload. Intrusion systems comprises of Pi camera for live feed with additional features like offline recording and motion detection. This is coupled with facial recognition to provide intruder notification and image processing for reading the electrical meters.

FUTURE SCOPE

Just like the world we live in even the technology is constantly evolving. The above listed features could be implemented in the time frame provided leading to a way for many improvements and future work.

1. Efficient modes for switches like standby mode or sleep mode can be implemented to better use the energy when they are rarely used or when the devices being operated draw current in the range below the normal threshold.
2. To have a better secured network, instead of integrating the camera modules at just the entryways, each switchboard can be integrated with a camera module which will be a part of network of camera modules finally connected to one central server.
3. The manual intervention in reading the meter reading and bill generation can be minimised by logging the power consumption levels remotely and then intimating the user about their monthly electricity bill.
4. The project can be made more user friendly by provision of touchscreen interface for setting the timer modes and provision of control of regulators to adjust the fan speed, light intensity can also be given.
5. With these improvements, the product may easily be suitable for customers through mass production.

REFERENCES

Ghosh, A. (2016). Intelligent appliances controller using Raspberry Pi, *2016 IEEE 7th Annual Information Technology, Electronics and Mobile Communication Conference (IEMCON)*, Vancouver, Canada, pp. 1-5.

GitHub. (2018). ADC Values Inaccurate and different from NodeMCU Issue #2672 esp8266/Arduino, [online]. Available at https://github.com/esp8266/Arduino/issues/2672

GitHub. (2018). esp8266/Arduino, [online]. Available at https://github.com/esp8266/Arduino

Pawar, P., & Vittal, K. P. (2017). Design of smart socket for power optimization in home energy management system, *2017 2nd IEEE International Conference on Recent Trends in Electronics, Information & Communication Technology (RTEICT),* Bangalore, pp. 1739-1744. 10.1109/RTEICT.2017.8256897

Santoro, P., Calderaro, V., Galdi, V., & Piccolo, A. (2016). Active smart socket design to perform local control of power demand in residential units, *8th IET International Conference on Power Electronics, Machines and Drives (PEMD 2016)*, Glasgow, UK, pp. 1-5. 10.1049/cp.2016.0380

Shie, M., Lin, P., Su, T., Chen, P., & Hutahaean, A. (2014). Intelligent energy monitoring system based on ZigBee-equipped smart sockets, *2014 International Conference on Intelligent Green Building and Smart Grid (IGBSG)*, Taipei, Taiwan, pp. 1-5. 10.1109/IGBSG.2014.6835281

Tong, Y., & Li, Z. Design of Intelligent Socket Based on WiFi, *2017 4th International Conference on Information Science and Control Engineering (ICISCE),* Changsha, China, pp. 952-955. 10.1109/ICISCE.2017.201

Xiao, M., Liu, Y. H., & Hu, Q. (2016). Design and Implementation of Socket-Based-Network Connections Smart Home System, *2016 Sixth International Conference on Instrumentation & Measurement, Computer, Communication and Control (IMCCC)*, Harbin, China, pp. 756-760. 10.1109/IMCCC.2016.113

Xu, A., & He, S. (2017). The wireless smart socket control system design, *2017 2nd International Conference on Advanced Robotics and Mechatronics (ICARM),* Hefei, China, pp. 698-703. 10.1109/ICARM.2017.8273247

Chapter 17
Advanced Encryption Standard With Randomized Round Keys for Communication Security in IoT Networks

Ishpal Singh Gill
Namibia University of Science and Technology, Namibia

Dharm Singh Jat
Namibia University of Science and Technology, Namibia

ABSTRACT

Internet of things (IoT) is a rapidly emerging architecture connecting smart devices all across the world in various fields like smart homes, smart cities, health sector, security, etc. Security is a very important aspect of IoT. As more and more devices are connecting to the Internet, it becomes a lucrative target for hackers. The communication between the various devices, nodes, and between nodes and the cloud, needs to be secured. A combination of public and private key cryptography systems is used to secure the IoT networks. The Advanced Encryption Standard (AES) is used for encrypting the data in transit. However, the AES is known to be prone to brute force attacks, side channel attacks, and other forms of cryptanalysis. This chapter proposes a more secure AES algorithm with randomised round keys, which provides better security with negligible overheads, and is ideal for use in IoT networks.

INTRODUCTION

Internet of Things is rapidly emerging architecture in all fields in the world. It has tremendous benefits in the health, medical, traffic management, smart homes, smart cities (Zanella, Bui, Castellani, Vangelista, & Zorzi, 2014) etc sectors. The basic architecture of IoT is a three layered architecture, Layer 1 consists of the various smart devices, sensors etc., Layer 2 consists of the network protocols which help these devices to communicate with the Internet and the third layer is the applications and processes which are the deliverables to the end user. The IoT networks also require security for confidentiality,

DOI: 10.4018/978-1-5225-9246-4.ch017

integrity, authentication and availability. A combination of public and private cryptography is ideal for this purpose, with public key cryptography being used for providing authentication, integrity, key distribution and private key cryptography algorithms like AES being used for encryption and confidentiality of data communication. The combination is based on the utilizing the inherent strength of both types of crypto algorithms and averting the disadvantages of each. The symmetric algorithms or the private key algorithms are inherently much stronger but have the problem of key distribution and management, whereas the asymmetric algorithms or the public-private key algorithms, although not as strong as symmetric algorithms, provide for easier key management and secure distribution over networks. (Stallings, Cryptography and network security principles and practices, fourth edition, 2005, Forouzan, 2007).

The Advanced Encryption Standard (AES) is a 128 bit block cipher with key sizes of 128, 192, 256 bits (Kak, 2018). It is the FIPS standard for symmetric algorithms based on the design of the Rijndael algorithm, and is the strongest known symmetric algorithm. The only known vulnerabilities of AES are brute force and side channel attacks. A number of modifications of the AES have been attempted in order to make it more complex and secure. However, none of them address the use of randomised round keys to strengthen the security.

The only known unbreakable cipher is the One Time Pad (OTP), which uses a truly random key stream equal in length to the text size to encrypt the text. The output is a truly random encrypted text which is impossible to decrypt without the key stream and the key stream cannot be obtained through any known cryptanalysis technique. However, this system has the disadvantage of manual key distribution which involves huge administrative and logistics effort and cost (Schneier, 1996; Menezes, Oorschot, & Vanstone, 1997).

This chapter studies and analyses the various earlier usage of AES and AES modifications to secure IoT networks. A number of such modifications have tried to enhance the security of the AES algorithm. However, none address the use of randomised keys to enhance security., This chapter proposes a modified AES algorithm with randomized round keys, which will make AES more complex with randomize round keys, thus making it more difficult to break with negligible effect on efficiency, and thus ideally suited for security of IoT networks.

LITERATURE REVIEW

Advanced Encryption Standard

The Advanced Encryption Standard (AES) is a symmetric cipher which works on 128 bit block data with three key size variants of 128, 192 and 256 bits. The encryption consists of repeated iterations of four operations performed on the data over 10, 12 and 14 rounds for the 128, 192 and 256 bit key variants respectively. The four operations are Sub Bytes, Shift Rows, Mix Columns and Add Round Key. For the 128 bit key version, the single 128 bit key is broken down into 11 keys of 128 bits. One key is used in the initially to XOR with the text/data and thereafter the balance 10 keys are used for each of the 10 rounds. The last round has three operations without the Mix Columns operation. The decryption is the exact reverse of the encryption process *(FIPS Publication 197, 2001;* Daemen & Rijmen, 1999, 2013). The AES algorithm is the strongest known symmetric algorithm. The only known attacks against AES are the brute force attack, wherein the attacker obtains the key by trying out all possible bit combinations of the key length and the side channel attacks which makes use of the incorrect or flawed implementation of the algoithm.

The high level algorithm for AES encryption is given below:-

```
AES_Round (Data, RoundKey[j])
{
SubBytes (Data);
ShiftRows (Data);
MixColns (Data);
AddRoundKey (Data, RoundKey[j]);
}
AES_RoundLast (Data, RoundKey[NumofRounds])
{
SubBytes (Data);
ShiftRows (Data);
AddRoundKey (Data, RoundKey[Numof Rounds]);
}
```

Security for IoT

This paper analyses the security issues of IoT networks and various cryptographic techniques to ensure Confidentiality, Integrity and Availability. It proposes set of security measures that are best suited for IoT networks. It finally concludes that a decentralized architecture with AES and ECC is best suited for IoT networks like smart homes etc. (Hole, 2017)

Secure Data Transmission using AES in IoT

This paper proposes a secure data transmission using plain AES to enhance security. The solution is implemented in MATLAB and analyses the execution time (Khambra & Dabas, 2017). The results are depicted in Table 1.

The solution increases the time, which is inevitable when encryption is used. However, the solution does not strengthen the plain AES and remains vulnerable to brute force and side channel attacks of AES.

Table 1. Execution Time using AES

No of Nodes	Execution Time in Millisecs Without AES	Execution Time in Millisecs With AES
100	10.2239	76.5304
200	13.2843	80.8851
300	18.8278	85.7570

IoT and Enhanced Security

This paper carries out a detailed survey of IoT and its security issues. It also proposed a model wherein data is encrypted using firstly the AES and then hiding it using the Steganography technique. This model has been implemented using the MATLAB (Srivastava, Agarwal, & Mathur, 2015). However, the use of two cryptographic techniques will lead to increased execution times and reduced efficiency. It is better to use one strong symmetric algorithm like AES but with increased strength.

Preventing DoS Attacks in IoT

This paper analyses the various attacks that can take place in IoT networks. It also proposes a modified AES for security in IoT networks. The modified AES consists of a white box in lieu of the S box and doubles the encryption process to strengthen it (Javed, Khan, Qahar, & Abdullah, 2018). The proposal does strengthen the AES but it entails an increase in the execution time due to double encryption. Thus double encryption is a very inefficient way of enhancing security with almost doubled execution time also.

Implementation of AES With Time Complexity

This paper highlights the vulnerability of the AES algorithm and time constraints to data inputs other than text. The paper gives experimental results of AES operating on text, video and audio files. It is evident that the audio file takes maximum time for encryption and decryption. It also analyses a side channel attack on AES. However the paper does not propose any solution to strengthen the AES algorithm against brute force and side channel attacks (More & Bansode, 2015).

Secure Military Communication Based on AES

This paper pertains to proposing a solution to implement transmission of secure messages between remotely located terminals in military applications. The solution uses plain AES to implement a model for the management of secret keys for secure storage, retrieval and management of these keys. However, the paper does not give a solution to strengthen the AES. It also remains silent on the important issue of secure and efficient key distribution. (Bardis, Doukas, & Ntaikos, 2008).

Optimised Architecture for AES

This paper proposes an optimised architecture for implementation of AES. It proposes 10 levels of pipelining and use of Block RAM to decrease slice utilisation, which results in 36% higher throughput as compared too other designs. The solution is good for increasing efficiency in execution time especially for near real time applications. However, the paper does not give a solution to strengthen the AES (S, Goswami, Tadi, & Pandey).

Enhancing AES Algorithm With Arithmetic Coding

This paper proposes a technique to increase the throughput by first compressing the data using arithmetic coding and encrypting it using AES as AES accepts only 128 bit input data at one time. The process is reversed at the receiver end. The same has been implemented in MATLAB. This technique increases the efficiency as it will decrease the overall execution time for encryption and decryption, but does not strengthen the AES algorithm per se. Arithmetic coding is a simple technique to compress data and not a security technique or algorithm (Mukesh, Pandya, & Pathak, 2013)

Enhanced AES Algorithm for Next Generation IoT

This paper suggests encrypting data blocks of 200 bits using 400 bit keys. The key is broken down into two sections of 200 bit each and used for different cascaded AES comprising of five rounds each. The result show an improvement in the efficiency of the AES algorithm. It also claims to enhance security as using cascaded rounds with different keys (Bhara, Gupta, & Jaiswal, 2017). However, this proposal changes the basic mathematical structure of the AES and thus cannot be more secure than the plain AES as there is no mathematical foundation given to verify the claim of enhanced security. It has been well established that the strongest known symmetric algorithm is AES and any changes to its mathematical foundation will make it less secure (FIPS publication 197, 2001). Hence the enhancement will be only in terms of efficiency but at lower security.

ISSUES AND PROBLEMS

Most modifications discussed above either introduce a degree of complexity in the AES, which makes it difficult for an attacker to identify a pattern and hence make the AES algorithm stronger or primarily focus on increasing the efficiency of the AES implementation to reduce overheads and get speedier execution time for both encryption and decryption. Some modifications also propose a change in the mathematical structure of AES, which makes it appear more complex but actually reduces the basic security of the algorithm. The point to be understood is that the AES standard has been chosen with due diligence of the underlying mathematics and complexity and any attempt to tinker or change the basic mathematical structure or foundation will reduce the strength. Certain other modifications (not discussed in the chapter), use various methods like use of hash functions, increased key sizes, increase in a number of rounds, use of MAC addresses etc. to increase the complexity of AES. However, none of the earlier proposed designs and implementations try to strengthen AES by addressing both the complexity and introducing a degree of randomness, like the OTP system. As mentioned earlier, the strongest algorithm is the OTP, which is truly random in nature but it is not possible to use such an algorithm on networks because of huge logistics and administrative issues for key management. True randomness cannot be replicated on a computer system. However, the use of a near random or pseudo random algorithm is possible to implement on the computer. The other important issue is that IoT applications and devices require near real time communications and hence it is very important and critical that the security algorithm is efficient. Hence there is an inescapable need to strengthen the algorithm with negligible overheads of additional execution time.

SOLUTIONS AND RECOMMENDATIONS

Proposed AES Algorithm With Randomised Keys

The basic logic of the proposed algorithm is to randomise the keys much like the random keys used by OTP system of encryption which makes it the most secure algorithm in the world. The proposed AES algorithm will make use of a 1408 bit random number generator to randomise the 11 X 128 bit keys required for the initial add key operation and the Add Round Key operation in each of the 10 rounds of the AES-128 bit algorithm. This will increase the randomness and complexity of the AES and make it stronger and more secure. It will also have negligible effect on the execution time as the key generation is done only once during the encryption and decryption process irrespective of the length of data being encrypted or decrypted. The pseudo code for the same is depicted below:

```
Create Expanded keys (11 X 128 bit from 128 bit key)
w = key_expansion (key, s_box, rcon, 1);
Generate randomn number matrix r of size 44 X 4
r = randi (255,44,4,'double');
XOR r with w to get new key matrix of 11 X 128 bit keys
w = bitxor (w,r);
```

RESULTS AND DISCUSSIONS

The performance of the new modified AES with randomised round keys will be tested for two performance metrics namely complexity and efficiency or execution time.

Complexity

The complexity of an algorithm for a brute force attack, which is the only known vulnerability against the AES, is measured in terms of the number of possible combinations of the 128 bit key, which in case of plain AES is 2^{128}, which is a measure of the total key combinations possible for a 128 bit key. However, in case of the proposed AES, for a brute force attack to work, the attacker will need to try out not only the 2^{128} possible combinations to get the right key but also the 1408 bits of the randomized number to get the randomized round keys. Hence the strength of the proposed AES against a brute force attack will become:

Complexity $= 2^{128} + 2^{1408}$

Thus the AES with randomized round keys will be more secure and will provide higher security than any of the earlier attempts to use AES for security of IoT networks.

Execution Time

The execution timings will be similar to the normal plain AES. The random number generation (pseudo code given at para 3 above), will be done only once during the initial key expansion into the 11 X 128 bit keys for the rounds. Once the round keys are randomized, the code has no effect on the overall encryption and decryption process, irrespective of the length or size of data encrypted or decrypted. The slight overhead of the random number generation thus becomes negligible for large data sizes. The execution time would be comparable to the normal plain AES. Hence we are able to achieve much higher security but with negligible overheads, making this new proposed AES algorithm is ideally suited for IoT networks and applications.

APPLICATION OF PROPOSED AES ALGORITHM FOR IOT NETWORKS

The section explains the use of the enhanced and more secure AES for a smart home IoT network. As depicted in Fig. 1 below, the enhanced AES will be used to provide secure data communication between the various IoT nodes, devices to a central home gateway, which in turn provides secure communication to the Internet again using the enhanced AES.

The proposed AES could also be used to encrypt and store user data and information like profiles, preferences, history etc. on smart IoT devices.

FUTURE RESEARCH DIRECTIONS

Future research on the subject could be on conversion of the AES algorithm from a block cipher to a stream cipher with randomized keys being used. The stream cipher would ensure better efficiency and execution speed. Research could also be done in the field of algorithms for efficient generation of random numbers, which generate near true random numbers.

Figure 1. Secure Data Communication in Smart Home Network using AES with Randomised Round Keys

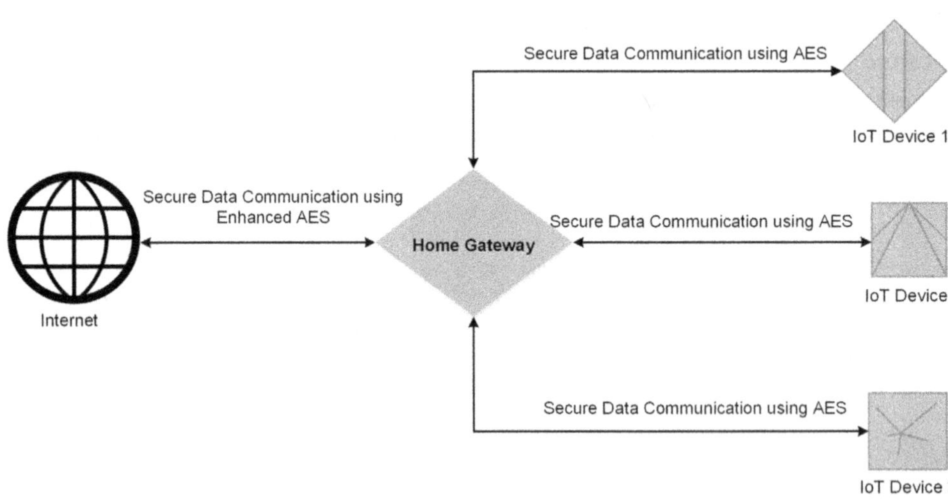

CONCLUSION

Security in IoT networks is a very important and critical aspect. The AES is a popular symmetric key standard for securing IoT networks and applications. However, it is susceptible to brute force and side channel attacks. This chapter proposes the enhanced AES for data security in IoT networks and applications. The enhanced AES provides a much higher degree of security with negligible overheads.

ACKNOWLEDGMENT

This research has been supported by the Namibian Ministry of Defence.

REFERENCES

Bardis, N. G., Doukas, N., & Ntaikos, K. (2008). Design and Development of a Secure Military Communication based on AES Prototype Crypto Algorithm and Advanced Key Management Scheme. *WSEAS Transactions on Information Science & Applications.*

Bhara, R., Gupta, A., & Jaiswal, M. (2017). An Enhanced AES Algorithm Using Cascading Method On 400 Bits Key Size Used In Enhancing The Safety Of Next Generation Internet Of Things (IoT). *International Conference on Computing, Communication and Automation (ICCCA2017).*

Daemen, J., & Rijmen, V. (1999). *AES proposal: Rijndael.* AES Algorithm Submission.

Daemen, J., & Rijmen, V. (2013). *The design of rijndael: AES.* Springer Science and Business Media. *FIPS publication 197.* (2001, Nov 26). Retrieved from Advanced Encryption Standard: https://nvlpubs.nist.gov/nistpubs/fips/nist.fips.197.pdf

Forouzan, B. A. (2007). *Cryptography and network security, special indian edition.* Tata McGraw Hill Pvt Ltd.

Hole, C. E. (2017). *Analysis of Security for IoT.*

Javed, Y., Khan, A. S., Qahar, A., & Abdullah, J. (2018). Preventing DoS Attacks in IoT Using AES.

Kak, A. (2018). Lecture 8 AES: The Advanced Encryption Standard. In Lecture Notes on "Computer and Network Security", Purdue University.

Khambra, D., & Dabas, P. (2017). *Secure Data Transmission using AES in IoT. International Journal of Application or Innovation in Engineering & Management.* IJAIEM.

More, S., & Bansode, R. (2015). *Implementation of AES with Time Complexity Measurement for Various Input. Global Journal of Computer Science and Technology, 15(4).*

Mukesh, P. S., Pandya, M. S., & Pathak, S. (2013). Enhancing AES Algorithm with Arithmetic Coding. *International Conference on Green Computing, Communication and Conservation of Energy (ICGCE).* 10.1109/ICGCE.2013.6823404

S. A. P., Goswami, D., Tadi, S., & Pandey, K. (n.d.). *Optimized Architecture for AES.* Allahabad, India.

Srivastava, A. K., Agarwal, A., & Mathur, A. (2015). *Internet of Things and its enhanced data security. International Journal of Engineering and Applied Sciences. IJEAS.*

Stallings, W. (2005). *Cryptography and network security principles and practices* (4th ed.). Prentice Hall. Retrieved from http://www.inf.ufsc.br/~bosco.sobral/ensino/ine5680/material-cripto-seg/2014-1/Stallings/Stallings_Cryptography_and_Network_Security.pdf

Stallings, W. (2011). Network security essentials: Applications and standards (4th ed.). London, UK: Pearson Education.

Vanishreeprasad, S., & Pushpalatha, M. N. (2015, May-June). Design and implementation of hybrid cryptosystem using AES and hash function. *IOSR Journal of Electronics and Communication Engineering*, *10*(3), 18–24. doi:10.9790/2834-10321824

Zanella, A., Bui, N., Castellani, A., Vangelista, L., & Zorzi, M. (2014). *Internet of Things for Smart Cities. IEEE INTERNET OF THINGS JOURNAL.*

ADDITIONAL READING

Daemen, J., & Rijmen, V. (1999). *AES proposal: Rijndael.* AES Algorithm Submission.

Stallings, W. (2005). Cryptography and network security principles and practices, fourth edition. FIPS publication 197. (2001, Nov 26). Retrieved Apr 18, 2018, from Advanced Encryption Standard

Chapter 18
Security for Smart Vehicle in IOT

Deepa Sethuramalingam
PSG College of Technology, India

Brindha N.
PSG College of Technology, India

S. Balamurugan
https://orcid.org/0000-0003-3428-3849
QUANTS IS & Consultancy Services, India

ABSTRACT

The smart city is not possible without a smart road. It can provide citizens with smart mobility. In order to overcome the complications handled by the parking system, smart parking has been developed. A model IoT-based parking system that uses a unified component called parking meter to address the issues as well as to provide smart parking management throughout the city is proposed in this chapter.

INTRODUCTION

The IoT is in simple contrast to the internet of people. Instead of people accessing data and interactive with one another, the IoT does the process. The IoT connects not just computers, smart phones and tablets, but also lots of additional things.

This type of networks is developed as part of the Intelligent Transportation Systems **(ITS)** to bring important development to the transportation systems performance. The goal of ITSs is to improve safety on the roads and to reduce traffic congestion, waiting times and fuel consumptions. The integration of the surrounded computers, sensing devices, navigation systems, digital maps and the wireless communication devices along with intelligent algorithms will help to develop frequent types of applications for the ITS to improve safety on the roads (Miller. 2018).

The types of communication in IoT are satellite, Wi-Fi, Radio Frequency (RF), RFID, Bluetooth and NFC. Satellite allows the devices such as mobile phones to send and receive the data through cellular network. It provides stable connection and universal compatibility, but it has no direct communication

DOI: 10.4018/978-1-5225-9246-4.ch018

(Greengard, 2015). Wi-Fi a wireless local area network (WLAN) which offers internet access to devices that is within the range of about 66 feet from AP (access point) whichtransfers and receives data. Radio Frequency (RF) provides low energy and simplicity for its technology which is not dependent on the new functionality of phones. Radio frequency identification (RFID) is the use of electromagnetic fields to recognize objects which does not require power but highly insecure. Its range is from 10cm to 200cm. Bluetooth is also wireless technology standard for substituting data over short distances (Mangwani, 2016). NFC is a Near-field communication which uses electromagnetic induction within each other's near field.

IoT has become an active area of research, standardization and development because it has incredible potential to improve vehicle and road safety, traffic efficiency and convenience as well as comfort to both drivers and passengers. A lot of IoT research works have motivated on specific areas including routing, broadcasting, quality of service (QoS), safety and security.

The various applications of IoT are smart home, IoT in Poultry and Farming, Smart Wearable's, smart city etc. Over the last few years, the need of the term "smart" has turn into a widespread method at all levels. Presently everyone use smartphones, there is hardly a city that does not seek to be a Smart City, technology tends to become smart in numerous areas, new buildings are smart buildings and even energy is smart. Therefore, there is a need to convert the roads into smart roads in order to develop journey planning with real time data and satellite navigation through traffic management and other measurements like public transport priority measurement, speed and access control management. Traffic management is a critical challenge in most of the metropolitan cities. To calm the problems on road, the smart road concept is needful in the current era. The smart road concept is implemented to overcome the crisis on road accidents and parking. The various security measures for using IoT are helmet detection, accident detection, weather detection and congestion detection.

EXISTING TECHNOLOGY AND ITS REVIEW

The major source of road accidents is humanoid error in the existing system. Even with today's greater prominence on automobile safety, accidents continue to occur. According to survey done by the National Highway Traffic Safety Administration (NHTSA) some of the causes of accidents prevailing in the existing system are listed below:

- **Over Speeding:** The major factor contributing to the increased number of road accidents is speeding. This has resulted in 41% of the total deaths due to road accidents in India in 2016.
- **Drunken Driving:** Even though driving under the influence of alcohol is strictly prohibited, many flaunt this rule, which at times results in road accidents.
- **Ambulance Delay:** At present criteria, detection of accident zone automatically is not existent, leading to the death of an individual by ambulance delay.
- **Inefficient Parking:** Spending excessive time in searching for parking spaces leads to parking at inappropriate places.

METHODOLOGY AND RESEARCH DESIGN

In order to provide road safety, the proposed system deals with two major categories. They are Smart Roads and Smart Parking. Over the last few years, the need for the term "smart" has turned into a pervasive method at all levels. There is a need to transform the roads into smart roads in order to improve journey planning with real-time information and satellite navigation through traffic management and other measurements like public transport priority measurement, speed and access control management. Nowadays traffic management is a critical challenge in most of the metropolitan cities (Mangwani, 2016). To placate the problems on road, the smart road concept is needing in the current era which is implemented to overcome the crisis in road accidents and parking. The various security measures for using IOT are Helmet detection, accident detection, weather detection and congestion detection.

The Helmet Detection Sensor (HDS) is used to enrich the security of the rider. A Force Sensing Resistor (FSR) and BLDC Fan are used for discovery of the rider's head and recognition of bike's speed separately. The smart helmet that is fixed with sensors which go about as detectors for rider's head and the seat strap itself. The motor of the bike can begin just if the rider has locked in its helmet belt. The second security strategy is existing with another sensor which goes about as an alert to the rider when the cruiser speed exceeds 100km/h. It uses the sensors like the Force Sensing sensor which is kept inside the helmet as a material sensor to identify whether the rider is wearing the helmet. It comprises of a Polymer Thick Film (PTF) gadget which shows a shrinking in obstruction with an expansion in the power connected to the dynamic surface. HDS uses a Brushless Direct Current (BLDC) Fan engine which is broadly used in electronics as a cooling gadget whereas here it acts as a speed sensor. This is on the grounds that the attributes of BLDC Fan are high effectiveness, minimal effort, straightforward mechanical development and support free. The stage twisting of a brushless changeless magnet engine can be sorted as single-stage, two-stage or three-stage, their motion appropriation can be either sinusoidal or trapezoidal. This BLDC engine is single-stage which the trapezoidal transition is adopted. HDS uses IC LM311 (Comparator IC) which is a gadget looking at two voltages in the meantime and give HIGH or LOW in the yield. A reference voltage is settled while the information voltage is variable from zero to the supply voltage. Hypothetically, the reference and information voltages can be used between any sort of qualities from zero and the supply voltage, yet there are down to earth constraints on the genuine range contingent upon the specific gadget used. HDS also uses IC Timer 555 which is a stable circuit utilizing IC clock 555 creates the square wave. This circuit delivered 'blare signal' sound for the ringer. This square wave which is in the advanced frame will change LOW (0V) to HIGH (+Vs). The span of the LOW and HIGH state may differ because of circuit and estimation of the segment.

Weather detection is the IoT application which is implemented in automatic vehicles for improving smart road packages. Driving relief functions are core technologies of automatic vehicles. Eminent applications are lane keeping, traffic sign detection and interpretation, light assistant and adaptive cruise control. All of those methods are capable to scan visual images and to find lanes, signals and lights of preceding or opposite traffic. The identification of the driving situation by means of the surrounding weather situation is real and can provide important information for many other driving applications (Smart Parking, n.d.). Weather condition information obtained by this combination is a label with both quantitative and qualitative methods. LIDAR technology enables advanced object recognition used for weather detection. LIDAR is optimistic technology for weather detection, due to its very high sensitivity and spatial resolution, this active detection technique enables efficient location of weather effects during both day and night and over a considerable range. LIDAR (Light Detection and Ranging) is based on laser technology and eliminates most of the common disadvantages of cameras (Chinrungrueng and Sunantachaikul, 2007). Vehicle indicating mounted cameras capture the visual signal in which the driver needs to drive safely. Processing this in real time allows identifying pavement markings, road signs or hazards such as obstacles, in order to assess time to lane crossing, time to collision or other risk indicators. It is also possible to find reduced visibility conditions such as fog, rain, or glare, and even to quantify their impact on sight distance or target visibility, in order to give speed warning, distance warning to vehicles ahead or to control light beams. In some cases, visibility can even be restored, up to a point, by cleave and in-painting techniques for Head-Up-Displays.

Camera-based fog detection systems can analyze distinct objects in the image or image regions like the road region or the horizon. These approaches are not consistent with everyday use. A new method has been proposed to use the only reliable visible attribute of foggy weather conditions like a decrease in divergence and muddling in the whole image. The power spectrum is the squared magnitude of the Fourier transform for the image that holds information about the frequencies in the image. The matter of neglect spatial information and its decision-informing potential will be analyzed. From the power spectrum, one builds the image features that can be then fed to the classifier that has been trained on fog and fog-free images. It turns out, that in the case of an observed fog scene, the frequency components are concentrated at the zero frequency whereas in a scene without fog one finds a broadly spread spectrum caused by the contrast attenuation and muddling in the image aggravated by the fog. Sharp Edges are modeled by many different low and high frequencies, whereas smooth edges created by only low frequencies.

The principle behind LIDAR is slightly simple which during Weather such as fog, rain, and snow the backscatter leads to the perception of an atmospheric veil. For object and obstacle detection for Adaptive Cruise Control (ACC) or highly automated driving the weather, influence is separated out. For obtaining recognized weather object information the LIDAR reflections have to be particle filtered via Kalman filters. Estimates for fog intensity are statistically calculated using a Kalman filter algorithm in a common error correcting procedures to produce a most probable estimate value for fog intensity. This process has become known as the Ensemble Kalman Filter (EnKF) process. Estimated value is provided to the algorithm as an initial prior that is sensitized by intervention values from the sensor data to provide a posterior value, which recursively becomes the new preceding for the next iteration of the filter. The Kalman filter corresponds, the historical data of previous sample data with the value of the new preceding, producing an error value which becomes the estimate. The set of weather condition data comprises of Rain, Snow, Fog, Dust, Mist and Haze.

Nowadays huge traffic congestion takes place due to overpopulated vehicles. Congestion happens when the vehicle tends to move slowly due to traffic. This leads to an increase in traveling time, delay in trips, rise of air pollution, fuel consumption of vehicles etc. To reduce this, traffic congestion detection technology based on IoT is used to detect and controls the traffic. To detect congestion array of ultrasonic sensors are used. They are fixed at roadsides in order to monitor the different traffic levels. The data from the sensor is sent to the controller to estimate the total level of traffic. If any lane is detected with high traffic then more signal timing is provided to that lane and if any lane is detected to have fewer vehicles then that lane is provided with less amount of time. The data about the traffic levels will be sent to the server and is analyzed by IoT analytics and gets stored in a server database.

The system consists of ultrasonic sensors (HC-SR), AR controller (LPC2138), 16*2 Liquid Crystal Display (LCD) and a Wi-Fi module (ESP8266). An Ultrasonic sensor is used to detect the level of traffic. These sensors are transducers which converts ultrasound waves to electrical waves and vice versa.

These sensors are used to detect real-time traffic levels at lanes. It is operated at 40 KHz frequency and detects objects within the range of 2cm to 4cm.AR controller for converting analog to digital data. It is a microcontroller with flash ROM on-chip, 32Kb RAM, two I2C serial interfaces, two 8 channel 10 bit analog to digital converter, CPU clock of range up to 60MHz, a real-time clock having optional battery backup. It is a low power consumption controller which is minimal in size. LCD for displaying traffic levels and signal time (Road accidents, 2017). It is used at a road intersection to display signals by collecting data from ultrasonic sensors. Wi-Fi module to transfer data to the server. It is a set of high performance and high integration wireless system-on-a-chip. It is used to embed the Wi-Fi capacities to other systems. This application requires less space and can be made with low cost. It is used to transfer data from the controller side to the server.

Accident detection is another interesting and important provision in this fast era. The fame of vehicles has moreover extended the traffic risks and road accidents. In order to reduce such incidents, a modified alert contraption for vehicle disasters is displayed. This method can identify accidents in very minimal time and sends an alert message to the nearby hospital. Global Positioning System (GPS) module is used to find the location where the accident occurred. The alert message is sent by using the Global System for Mobile Communication (GSM) module. The accident can be recognized completely with the help of both Micro electromechanical structure (MEMS) sensor and vibration sensor. The model of programmed vehicle accident detection and informing using GSM and GPS modem utilizes ARM7 which accompanies the following advantages (Chinrungrueng and Sunantachaikul, 2007).

The entire layout of the complete setup will be drawn in the form of a block diagram. Then the accident will be detected using a piezoelectric sensor and it delivers the output to the microcontroller. Finally, the location of the vehicle is identified using GPS and the location details will be sent to the phone number which is been already saved in the EEPROM through GSM (Mangwani, 2016). GPS is used in vehicles for tracking and navigating purpose. The purpose of using the tracking system is to track the vehicle without the intervention of the driver and the navigation system causes the driver to achieve the destination. When an accident happens the GPS locates the vehicle and sends the emergency message to any hospital which is near to that geographic location through GSM which is used for communication between devices. It has been designed with reliable structure and easy to work. But the main drawback is cannot be used without a network. This technique can be implemented in vehicle theft recovery and shipment tracking.

Smart Parking

In recent times the development of cities into smart cities has become realizable using developments in the field of Internet of Things. IOT technology raises in several fields of smart applications. Due to the growth in population and vehicular usage that is quickly increasing, it becomes problematic for the drivers to find a parking lot at peak hours. To overwhelmed these issues smart parking is implemented using token rendering system. IOT system could perform real-time processing of data thus facilitating highly responsive technologies. As the population is rapidly growing in this era, an appropriately planned and easy method is provided by Smart Parking Technology (Greengard, 2015). The number of parking places available by is less when associated with the number of vehicles that are on road. Whenever a driver wants to search a parking lot it takes more time to identify the free space by which gradually reducing the speed of the vehicle. It consumes more fuel also since moving increases fuel consumption. It affects others in terms of speed and fuel. It tests one's temper and paves way for more chaos leading to traffic congestion. There are few more issues in parking a vehicle such as insufficient information about parking space availability, inconvenient parking spaces, lack of importance about parking issues etc. Smart Parking Technology focusses on efficiently using the available space and providing a hassle-free environment.

The Smart Parking Technology is designed in such a way that it is appropriate for mall parking on a real-time basis and is designed by creating use of some IOT supportable hardware such as motion detection sensors (Mangwani, 2016). It detects any physical movement in a particular area. These sensors convert motion into an electric signal.

A token rendering machine is present at the entrance of the parking gate. When a parker wants to park a vehicle, just pushes a button and a ticket (token) is generated. As the parker accepts the token the barrier gate opens. The token which has the information about the specified lot. When the car moves to the particular space the sensors placed in the parking location will detect that the car has reached its parking space and it is made non-available to others. IR sensors are used for sensing parking slot along with a dc motor to simulate as gate opener motors. Here this system uses a WIFI modem for internet connectivity and an AVR microcontroller for operating the system also uses IoT Gecko for online connectivity and IOT management GUI design. The system senses if parking slots are engaged or not using IR sensors. Also, it uses IR technology to sense if a vehicle has reached on the gate for automated gate opening. The system can read the number of parking slots available and updates the data with the cloud server to allow for checking the slot availability through online. This allows users to check for available parking spaces through online from anywhere and avail hassle free parking. Thus the system resolves the parking issue for cities and gets the users an efficient IoT based parking management system.

CONCLUSION

It has proved that the motorcycle's engine will only start when the helmet is worn and when the helmet belt has been locked. Therefore, it will reduce effect from the accident and motorcycles cannot be stolen. If the motorcycle go beyond 100km / hour, then the LED light flashes as an, alarm because over speed limit signal which has been given to alert the rider. Increasingly the automotive industry is developing and producing innovative driving intervention functions to support vehicle drivers. In this context, it is an important functionality to capture local weather in the vicinity of the car. It is clear that the detection of weather (road conditions) is a major element and data source in this quest. Sharing information from vehicles with the sensors labelled in this process is regarded as an essential benefit for driving safely. Smart parking technology is easy to manage time, fuel and reduce traffic congestion at peak hours. As a result of IoT requires gathering, accessing and processing information about available parking spaces. This method senses vehicle accidents and sends an alert message automatically. It is very useful because the alert message will be delivered to the hospital with the emergency alert and the location of the accident can be perfectly identified by the help of latitudinal and longitudinal position which is very beneficial.

REFERENCES

Auto Car Pro. (n.d.). *Economic Loss Road Accidents India Estimated.* Retrieved from http://www.autocarpro.in/news-national/economic-loss-road-accidents-india-estimated-gdp

Bahga, A., & Madiseti, V. (2017). *Internet of Things: A hands on approach*. Hyderabad: Universities press.

Balachandran, M. (2016, April 23). *Accidents on India's deadly roads cost the economy over $8 billion every year*. Retrieved from https://qz.com/689860/accidents-on-indias-deadly-roads-cost-the-economy-over-8-billio every-year/

Checkoway, S., McCoy, D., Anderson, D., & Kantor, B. (2011, August). *Comprehensive Experimental Analyses of Automotive Attack Surfaces*. Retrieved from https://www.researchgate.net/publication/259753269_Comprehensive_Experimental_Analyses_of_Automotive_Attack_Surfaces

Chinrungrueng, J., & Sunantachaikul, U. (2007). Smart Parking: an Application of optical Wireless Sensor Network. *IEEE Proceedings of the 2007 International Symposium on Applications and the Internet Workshops (SAINTW'07).*

Greengard, S. (2015). *Internet Of Things*. London: MIT Press. doi:10.7551/mitpress/10277.001.0001

Gupta, M. D. (2017, September 6). *Road accidents killed 17 people every hour in India in 2016, Delhi most unsafe*. Retrieved from http://www.hindustantimes.com/india-news/road-accidents-claimed-nearly-400-lives-every-day-in-india-in-2016/story-7DlmtdnvMYLLZVGxXKOaJN.html

Hanche, S.C., Munot, P., Bagal, P., Sonawane, K. & Pise, P. (2013). Automated Vehicle Parking System using RFID, *ITSI Transactions on Electrical and Electronics Engineering (ITSI-TEEE), 1(2).*

International Parking & Mobility Institute. (n.d.). Retrieved from http://www.parking.org/

Mangwani, P. (2016). Smart Parking Assist System using Internet of Things (IoT), International Journal of Control Theory and Applications, Volume 9-Number 40.

Miller, M. (2018). *Internet Of Things: How smart TVs, smart cars, smart homes, and smart cities are changing the world*. Chennai, India: Pearson.

Olariu, S., & Weigle, M. C. (2009). *Vehicular Networks: from theory to practice*. CRC Press. doi:10.1201/9781420085891

Raut, S. H., & Ambulgekar, H. P. (2013, March 4). *Proactive and Reactive Routing Protocols in Multihop Mobile Ad hoc Network*. Retrieved from https://pdfs.semanticscholar.org/4704/972232ae2bf64df 8e4fd3dcc93b414fbebdb.pdf

Road Accidents in India 2015. (2015). Government of India Ministry of Road Transport & Highways Transport Research Wing.

Road Accidents in India 2016: 17 deaths on roads every hour, Chennai and Delhi most dangerous. (2017, September 11). Retrieved from http://indianexpress.com/article/india/road-accidents-in-india-2016-17-deaths-on-roads-every-hour-chennai-and-delhi-most-dangerous-4837832/

Routing Information Protocol. (2019, June 12). Retrieved from https://en.wikipedia.org/wiki/Routing_Information_Protocol

Smart Parking. (n.d.). *Happiest Minds Technology*. Retrieved from https://www.happiestminds.com/whitepapers/smart-parking.pdf

Wang, H., & He, W. (2011). A Reservation-based Smart Parking System. *The First International Workshop on Cyber-Physical Networking Systems*.

Zhou, F., & Li, Q. (2014). Parking Guidance System Based on ZigBee and Geomagnetic Sensor Technology. In *Proceedings of 13th International Conference on Distributed Computing and Applications to Business, Engineering and Science (DCABES)* (pp. 268-271). IEEE.

Compilation of References

Aarthi, R., Shaik, A., & Khadir, A. (2015). An efficient method of irrigation using sensors. *International Journal of advanced research in computer and communication engineering, 4*(7).

Abbasi, A., Adjeroh, D., Dredze, M., Paul, M. J., Zahedi, F. M., Zhao, H., & Huesch, M. D. (2014). Social media analytics for smart health. *IEEE Intelligent Systems, 29*(2), 60–80. doi:10.1109/MIS.2014.29

ABI Research. (2018, Jan. 31). *Smart Footwear to Provide Unique Data Opportunities for Enterprises and Healthcare Providers.* Retrieved from https://www.prnewswire.com/news-releases/smart-footwear-to-provide-unique-data-opportunities-for-enterprises-and-healthcare-providers-300591076.html

Abraham, S. C. (2016). *Internet of Things (IoT) with Cloud Computing and Machine-to- Machine (M2M).* Communication. doi:10.18535/ijetst/v3i09.13

Abramson, N. (1970). The ALOHA system-Another alternative for computer communications, In *Proceedings of Fall Joint Computer Conference,* (pp. 281-285). ACM.

Agrawal, N., & Singhal, S. (2015). Smart drip irrigation system using raspberry pi and Arduino.

Ahlgren, B., Hidell, M., & Ngai, E. C.-H. (2016). Internet of Things for Smart Cities: Interoperability and Open Data. *IEEE Internet Computing, 20*(6), 52–56. doi:10.1109/MIC.2016.124

Akkineni, H., & Junapudi, V. (2010). Agriculture wrapped with social networks, data mining, and mobile computing to boost up crop productivity. *International Journal of Biological Sciences and Engineering. 1*(1), pp. 43:48.

Akyildiz, I., & Jornet, J. (2010). The Internet of nano-things. *IEEE Wireless Communications, 17*(6), 58–63. doi:10.1109/MWC.2010.5675779

Al Mamun, M. A., Hannan, M. A., & Hussain, A. (2014). A Novel Prototype and Simulation Model for Real Time Solid Waste Bin Monitoring System. *Jurnal Kejuruteraan, 26,* 15–19. doi:10.17576/jkukm-2014-26-02

Al-Ali, A. R., Qasaimeh, M., Mamoun, A. M., Radder, S., & Zualkernan, A. (2015). ZigBee-based irrigation system for home gardens. *Communications.*

Al-Dowaihi, D., Al-Ajlan, M., Al-Zahrani, N., Al-Quwayfili, N., al-Jwiser, N., & Kanjo, E. (2013, January). *Mbreath: Asthma monitoring system on the go.* Paper presented at the International Conference on Computer Medical Applications (ICCMA).

Alemdar, H., & Ersoy, C. (2010). Wireless sensor networks for healthcare: A survey. *Computer Networks, 54*(15), 2688–2710. doi:10.1016/j.comnet.2010.05.003

AlEnezi, A., AlMeraj, Z., & Manuel, P. (2018, April). Challenges of IoT Based Smart-Government Development. In *Green Technologies Conference (GreenTech),* 2018 (pp. 155-160). IEEE.

Al-Maaded, M., & Madi, N. K., Kahraman, R., Hodzic, A., & Ozerkan, N. G. (2012, March). An Overview of Solid Waste Management and Plastic Recycling in Qatar. *Journal of Polymers and the Environment*, *20*(1), 186–194. doi:10.100710924-011-0332-2

Almishari, S., Ababtein, N., Dash, P., & Naik, K. (2017). An Energy Efficient Real-time Vehicle Tracking System. *IEEE Pacific Rim Conference on Communications, Computers and Signal Processing*. 10.1109/PACRIM.2017.8121884

Almusalli, F. A., Zaman, N., & Rasool, R. (2017). *Energy efficient middleware: Design and development for mobile applications.* Paper presented at the 2017 19th International Conference on Advanced Communication Technology (ICACT). 10.23919/ICACT.2017.7890149

Al-Qadi, I. L., Loulizi, A., Elseifi, M., & Lahouar, S. (2004). The Virginia Smart Road: The impact of pavement instrumentation on understanding pavement performance. *Electronic Journal of the Association of Asphalt Paving Technologists*, *73*, 427–466.

Alsayid, B., Jallad, J., Dradi, M., & Al-Qasem, O. *(2013). Automatic irrigation system with pv solar tracking. Int. J Latest Trends.*

Al-Tamimi, N., Slater, N., Kayyali, R., & ElShaer, A. (in press). Perceptions by adult patients with type 1 and 2 diabetes of current and advanced technologies of blood glucose monitoring: A prospective study. *Canadian Journal of Diabetes*. PMID:30026045

Altun, K., Barshan, B., & Tunçel, O. (2010). Comparative study on classifying human activities with miniature inertial and magnetic sensors. *Pattern Recognition*, *43*(10), 3605–3620. doi:10.1016/j.patcog.2010.04.019

Alvares, D., Wieczorek, L., Raguse, B., Ladouceur, F., & Lovell, N. H. (2013). Development of nanoparticle film-based multiaxial Tactile sensors for biomedical applications. *Sensors and Actuators. A, Physical*, *196*(1), 38–47. doi:10.1016/j.sna.2013.03.021

Aly, W. H. F. (2013). Wireless Sensor Networks for Water Management that supports Differentiated Services. *International Journal of Scientific and Engineering Research.*

Amplify. (2018). *Amplifyandroid.* Available at: http://amplifyandroid.com/

Anastasova, S., Crewther, B., Bembnowicz, P., Curto, V., Ip, H. M. D., Rosa, B., & Yang, G. (2017). A wearable sensing patch for continuous sweat monitoring. *Biosensors & Bioelectronics*, *93*, 139–145. doi:10.1016/j.bios.2016.09.038 PMID:27743863

Anderson, J. G. (n.d.). Social, ethical and legal barriers to e-health. - PubMed - NCBI. Retrieved from https://www.ncbi.nlm.nih.gov/pubmed/17064955

Angeles, R. (2005). RFID technology: Supply-chain applications and implementation issues. *Information Systems Management*, *22*(1), 51–65. doi:10.1201/1078/44912.22.1.20051201/85739.7

Aniket, D., Mayur, D., Mayur, P., & Aakash, R. (2017). E-Agriculture Information Monitoring System using Data Mining. *International Journal of Advanced Research in Computer and Communication Engineering*, *6*(5).

Animaw, W., & Seyoum, Y. (2017). Increasing prevalence of diabetes mellitus in a developing country and its related factors. *PLoS One*, *12*(11), 1–11. doi:10.1371/journal.pone.0187670 PMID:29112962

Ankrum, D. R. (1992). Ivhs-smart vehicles, smart roads. Traffic Safety (Chicago), 92(3).

Anoop, I., Jain, A., Pathak, S., & Yadav, G. (2017). IOT based Smart Waste Management. *International Journal of Advanced Research in Computer and Communication Engineering, 6*(1).

Antolín, D., Medrano, N., Calvo, B., & Pérez, F. (2017). A Wearable Wireless Sensor Network for Indoor Smart Environment Monitoring in Safety Applications. *Sensors (Basel)*, *17*(2), 365. doi:10.339017020365 PMID:28216556

Appea, A. K. (2003). *Validation of FWD testing results at the Virginia Smart Road: Theoretically and by instrument responses* (Doctoral dissertation, Virginia Tech).

Arakawa, T., Kuroki, Y., Nitta, H., Toma, K., Mitsubayashi, K., Takeuchi, S., . . . Minakuchi, S. (2015, December). *Mouth guard type biosensor "cavitous sensor" for monitoring of saliva glucose with telemetry system.* Paper presented at 2015 9th International Conference on Sensing Technology, Auckland, New Zealand.

Arduino. (2018). Mega 2560 Reference Design.

Arora, Y., & Goyal, D. (2015). Big Data Technologies: Brief Overview. *International Journal of Computer Applications, 131*(9), 1-6.

Arora, Y., & Goyal, D. (2016). Big Data: A Review of analytics methods & Techniques. *Proceedings of 2nd International Conferences on Contemporary Computing and Informatics, 225-230.* IEEE. 10.1109/IC3I.2016.7917965

Årsand, E., Muzny, M., Bradway, M., Muzik, J., & Hartvigsen, G. (2015). Performance of the first combined smartwatch and smartphone diabetes diary application study. *Journal of Diabetes Science and Technology*, *9*(3), 556–563. doi:10.1177/1932296814567708 PMID:25591859

Asif, Z., & Mandviwalla, M. (2005). Integrating the supply chain with RFID: A technological and business analysis. *Communications of the Association for Information Systems*, *15*(24), 393–427.

Atallah, L., Jones, G. G., Ali, R., Leong, J. J., Lo, B., & Yang, G. (2011). Observing Recovery from Knee-Replacement Surgery by Using Wearable Sensors. *2011 International Conference on Body Sensor Networks.* 10.1109/BSN.2011.10

Atul (2016). Smart Greenhouse: The future of agriculture. Retrieved from https://www.hackster.io/synergy-flynn-9ffb33/smart-greenhouse-the-future-of-agriculture-5d0e68

Atulker, D. (2009). Soci-Economic status of Agricultural Landless labors in Indore Block of Indore District of Madhya Pradesh. (MS Thesis Jawaharlal Nehru Agriculture University). Retrieved from http://krishikosh.egranth.ac.in/bitstream/1/5810019780/1/T-82174.pdf

Atzori, L., Iera, A., & Morabito, G. (2010). The Internet of Things: A survey. *Computer Networks*, *54*(15), 2787–2805. doi:10.1016/j.comnet.2010.05.010

Auto Car Pro. (n.d.). *Economic Loss Road Accidents India Estimated.* Retrieved from http://www.autocarpro.in/news-national/economic-loss-road-accidents-india-estimated-gdp

Bahga, A., & Madiseti, V. (2017). *Internet of Things: A hands on approach.* Hyderabad: Universities press.

Balachandran, M. (2016, April 23). *Accidents on India's deadly roads cost the economy over $8 billion every year.* Retrieved from https://qz.com/689860/accidents-on-indias-deadly-roads-cost-the-economy-over-8-billio every-year/

Balasubramanian, A. (2013). Technical Report on Agro-Ecological zones of India. Retrieved from https://www.researchgate.net/publication/314206350

Banaeianjahromi, N., & Smolander, K. (2016). What do we know about the role of enterprise architecture in enterprise integration? A systematic mapping study. *Journal of Enterprise Information Management*, *29*(1), 140–164. doi:10.1108/JEIM-12-2014-0114

Bandodkar, A. J., Jia, W., Yardımcı, C., Wang, X., Ramirez, J., & Wang, J. (2014a). Tattoo-based noninvasive glucose monitoring: A proof-of-concept study. *Analytical Chemistry*, *12*(87), 394–398. PMID:25496376

Bandodkar, A. J., Molinnus, D., Mirza, O., Guinovart, T., Windmiller, J. R., Valdes-Ramirez, G., ... Wang, J. (2014b). Epidermal tattoo potentiometric sodium sensors with wireless signal transduction for continuous non-invasive sweat monitoring. *Biosensors & Bioelectronics*, *87*(1), 603–609. doi:10.1016/j.bios.2013.11.039 PMID:24333582

Bandyopadhyay, D., & Sen, J. (2011). Internet of things: Applications and challenges in technology and standardization. *Wireless Personal Communications*, *58*(1), 49–69. doi:10.100711277-011-0288-5

Bansal, R. (2003). Coming Soon to a Wal-Mart Near You. *IEEE Antennas & Propagation Magazine*, *45*(6), 105–106. doi:10.1109/MAP.2003.1282186

Bardis, N. G., Doukas, N., & Ntaikos, K. (2008). Design and Development of a Secure Military Communication based on AES Prototype Crypto Algorithm and Advanced Key Management Scheme. *WSEAS Transactions on Information Science & Applications.*

Barreiros, E., Almeida, A., Saraiva, J., & Soares, S. (2011, September). *A systematic mapping study on software engineering testbeds.* Paper presented at 2011 International Symposium on Empirical Software Engineering and Measurement, Alberta, Canada. 10.1109/ESEM.2011.19

Belli, L., Cirani, S., Davoli, L., Gorrieri, A., Mancin, M., Picone, M., & Ferrari, G. (2015). Design and Deployment of an IoT Application-Oriented Testbed. *Computer*, *48*(9), 32–40. doi:10.1109/MC.2015.253

Bengio, Y., Courville, A., & Vincent, P. (2013). Representation learning: A review and new perspectives. *IEEE Transactions on Pattern Analysis and Machine Intelligence*, *35*(8), 1798–1828. doi:10.1109/TPAMI.2013.50 PMID:23787338

Berard, A. J., Mentzer, J. L., & Nixon, D. C. (1996). U.S. Patent No. 5,515,043. Washington, DC: U.S. Patent and Trademark Office.

Bergmo, T. S. (2010). Economic evaluation in telemedicine – still room for improvement. *Journal of Telemedicine and Telecare*, *16*(5), 229–231. doi:10.1258/jtt.2010.009008 PMID:20501629

Bertolotti, G. M., Cristiani, A. M., Colagiorgio, P., Romano, F., Bassani, E., Caramia, N., & Ramat, S. (2016). A Wearable and Modular Inertial Unit for Measuring Limb Movements and Balance Control Abilities. *IEEE Sensors Journal*, *16*(3), 790–797. doi:10.1109/JSEN.2015.2489381

Bhara, R., Gupta, A., & Jaiswal, M. (2017). An Enhanced AES Algorithm Using Cascading Method On 400 Bits Key Size Used In Enhancing The Safety Of Next Generation Internet Of Things (IoT). *International Conference on Computing, Communication and Automation (ICCCA2017).*

Biao, L. Ai qun, H., & Zhong-Yuan, Q. (2006, June). Trends and Brief Comments on Anti-collision Techniques in Radio Frequency Identification System. In 6th International Conference on ITS Telecommunications Proceedings (pp. 241-245). IEEE.

Bichler, M. (2006). Design science in information systems research. *Wirtschaftsinformatik*, *48*(2), 133–135. doi:10.100711576-006-0028-8

Bircher, S., Skou, N., Jensen, K. H., Walker, J. P., & Rasmussen, L. (2012). A soil moisture and temperature network for SMOS validation in Western Denmark. In 2015 International Conference Computing, Communication & Automation (ICCCA), Denmark. *Hydrology and Earth System Sciences*, *16*(5), 1445–1463. doi:10.5194/hess-16-1445-2012

Blobel, B. G. (2007). Educational Challenge of Health Information Systems' Interoperability. *Methods of Information in Medicine*, *46*(01), 52–56. doi:10.1055-0038-1628132 PMID:17224981

Blue Stream Consultancy. (2018). *Smart healthcare.* Available at http://www.bluestream.sg/smart-healthcare

Bojanova, I., Hurlburt, G., & Voas, J. (2014). Imagineering an Internet of Anything. *Computer*, *47*(6), 72–77. doi:10.1109/MC.2014.150

Bolinder, J., Antuna, R., Geelhoed-Duijvestijn, P., Kröger, J., & Weitgasser, R. (2016). Novel glucose-sensing technology and hypoglycaemia in type 1 diabetes: A multicentre, non-masked, randomised controlled trial. *Lancet*, *388*(10057), 2254–2263. doi:10.1016/S0140-6736(16)31535-5 PMID:27634581

Bonomi, F., Milito, R., Zhu, J., & Addepalli, S. (2012). Fog computing and its role in the internet of things. *Proceedings of the first edition of the MCC workshop on Mobile cloud computing - MCC '12*. 10.1145/2342509.2342513

Bonuccelli, M. A., Lonetti, F., & Martelli, F. (2006, June). Tree slotted ALOHA: a new protocol for tag identification in RFID networks. In *Proceedings of IEEE International Symposium on a World of Wireless, Mobile and Multimedia Networks* (pp. 603-608). IEEE Computer Society.

Bormann, C., Ersue, M., & Keranen, A. (2014). *Terminology for Constrained-Node Networks*. doi:10.17487/rfc7228

Boudriga, N. (2009). Security of mobile communications. Boca Raton, FL: Auerbach Publications. doi:10.1201/9780849379420

Brazeal, M. (2009). *RFID: Improving the Customer Experience*. New York, NY: Paramount Marketing Publishing.

Brennan, N., Barnes, R., Calnan, M., Corrigan, O., Dieppe, P., & Entwistle, V. (2013). Trust in the health-care provider–patient relationship: A systematic mapping review of the evidence base. *International Journal for Quality in Health Care*, *25*(6), 682–688. doi:10.1093/intqhc/mzt063 PMID:24068242

Briscoe, G., & De Wilde, P. (2006). Digital ecosystems: Evolving service-orientated architectures. Paper presented at *1st international conference on Bio inspired models of network, information and computing systems,* Madonna di Campiglio, Trenito, Italy.

Brown, A. (2015). *Technology that is Flexible, Sticky, and Smart = Wearable Patches*. Retrieved from https://www.wearable-technologies.com/2015/08/technology-that-is-flexible-sticky-and-smart-wearable-patches/

Brown, M. (2018). *Smart Farming—Automated and Connected Agriculture,* engineering.com. Retrieved from https://www.engineering.com/DesignerEdge/DesignerEdgeArticles/ArticleID/16653/Smart-FarmingAutomated-and-Connected-Agriculture.aspx

Bruen, D., Delaney, C., Florea, L., & Diamond, D. (2017). Glucose sensing for diabetes monitoring: Recent developments. *Sensors (Basel)*, *17*(8), 1–21. doi:10.339017081866 PMID:28805693

Bui, N., & Zorzi, M. (2011). Health care applications. *Proceedings of the 4th International Symposium on Applied Sciences in Biomedical and Communication Technologies - ISABEL '11*. 10.1145/2093698.2093829

Business Insider Intelligence. (2015, May 21). *THE WEARABLES REPORT: Growth trends, consumer attitudes, and why smartwatches will dominate*. Retrieved from http://bit.ly/mpo0617tp3

Butean, A., David, A., Buduleci, C., & Daian, A. (2015). Auxilum Medicine: A Cloud Based Platform for Real-Time Monitoring Medical Devices. *2015 20th International Conference on Control Systems and Computer Science*. doi:10.1109/cscs.2015.135

Caduff, A., Zanon, M., Zakharov, P., Mueller, M., Talary, M., Krebs, A., ... Donath, M. (2018). First experiences with a wearable multisensor in an outpatient glucose monitoring study, Part I: The users' view. *Journal of Diabetes Science and Technology*, *12*(3), 562–568. doi:10.1177/1932296817750932 PMID:29332423

Cappiello, C., Francalanci, C., & Pernici, B. (2004). Time-related factors of data quality in multichannel information systems, *20*(3), 71-91.

Cappon, G., Vettoretti, M., Marturano, F., Facchinetti, A., & Sparacino, G. (2018). A neural-network-based approach to personalize insulin bolus calculation using continuous glucose monitoring. *Journal of Diabetes Science and Technology*, *12*(2), 265–272. doi:10.1177/1932296818759558 PMID:29493356

Cardiovascular diseases. (n.d.). Retrieved from http:/ /www.searo.who.int/ india/ topics/cardiovascular_diseases

Cha, J. R., & Kim, J. H. (2005, July). Novel anti-collision algorithms for fast object identification in RFID system. In *ICPADS (2)* (pp. 63-67).

Chakraborty, P., & Chakrabarty, D. K. (2008). A brief survey of computerized expert systems for crop protection being used in India. *Progress in Natural Science, 18,* 469-473.

Chakraborty, P., & Chakrabarty, D. K. (2008). *An example of agricultural expert systems being used in India. Georgian Electronic Scientific Journal of Computer Science and Telecommunications, 1,* 15.

Chang, E., & West, M. (2006). Digital Ecosystems a Generation of the Collaborative Environment. Paper presented at Eighth International Conference on Information Integration and Webbased Applications, Yogyakarta, Java, Indonesia.

Chawhan, S., Ghodichor, P., & Paunikar, K. (2017). Survey on Data Mining for Increasing Agriculture Productivity. *International Journal for Research in Applied Science and Engineering Technology*, *5*(3).

Checkoway, S., McCoy, D., Anderson, D., & Kantor, B. (2011, August). *Comprehensive Experimental Analyses of Automotive Attack Surfaces*. Retrieved from https://www.researchgate.net/publication/259753269_Comprehensive_Experimental_Analyses_of_Automotive_Attack_Surfaces

Chen, X., & Zong, Z. (2016). *Android app energy efficiency: The impact of language, runtime, compiler, and implementation.* Paper presented at the 2016 IEEE International Conferences on Big Data and Cloud Computing, Social Computing and Networking, Sustainable Computing and Communications. 10.1109/BDCloud-SocialCom-SustainCom.2016.77

Ching, K. W., & Singh, M. M. WEARABLE TECHNOLOGY DEVICES SECURITY AND PRIVACY VULNERABILITY ANALYSIS. Retrieved from https://www.scribd.com/document/315235407/WEARABLE-TECHNOLOGY-DEVICES-SECURITY-AND-PRIVACY-VULNERABILITY-ANALYSIS

Chinrungrueng, J., & Sunantachaikul, U. (2007). Smart Parking: an Application of optical Wireless Sensor Network. *IEEE Proceedings of the 2007 International Symposium on Applications and the Internet Workshops (SAINTW'07).*

Choi, T. M., Yeunge, W. K., Cheng, T. C. E., & Yue, X. (2018). Optimal Scheduling, Coordination, and the Value of RFID technology in Garment Manufacturing Supply Chains. *IEEE Transactions on Engineering Management*, *65*(1), 1. doi:10.1109/TEM.2017.2739799

Chuang, M. L., & Shaw, W. H. (2005). How RFID will Impact Supply Chain Networks, *IEEE International Engineering Management Conference*, Newfoundland, Canada, 231-235.

Chu, M. K., Iguchi, S., Miyajima, K., Arakawa, T., Kudo, H., & Mitsubayashi, K. (2011). Development of a soft contact-lens biosensor for in-vivo tear glucose monitoring. In *Proceedings of 5th European Conference of the International Federation for Medical and Biological Engineering* (vol 37. pp. 1007-1010). Berlin, Germany: Springer 10.1007/978-3-642-23508-5_262

Cicero, M. X., Walsh, B., Solad, Y., Whitfill, T., Paesano, G., Kim, K., ... Cone, D. C. (2015). Do You See What I See? Insights from Using Google Glass for Disaster Telemedicine Triage. *Prehospital and Disaster Medicine*, *30*(01), 4–8. doi:10.1017/S1049023X1400140X PMID:25571779

Cirani, S., Davoli, L., Ferrari, G., Leone, R., Medagliani, P., Picone, M., & Veltri, L. (2014). A Scalable and Self-Configuring Architecture for Service Discovery in the Internet of Things. *IEEE Internet of Things Journal*, *1*(5), 508–521. doi:10.1109/JIOT.2014.2358296

Copacino, W., & Anderson, D. (2003). Connecting with the Bottom Line: A Global Study of Supply Chain Leadership and its Contribution to the High Performance Business, *Accenture,* p.1.

Coughlin, J. F. (2010). Disruptive Demographics, Design, and the Future of Everyday Environments. *Design Management Review*, *18*(2), 53–59. doi:10.1111/j.1948-7169.2007.tb00083.x

Crisp, B. R. (2017). *The Routledge handbook of religion, spirituality and social work.* UK: Taylor & Francis. doi:10.4324/9781315679853

D'Amato, G., Vitale, C., Molino, A., Stanziola, A., Sanduzzi, A., Vatrella, A., & D'Amato, M. (2016). Asthma-related deaths. *Multidisciplinary Respiratory Medicine*, *11*(1), 37. doi:10.118640248-016-0073-0 PMID:27752310

Daemen, J., & Rijmen, V. (2013). *The design of rijndael: AES.* Springer Science and Business Media. *FIPS publication 197.* (2001, Nov 26). Retrieved from Advanced Encryption Standard: https://nvlpubs.nist.gov/nistpubs/fips/nist.fips.197.pdf

Daemen, J., & Rijmen, V. (1999). *AES proposal: Rijndael.* AES Algorithm Submission.

Daniel, K. M., Cason, C. L., & Ferrell, S. (2009). Emerging Technologies to Enhance the Safety of Older People in Their Homes. *Geriatric Nursing*, *30*(6), 384–389. doi:10.1016/j.gerinurse.2009.08.010 PMID:19963147

Data.gov. (n.d.). Retrieved from http://www.data.gov

dataworldbank.org. (n.d.). Retrieved from http://www.dataworldbank.org

Datta, S. K., Bonnet, C., & Nikaein, N. (2014). *Self-adaptive battery and context aware mobile application development.* Paper presented at the IWCMC. 10.1109/IWCMC.2014.6906452

Deepiga, M. T., & Sivasankari, A. (2015). Smart water monitoring system using wireless sensor network at Home/Office. *International Research Journal of Engineering and Technology*, *2*(4), 1305–1314.

Del Favero, S., Place, J., Kropff, J., Messori, M., Keith-Hynes, P., Visentin, R., & Di Palma, F. (2015). Multicenter outpatient dinner/overnight reduction of hypoglycemia and increased time of glucose in target with a wearable artificial pancreas using modular model predictive control in adults with type 1 diabetes. *Diabetes, Obesity & Metabolism*, *17*(5), 468–476. doi:10.1111/dom.12440 PMID:25600304

Delen, D., Hardgrave, B. C., & Sharda, R. (2007). RFID for better supply management through enhanced information visibility. *Production and Operations Management*, *16*(5), 613–624. doi:10.1111/j.1937-5956.2007.tb00284.x

Deloitte. (2018). *A journey towards smart health. The impact of digitalization on patient experience.* Available at https://www2.deloitte.com

Dempsey, J. S. (2010). Introduction to private security. Boston, MA: Cengage Learning.

Dicheva, D., Dichev, C., Agre, G., & Angelova, G. (2015). Gamification in education: A systematic mapping study. *Journal of Educational Technology & Society*, *18*(3), 75–88.

Dickens, B., & Cook, R. (2006). Legal and ethical issues in telemedicine and robotics. *International Journal of Gynaecology and Obstetrics: the Official Organ of the International Federation of Gynaecology and Obstetrics*, *94*(1), 73–78. doi:10.1016/j.ijgo.2006.04.023 PMID:16777109

Ding, Y., Jin, Y., Ren, L., & Hao, K. (2013). An Intelligent Self-Organization Scheme for the Internet of Things. *IEEE Computational Intelligence Magazine*, *8*(3), 41–53. doi:10.1109/MCI.2013.2264251

Dirkx, E., Hager, C., Tadross, M., Bethune, S., & Curtis, B. (2008). Climate change vulnerability and adaptation assessment Namibia. Final Report. March.

Divya, J., & Sreekumar, K. (2014). A Survey on Expert System in Agriculture. *International Journal of Computer Science and Information Technology* *5*(6). 7860-7864.

Djeddou, M., Khelladi, R., & Benssalah, M. (2013). Improved RFID anti-collision algorithm. *International Journal of Electronics and Communications*, *67*(3), 256–262. doi:10.1016/j.aeue.2012.08.009

Do, Q., Tran, S., & Robinson, K. (2015, December). Big data and mHealth drive asthma self-management. Paper presented at International Conference on Computational Science and Computational Intelligence, Las Vegas, Nevada. 10.1109/CSCI.2015.129

Doukas, C., & Maglogiannis, I. (2012). Bringing IoT and Cloud Computing towards Pervasive Healthcare. *2012 Sixth International Conference on Innovative Mobile and Internet Services in Ubiquitous Computing*. 10.1109/IMIS.2012.26

Dufour, J. Y. (Ed.). (2012). *Intelligent video surveillance systems*. Hoboken, NJ: John Wiley & Sons. doi:10.1002/9781118577851

Du, H. T., Xu, K. L., & Wang, W. L. (2006). An anti-collision algorithm based on binary-tree searching of backtracking. *Journal of Yunnan University*, *28*, 133–136.

Elsherif, M., Hasan, M. U., Yetisen, A. K., & Butt, H. (2018). Wearable contact lens biosensors for continuous glucose monitoring using smartphones. *ACS Nano*, *12*(6), 5452–5462. doi:10.1021/acsnano.8b00829 PMID:29750502

Eltaieb, A. A. M., & Min, Z. J. (2015). Automatic Water Level Control System. *International Journal of Science and Research*, *4*(12), 1505–1509.

Endeavour Partners. (n.d.). Retrieved from https://endeavour.partners

EPC. (2019). Available at https://www.gs1.org/epcglobal

Eramo, L. (2018). *Three ways asthma treatment is getting connected.* Available at https://www.futurehealthindex.com

Essén, A., & Conrick, M. (2008). New e-service development in the homecare sector: Beyond implementing a radical technology. *International Journal of Medical Informatics*, *77*(10), 679–688. doi:10.1016/j.ijmedinf.2008.02.001 PMID:18514021

Fatima, A. (2011). E-Banking Security Issues-Is There A Solution in Biometrics? *Journal of Internet Banking and Commerce*, *16*(2), 1.

Faustine, A., Mvuma, A. N., Mongi, H. J., Gabriel, M. C., Tenge, A. J., & Kucel, S. B. (2014). Wireless sensor networks for water quality monitoring and control within Lake Victoria basin: Prototype development. *Wireless Sensor Network*, *6*(12), 281–290. doi:10.4236/wsn.2014.612027

Ferraro, R., & Aktihanoglu, M. (2011). *Location-aware applications*. Shelter Island, NY: Manning Publications. Retrieved from https://www.linkedin.com/pulse/what-next-iot-ahmed-banafa?trk=mp-author-card

Ferrer, G., Dew, N., & Apte, U. (2010). When is RFID right for your service? *International Journal of Production Economics*, *124*(2), 414–425. doi:10.1016/j.ijpe.2009.12.004

Ferris, B., Watkins, K., & Borning, A. (2010, April). OneBusAway: results from providing real-time arrival information for public transit. *Proc. SIGCHI Conf. Hum. Comput. Interact.,* (pp. 1807–1816). ACM. 10.1145/1753326.1753597

Fettweis, G. P. (2014). The Tactile Internet: Applications and Challenges. *IEEE Vehicular Technology Magazine*, *9*(1), 64–70. doi:10.1109/MVT.2013.2295069

Fichman, R. G. (2004). Going Beyond the Dominating for Information Technology Innovation Research: Emerging Concepts and Methods. *Journal of the Association for Information Systems*, *5*(8), 314–355. doi:10.17705/1jais.00054

Finkenzeller, K. (2003). *RFID Handbook: Fundamentals and Applications in Contactless Smart Cards and Identification* (2nd ed.). New York, NY: John Wiley. doi:10.1002/0470868023

Foerster, F., Smeja, M., & Fahrenberg, J. (1999). Detection of posture and motion by accelerometry: A validation study in ambulatory monitoring. *Computers in Human Behavior*, *15*(5), 571–583. doi:10.1016/S0747-5632(99)00037-0

Food and Agriculture Organization of the United Nations. (2015). Retrieved from http://www.fao.org/nr/water/aquastat/countries_regions/IND

Forouzan, B. A. (2007). *Cryptography and network security, special indian edition.* Tata McGraw Hill Pvt Ltd.

Frodigh, M., Johansson, P., & Larsson, P. (2000). Wireless ad hoc networking: the art of networking without a network. *Ericsson review, 4*(4), 249.

Fuentes, J. S., Gonzalez-Manzano, L., Solanas, A., & Veseli, F. (2018). Attribute-Based Credentials for Privacy-Aware Smart Health Services in IoT-Based Smart Cities. *Computer*, *51*(7), 44–53. doi:10.1109/MC.2018.3011042

Future Market Insights. (n.d.). *Wearable Medical Devices Market - Global Industry Analysis. Size and Forecast, 2016 to 2026.* Retrieved from http://bit.ly/mpo0617tp4

Gaillard, M. (2017). "Invasive" and "Non-invasive" technologies in neuroscience communication. *Bioéthique Online*, *6*(11), 1–10.

Ganesan, K., & Vignesh, K. (2007). Automated parking slot allocation using RFID technology. (pp. 1-4). *9th International Symposium on Signal Processing and Its Applications.* 10.1109/ISSPA.2007.4555296

Gaukler, G. M., & Hausman, W. H. (2008). RFID in a mixed-model automotive assembly operation: Process and quality cost saving. *IIE Transactions*, *40*(11), 1083–1096. doi:10.1080/07408170802167654

Gayatri, M. K., Jayasakthi, J., & G.S., & Mala, G. A. (2015). Providing Smart Agriculture Solutions to Farmers for better yielding using IOT. *IEEE International Conference on Technology Innovations in ICT for Agriculture and Rural Development.* (pp 40-43). IEEE.

Geetha, M. C. S. (2015). A survey on Data mining Techniques in Agriculture. *International Journal of Innovative Research in Computer and Communication Engineering*, *3*(2), 887-892.

Geiger, G., Werner, T., & Matko, D. (2003): Leak Detection and Locating – A Survey. *35th Annual PSIG Meeting*, Bern, Switzerland. Pipeline Simulation Interest Group.

Geng, Y., & Cassandras, C. G. (2011, October). A new "smart parking" system based on optimal resource allocation and reservations. In *2011 14th International IEEE Conference on Intelligent Transportation Systems (ITSC),* (pp. 979-984). IEEE.

Geng, Y., & Cassandras, C. G. (2013). New" Smart Parking" System Based on Resource Allocation and Reservations. *IEEE Transactions on Intelligent Transportation Systems*, *14*(3), 1129–1139. doi:10.1109/TITS.2013.2252428

Gershenfeld, N., Krikorian, R., & Cohen, D. (2004). The Internet of Things. *Scientific American, 291*(4), 76–81. doi:10.1038cientificamerican1004-76 PMID:15487673

Ghosh, A. (2016). Intelligent appliances controller using Raspberry Pi, *2016 IEEE 7th Annual Information Technology, Electronics and Mobile Communication Conference (IEMCON)*, Vancouver, Canada, pp. 1-5.

Giffinger, R., Fertner, C., Kramar, H., Kalasek, R., Pichler-Milanović, N., & Meijers, E. (2007). Smart Cities: Ranking of European Medium-Sized Cities. Vienna, Austria: Centre of Regional Science (SRF), Vienna University Technology.

GitHub. (2018). ADC Values Inaccurate and different from NodeMCU Issue #2672 esp8266/Arduino, [online]. Available at https://github.com/esp8266/Arduino/issues/2672

GitHub. (2018). esp8266/Arduino, [online]. Available at https://github.com/esp8266/Arduino

Global Initiative for Asthma (GINA). (2014). *The Global Strategy for Asthma Management and Prevention.* Available at www.ginasthma.org

Global Position System. Retrieved from https://www.gps.gov/applications/agriculture/

Glouche, Y., & Couderc, P. (2013). A Smart Waste Management with Self-Describing objects. *The Second International Conference on Smart Systems, Devices and Technologies (SMART'13)*, Rome, Italy.

Glover, B., & Bhatt, H. (2006). RFID Essentials. Sebastopol, CA: O'Reilly Publishing.

Gogate, U., & Bakal, J. (2018). Healthcare monitoring system based on wireless sensor network for cardiac patients. *Biomedical & Pharmacology Journal, 11*(3), 1681–1688. doi:10.13005/bpj/1537

Goosem, M. (2009). 36 RETHINKING ROAD ECOLOGY. Living in a Dynamic Tropical Forest Landscape, 445.

Grandview Market research report. (2018). *Smart Cities Market Size, Share & Trends Analysis Report By Application (Education, Governance, Buildings, Mobility, Healthcare, Utilities), By Component (Services, Solutions), And Segment Forecasts 2018–2025.* Retrieved from https://www.grandviewresearch.com/industry-analysis/smart-cities-market

Grant, M. J., & Booth, A. (2009). A typology of reviews: An analysis of 14 review types and associated methodologies. *Health Information and Libraries Journal, 26*(2), 91–108. doi:10.1111/j.1471-1842.2009.00848.x PMID:19490148

Green Revolution. Retrieved from http://www.economicsdiscussion.net/essays/green-revolution-essays/essay-on-green-revolution-in-india/17559

Greengard, S. (2015). *Internet Of Things.* London: MIT Press. doi:10.7551/mitpress/10277.001.0001

Guarnieri, M., & Balmes, J. R. (2014). Outdoor air pollution and asthma. *Lancet, 383*(9928), 1581–1592. doi:10.1016/S0140-6736(14)60617-6 PMID:24792855

Gubbi, J., Buyya, R., Marusic, S., & Palaniswami, M. (2013). Internet of Things (IoT): A vision, architectural elements, and future directions. *Future Generation Computer Systems, 29*(7), 1645–1660. doi:10.1016/j.future.2013.01.010

Guinard, D., & Trifa, V. (2016). Building the web of things: with examples in node. js and raspberry pi. Shelter Island, NY: Manning Publications

Guinard, D., Trifa, V., Mattern, F., & Wilde, E. (2011). From the internet of things to the web of things: Resource-oriented architecture and best practices. In *Architecting the Internet of things* (pp. 97–129). Berlin, Germany: Springer. doi:10.1007/978-3-642-19157-2_5

Guo, Y., Wang, C., & Chen, X. (2017). Understanding Application-Battery Interactions on Smartphones: A Large-Scale Empirical Study. *IEEE Access: Practical Innovations, Open Solutions, 5*, 13387–13400. doi:10.1109/ACCESS.2017.2728620

Gupta, M. D. (2017, September 6). *Road accidents killed 17 people every hour in India in 2016, Delhi most unsafe.* Retrieved from http://www.hindustantimes.com/india-news/road-accidents-claimed-nearly-400-lives-every-day-in-india-in-2016/story-7DlmtdnvMYLLZVGxXKOaJN.html

Gupta, N., Kumar, S., Kumar, M., Mathur, D., & Jilani, E. (2016). *Wireless Water Level Controller Using Zigbee, V*(Iv), 79–81.

Hanche, S.C., Munot, P., Bagal, P., Sonawane, K. & Pise, P. (2013). Automated Vehicle Parking System using RFID, *ITSI Transactions on Electrical and Electronics Engineering (ITSI-TEEE), 1(2).*

Hari Baabu, V., Senthil Kumar, G., Deb, P., & Rai, A. (2016). Smart Parking Assist System using Internet of Things (2016). *International Journal of Control Theory and Applications, 9*(40), 2016.

Hartke, K. (2015). Observing Resources in the Constrained Application Protocol (CoAP). doi:10.17487/rfc7641

He, S., Liu, Y., & Zhou, H. (2015). *Optimizing smartphone power consumption through dynamic resolution scaling.* Paper presented at the Proceedings of the 21st Annual International Conference on Mobile Computing and Networking. 10.1145/2789168.2790117

Heggestuen, J. (2013, Dec. 15). One In Every 5 People In The World Own A Smartphone, One In Every 17 Own A Tablet [CHART]. Retrieved from http://www.businessinsider.com/smartphone-and-tablet-penetration-2013-10?IR=T

Heinemann, L. (2008). Finger pricking and pain: A never ending story. *Journal of Diabetes Science and Technology, 2*(5), 919–921. doi:10.1177/193229680800200526 PMID:19885279

Henning, A., Lauko, J., Grabmaier, A., & Wilson, C. (2014). Wireless tear glucose sensor. *Procedia Engineering, 87*, 66–69. doi:10.1016/j.proeng.2014.11.267

Hertleer, C., Tronquo, A., Rogier, H., Vallozzi, L., & Van Langenhove, L. (2007). An aperture-coupled patch antenna for integration into Wearable textile systems. *IEEE Antennas and Wireless Propagation Letters, 6*, 392–395. doi:10.1109/LAWP.2007.903498

Hevner, A. R. (2007). A three cycle view of design science research. *Scandinavian Journal of Information Systems, 19*(2), 4.

Hewett, T., Baecker, R., Card, S., Carey, T., Gasen, J., Mantei, M., ... Verplank, W. (1992). *ACM SIHCHI curricula for human–computer interaction. Association for Computing Machinery*. New York, NY: ACM. doi:10.1145/2594128

Hindy, J. (2018). *5 best battery saver apps for Android and other ways too!* Retrieved from https://www.androidauthority.com/best-battery-saver-android-apps-266980/

Hock, G. (2005). Highly available computer-based automation systems. *Computing & Control Engineering Journal, 16*(3), 17–20. doi:10.1049/cce:20050302

Ho, K., Borycki, E., Kushniruk, A., Juhra, C., & Hernandez, M. (2016). *The health perspective in using digital media for health and wellness. 2016 Digital Media Industry & Academic Forum.* DMIAF. doi:10.1109/dmiaf.2016.7574895

Holcombe, W. T. (2014). *U.S. Patent No. 8,754,775.* Washington, DC: U.S. Patent and Trademark Office.

Hole, C. E. (2017). *Analysis of Security for IoT.*

Hong, T. P., & Soh, C. Jaafar, & Ishak (2013). Real Time Monitoring System for Parking Space Management Services. System, Process & Control (ICSPC), IEEE.

Huang, C. T., Shen, C. L., Tang, C. F., & Chang, S. H. (2008). A wearable yarn-based piezo-resistive sensor. *Sensors and Actuators. A, Physical, 141*(2), 396–403. doi:10.1016/j.sna.2007.10.069

Hukeri, M. P. A., & Ghewari, M. P. (2017). Review paper on IoT based technology. AMGOI, Maharashtra, India.

Hung, K., & Zhang, Y. T. (2002, October). Usage of Bluetooth/sup TM/in wireless sensors for tele-healthcare. In *Proceedings of the Second Joint 24th Annual Conference and the Annual Fall Meeting of the Biomedical Engineering Society EMBS/BMES Conference, 3*(pp. 1881-1882). IEEE.

Hush, D. R., & Wood, C. (1998, August). Analysis of tree algorithms for RFID arbitration. In *Proceeding of IEEE International Symposium on Information Theory,* 107. IEEE.

IBGE. (2010). Census 2010 of the Brazilian Institute of Geography and Statistics.

Idris, M. Y. I., Leng, Y. Y., Tamil, E. M., Noor, N. M., & Razak, Z. (2009). Car park system: A review of smart parking system and its technology. *Information Technology Journal*, *8*(2), 101–113. doi:10.3923/itj.2009.101.113

International Parking & Mobility Institute. (n.d.). Retrieved from http://www.parking.org/

Isala, S. (2016). *The impact of green schemes on the livelihood of communities in the Kavango region, Namibia* (Doctoral dissertation, JKUAT).

Ishikawa, T., Hyodo, Y., Miyashita, K., Yoshifuji, K., Komoriya, Y., & Imai, Y. (2017). Wearable Motion Tolerant PPG Sensor for Instant Heart Rate in Daily Activity. *Proceedings of the 10th International Joint Conference on Biomedical Engineering Systems and Technologies*. 10.5220/0006109901260133

Iyawa, G. E., Herselman, M., & Botha, A. (2016a). Digital health innovation ecosystems: From systematic literature to conceptual framework. *Procedia Computer Science*, *100*, 244–252. doi:10.1016/j.procs.2016.09.149

Iyawa, G. E., Herselman, M., & Botha, A. (2016b). Identifying and defining the terms related to a Digital Health Innovation Ecosystem. In M. Herselman, & A. Botha (Eds.), *Strategies, Approaches and Experiences: Towards Building a South African Digital Health Innovation Ecosystem*. Pretoria, South Africa: CSIR.

Jamkar, R. G., & Chile, R. H. (2004). Microcontroller based Temperature Indicator and Controller. *Journal of Instrument. Society of India*, *34*(3), 180–186.

Jara, A. J., Alcolea, A. F., Zamora, M. A., Skarmeta, A. F., & Alsaedy, M. (2010). *Drugs interaction checker based on IoT. 2010 Internet of Things*. IOT. doi:10.1109/iot.2010.5678458

Javed, Y., Khan, A. S., Qahar, A., & Abdullah, J. (2018). Preventing DoS Attacks in IoT Using AES.

Jelicic, V., Magno, M., Brunelli, D., Paci, G., & Benini, L. (2013). Context-Adaptive Multimodal Wireless Sensor Network for Energy-Efficient Gas Monitoring. *IEEE Sensors Journal*, *13*(1), 328–338. doi:10.1109/JSEN.2012.2215733

Jia, L., Hu, Z., Song, Y., & Luo, Z. (2012, March). Optimal siting and sizing of electric vehicle charging stations. In *2012 IEEE International Electric Vehicle Conference (IEVC),* (pp. 1-6). IEEE. 10.1109/IEVC.2012.6183283

Jiang, L. F., Lu, G. Z., & Xin, Y. W. (2007). Research on anti-collision algorithm in radio frequency identification system. *Computer Engineering and Applications*, *15*, 29–32.

Jin, M., Bekiaris-Liberis, N., Weekly, K., Spanos, C. J., & Bayen, A. M. (2018). Occupancy detection via environmental sensing. *IEEE Transactions on Automation Science and Engineering*, *15*(2), 443–455. doi:10.1109/TASE.2016.2619720

Jin, M., Jia, R., & Spanos, C. J. (2017). Virtual occupancy sensing: Using smart meters to indicate your presence. *IEEE Transactions on Mobile Computing*, *16*(11), 3264–3277. doi:10.1109/TMC.2017.2684806

Johari, A., Wahab, M. H. A., Latif, N. S. A., Ayob, M. E., Ayob, M. I., Ayob, M. A., & Mohd, M. N. H. (2011). Tank water level monitoring system using GSM network. *International Journal of Computer Science and Information Technologies*, *2*(3), 1114–1115.

Johnson, D. (2002). RFID tags improve tracking, quality on Ford line in Mexico. *Control Engineering*, *49*(11), 16.

Jones, P., Clarke-Hill, C., Shears, P., Comfort, D., & Hillier, D. (2004a). Radio frequency identification in the UK: Opportunities and challenges. *International Journal of Retail & Distribution Management*, *32*(3), 164–171. doi:10.1108/09590550410524957

Jones, P., Clarke-Hill, C., Shears, P., Hillier, D., & Comfort, D. (2004b). Radio frequency identification and privacy and public policy issues. *Management Research News*, *27*(8/9), 46–56. doi:10.1108/01409170410784563

Juels, A., Rivest, R. L., & Szydlo, M. (2003, October). The blocker tag: selective blocking of RFID tags for consumer privacy. *Proceedings of the 8th ACM Conference on Computer and Communications Security,* ACM Press, Washington, DC, 103-110. 10.1145/948109.948126

Juhlin, O. (1994). *Information Technology hits the Automobile: rethinking road traffic as social interaction. Changing Large Technical Systems* (J. Summerton, Ed.). San Francisco, CA: Westview Press.

Kaewkannate, K., & Kim, S. (2018). *The Comparison of Wearable Fitness Devices*. Wearable Technologies. doi:10.5772/intechopen.76967

Kak, A. (2018). Lecture 8 AES: The Advanced Encryption Standard. In Lecture Notes on "Computer and Network Security", Purdue University.

Karkkainen, M. (2003). Increasing efficiency in the supply chain for short shelf life good using RFID tagging. *International Journal of Retail & Distribution Management*, *31*(10), 529–536. doi:10.1108/09590550310497058

Karpinski, M., Senart, A., & Cahill, V. (2006, March). Sensor networks for smart roads. In *Proceedings of the 4th annual IEEE international conference on Pervasive Computing and Communications Workshops* (pp. 306-310). IEEE.

Kascheev, N., Kozyrev, O., Leykin, M., & Vanyagin, A. (2017, September). *Non-invasive monitoring of blood glucose by means of wearable tracking technology*. Paper presented at 2017 IEEE East-West Design & Test Symposium, Novi Sad, Serbia. 10.1109/EWDTS.2017.8110137

Kasliwal Manasi, H., & Suryawanshi Smithkumar, B. (2016). A Novel approach to Garbage Management Using Internet of Things for smart cities. *International Journal of Current Trends in Engineering & Research, 2,* 348-53.

Kassem, A., Hamad, M., El-Moucary, C., Neghawi, E., Jaoude, G. B., & Merhej, C. (2013, September). *Asthma Care Apps*. Paper presented in Tripoli, Lebanon.

Katoch, M. (2017). *5 Farmers Who Prove Smart Agriculture Can Make you Rich.* Retrieved from https://www.thebetterindia.com/125477/kisan-diwas-successful-farmers-lucrative-business

Kenmotsu, M., Sun, W., Shibata, N., Yasumoto, K., & Mi-noru, I. (2012). Parking Navigation for alleviating Congestion in Multilevel Parking Facility. *Vehicular Technology Conference (VTC Fall). IEEE.* 10.1109/VTCFall.2012.6398911

Kennedy, T., Fink, P., Chu, A., & Studor, G. (2007). Potential space applications for body-centric wireless and e-textile antennas. In *Proc. IET Seminar Antennas and Propagation for Body-Centric Wireless Communications*, London, U.K., pp.77–83. 10.1049/ic:20070551

Khambra, D., & Dabas, P. (2017). *Secure Data Transmission using AES in IoT. International Journal of Application or Innovation in Engineering & Management*. IJAIEM.

Kianpisheh, A., Mustaffa, N., Limtrairut, P., & Keikhosrokiani, P. (2012). Smart parking system (SPS) architecture using ultrasonic detector. *International Journal of Software Engineering and Its Applications*, *6*(3), 55–58.

Kikidis, D., Konstantinos, V., Tzovaras, D., & Usmani, O. S. (2016). The digital asthma patient: The history and future of inhaler based health monitoring devices. *Journal of Aerosol Medicine and Pulmonary Drug Delivery*, *29*(3), 219–232. doi:10.1089/jamp.2015.1267 PMID:26919553

Kim, C., Yang, K. H., & Kim, J. (2008). A strategy for third-party logistics systems: A case analysis using the blue ocean strategy. *Omega*, *36*(4), 522–534. doi:10.1016/j.omega.2006.11.011

Kim, E. K., Kwak, S. H., Baek, S., Lee, S. L., Jang, H. C., Park, K. S., & Cho, Y. M. (2016). Feasibility of a patient-centered, smartphone-based, diabetes care system: A pilot study. *Diabetes & Metabolism Journal*, *40*(3), 192–201. doi:10.4093/dmj.2016.40.3.192 PMID:27098508

Kim, J., Araujo, W. R., Samek, I. A., Bandodkar, A. J., Jia, W., Brunetti, B., ... Wang, J. (2015). Wearable temporary tattoo sensor for real-time trace metal monitoring in human sweat. *Electrochemistry Communications*, *87*(1), 41–45. doi:10.1016/j.elecom.2014.11.024

Kim, J., Chu, C., & Shin, S. (2014). ISSAQ: An Integrated Sensing Systems for Real-Time Indoor Air Quality Monitoring. *IEEE Sensors Journal*, *14*(12), 4230–4244. doi:10.1109/JSEN.2014.2359832

Kim, J., Kim, M., Lee, M. S., Kim, K., Ji, S., Kim, Y. T., & Bien, F. (2017). Wearable smart sensor systems integrated on soft contact lenses for wireless ocular diagnostics. *Nature Communications*, *8*(14997), 1–8. PMID:28447604

Kingsley-Hughes, A. (2016). *The best battery saver app for Android: Greenify.* Available at https://www.zdnet.com/article/the-best-battery-saver-app-for-android-greenify/

Kinney, P. (2003, October). Zigbee technology: Wireless control that simply works. In Communications design conference (Vol. 2, pp. 1-7).

Kinsella, B. (2003). The Wal-Mart factors. *Industrial Engineering (American Institute of Industrial Engineers)*, *32–36*(November).

Kleinke, J. (2005). Dot-Gov: Market Failure And The Creation Of A National Health Information Technology System. *Health Affairs*, *24*(5), 1246–1262. doi:10.1377/hlthaff.24.5.1246 PMID:16162569

Ko, J. G., Lu, C., Srivastava, M. B., Stankovic, J. A., Terzis, A., & Welsh, M. (2010). Wireless sensor networks for healthcare. *Proceedings of the IEEE*, *98*(11), 1947–1960. doi:10.1109/JPROC.2010.2065210

Kokalki, S. A., Mali, A. R., Mundada, P. A., & Sontakke, R. H. (2017). *Smart health band using IoT.* Paper presented at International Conference on Power, Control, Signals and Instrumentation Engineering, Chennai, India.

Koli, R. P., & Jadhav, V. D. (2015). Agriculture Decision Support System as Android Application. *International Journal of Science and Research*, *4*(4), 903-906.

Kovatsch, M., Lanter, M., & Shelby, Z. (2014). Californium: Scalable cloud services for the Internet of Things with CoAP. *2014 International Conference on the Internet of Things (IOT).* 10.1109/IOT.2014.7030106

Kulkarni, A. A. (2013). Solar roadways. Rebuilding our Infrastructure and Economy. *International Journal of Engineering Research and Applications*, *3*(3), 1429–1436.

Kumar, S., Bhattacharyya, J. K., Vaidya, A. N., Chakrabarti, T., Devotta, S., & Akolkar, A. B. (2009). Assessment of the status of municipal solid waste management in metro cities, state capitals, class I cities, and class II towns in India: An insight. *Waste Management,* February 2009, Pages 883-895.

Kumar, S., Mahapatra, B., Kumar, R., & Turuk, A. K. (2018, March). *Security and privacy solution for I-RFID based smart infrastructure health monitoring*. Paper presented at 2018 Technologies for Smart-City Energy Security and Power. 10.1109/ICSESP.2018.8376733

Kumar, V., & Svensson, J. (Eds.). (2015). Promoting Social Change and Democracy Through Information Technology. Hershey, PA: IGI Global. doi:10.4018/978-1-4666-8502-4

Kumbhar, V., & Singh, T. P. (2013). A comprehensive study of application of decision support system in agriculture in Indian Context. *International Journal of Computer Application. 63*(14), 6-11.

Kushwaha, H. L., Sinha, J. P., Khura, T. K., Kushwaha, D. K., Ekka, U., Purushottam, M., & Singh, N. (2016, December). Status and Scope of Robotics in Agriculture. In *International Conference on Emerging Technologies in Agricultural and Food Engineering.* Agricultural and Food Engineering Department, IIT Kharagpur.

Kwak, C., Cho, Y., Ko, J. M., & Kim, C. O. (2011). Adaptive Product Tracking in RFID-Enabled Large-Scale Supply Chain. *Expert Systems with Applications*, *38*(3), 1583–1590. doi:10.1016/j.eswa.2010.07.077

Kwan, A. M., Fung, A. G., Jansen, P. A., Schivo, M., Kenyon, N. J., Delplanque, J. P., & Davis, C. E. (2015). Personal lung function monitoring devices for asthma patients. *IEEE Sensors Journal*, *15*(4), 2238–2247. doi:10.1109/JSEN.2014.2373134

Lajnef, N., Chatti, K., Chakrabartty, S., Rhimi, M., & Sarkar, P. (2013). *Smart pavement monitoring system (No. FHWA-HRT-12-072)*. USA: Federal Highway Administration.

Lasserre, J. A., Bishop, C. M., & Minka, T. P. (2006, June). Principled hybrids of generative and discriminative models. In *2006 IEEE Computer Society Conference on Computer Vision and Pattern Recognition,* (Vol. 1, pp. 87-94). IEEE. 10.1109/CVPR.2006.227

Lata, S. (2016). IoT Based Smart Poultry Farming using Commodity Hardware and Software. *Bonfring International Journal of Software Engineering and Soft Computing*, *6*(Special Issue), 171–175. doi:10.9756/BIJSESC.8269

Law, C., Lee, K., & Siu, K. Y. (2000). Efficient memoryless protocol for tag identification (extended abstract). In *Proceeding of 4th International workshop on Discrete Algorithms and methods for mobile computing and communications (DIALM 2000)*, New York, NY, 75-84. 10.1145/345848.345865

Lee, H. J., & Kim, M. (2018). The Internet of Things in a Smart Connected World. doi:. doi:10.5772/intechopen.76128

Lee, J. (2011). Smart health: Concepts and status of ubiquitous health with smartphone. Paper presented at International Conference on Information and Communication Technology Convergence, Seoul, South Korea. 10.1109/ICTC.2011.6082623

Lee, K. (2018). *Harmful Effects of the Green Revolution.* Retrieved from https://sciencing.com/harmful-effects-green-revolution-8587115.html

Lee, H., Song, C., Hong, Y. S., Kim, M. S., Cho, H. R., Kang, T., & Kim, D. H. (2017). Wearable/disposable sweat-based glucose monitoring device with multistage transdermal drug delivery module. *Science Advances*, *3*(3), 1–8. doi:10.1126ciadv.1601314 PMID:28345030

Lewalski, E. A. (2012). Sensor Telemetry Using WiFi Technology; *International Journal of Electrical & Computer Sciences IJECS-IJENS 5,* pp. 34-42.

Li, D., Lyu, Y., Gui, J., & Halfond, W. G. (2016). Automated energy optimization of http requests for mobile applications. In *Proceedings of the 38th international conference on software engineering*. 10.1145/2884781.2884867

Lin, T., Mayzel, Y., & Bahartan, K. (2018). The accuracy of a non-invasive glucose monitoring device does not depend on clinical characteristics of people with type 2 diabetes mellitus. *Journal of drug assessment, 7*(1), 1-7.

Liu, L., & Lai, S. (2006). ALOHA-Based Anti-Collision Algorithms Used in RFID System. In International Conference on Wireless Communications, Networking and Mobile Computing (pp.1-4), Wuhan, China. IEEE.

Liu, Z., & Halonen, Z. (2018). *ICT supporting healthcare for elderly in China: A systematic mapping study*. Paper presented at Bled Econference Digital Transformation: Meeting the Challenges, Bled, Slovenia. 10.18690/978-961-286-170-4.28

Liu, G., Ho, C., Slappey, N., Zhou, Z., Snelgrove, S. E., Brown, M., & Edwards, J. (2016). A wearable conductivity sensor for wireless real-time sweat monitoring. *Sensors and Actuators. B, Chemical*, *227*, 35–42. doi:10.1016/j.snb.2015.12.034

Liu, X., Mao, G., Ren, J., Li, R. Y. M., Guo, J., & Zhang, L. (2015). How might China achieve its 2020 emissions target? A scenario analysis of energy consumption and CO2 emissions using the system dynamics model. *Journal of Cleaner Production*, *103*, 401–410. doi:10.1016/j.jclepro.2014.12.080

Lmberis, A., & Dittmar, A. (2007). Advanced Wearable Health Systems and Applications - Research and Development Efforts in the European Union. *IEEE Engineering in Medicine and Biology Magazine*, *26*(3), 29–33. doi:10.1109/MEMB.2007.364926 PMID:17549917

Longhi, S., Marzioni, D., Alidori, E., Buò, G. D., Prist, M., Grisostomi, M., & Pirro, M. (2012, May). Solid waste management architecture using wireless sensor network technology. In 2012 5th International Conference on New Technologies, Mobility and Security (NTMS), (pp. 1-5). IEEE. 10.1109/NTMS.2012.6208764

Ma, X., Huang, P., Jin, X., Wang, P., Park, S., Shen, D., . . . Voelker, G. M. (2013). *Edoctor: Automatically diagnosing abnormal battery drain issues on smartphones*. Paper presented at the NSDI.

Mahajan, K. (2014, July). Waste Bin Monitoring System Using Integrated Technologies. *International Journal of Innovative Research in Science, Engineering and Technology, 3*(7).

Mahale, R. B., & Sonavane, S. S. (2016). Smart Poultry Farm Monitoring Using IOT and Wireless Sensor Networks. *International Journal of Advanced Research in Computer Science*, *7*(3), 187–190.

Mahale, R. B., & Sonavane, S. S. (2016). Smart Poultry Farm: An Integrated Solution Using WSN and GPRS Based Network, *International Journal of Advanced Research in Computer Engineering & Technology, 5*(6).

Mahmud, M. S. A., Buyamin, S., Mokji, M. M., & Abidin, M. S. Z. (2018). Internet of Things based Smart Environmental Monitoring for Mushroom Cultivation. *Indonesian Journal of Electrical Engineering and Computer Science, 10*(3), 847-852.

Mainetti, L., Palano, L., Patrono, L., Stefanizzi, M. L., & Vergallo, R. (2014, September). Integration of RFID and WSN technologies in a smart parking system. *In 2014 22nd International Conference on Software, Telecommunications and Computer Networks (SoftCOM),* (pp. 104-110). IEEE. 10.1109/SOFTCOM.2014.7039099

Majumdar, J., Naraseeyappa, S., & Ankalaki, S. (2017). Analysis of agriculture data using data mining techniques: application of big data. *Journal of Big Data, pp. 4*(1), 20.

Maksimović, M., Vujović, V., & Omanović-Miklić anin, E. (2015). Application of internet of things in food packaging and transportation. Int. J. Sustainable Agricultural Management and Informatics, 1(4), 333-350.

Manda, V. L. K., Kushal, V., & Ramasubramanian, N. (2018). An Elegant Home Automation System Using GSM and ARM-Based Architecture. *IEEE Potentials*, *37*(5), 43–48. doi:10.1109/MPOT.2016.2515644

Manes, G., Collodi, G., Fusco, R., Gelpi, L., & Manes, A. (2012). A Wireless Sensor Network for Precise Volatile Organic Compound Monitoring. *International Journal of Distributed Sensor Networks*, *8*(4), 820716. doi:10.1155/2012/820716

Mangwani, P. (2016). Smart Parking Assist System using Internet of Things (IoT), International Journal of Control Theory and Applications, Volume 9-Number 40.

Mano, L. Y., Faiçal, B. S., Nakamura, L. H., Gomes, P. H., Libralon, G. L., Meneguete, R. I., & Ueyama, J. (2016). Exploiting IoT technologies for enhancing Health Smart Homes through patient identification and emotion recognition. *Computer Communications*, *89*, 178–190. doi:10.1016/j.comcom.2016.03.010

Martins, M., Cappos, J., & Fonseca, R. (2015). Selectively Taming Background Android Apps to Improve Battery Lifetime. In *USENIX Annual Technical Conference*, (pp. 563-575).

Martin, T., & Raskovic, D. (2008). Issues in wearable computing for medical monitoring applications: A case study of a wearable ECG monitoring device. *Sensors and Actuators. A, Physical*, *141*(2), 396–403.

Marwaha, S. (2012). AGRIDAKSH- A Tool for developing the online expert system. *Agro-Informatics and Precision Agriculture (AIPA 2012)*. 17-23.

Mason, A. G. (2002). Cisco Secure Virtual Private Network (p. 7). Indianapolis, IN: Cisco Press.

Mathie, M. J., Coster, A. C., Lovell, N. H., & Celler, B. G. (2003). Detection of daily physical activities using a triaxial accelerometer. *Medical & Biological Engineering & Computing*, *41*(3), 296–301. doi:10.1007/BF02348434 PMID:12803294

Mattern, F., & Floerkemeier, C. (2010). From the Internet of Computers to the Internet of Things. *Lecture Notes in Computer Science*, *6462*, 242–259. doi:10.1007/978-3-642-17226-7_15

Mayagoitia, R. E., Nene, A. V., & Veltink, P. H. (2002). Accelerometer and rate gyroscope measurement of kinematics: An inexpensive alternative to optical motion analysis systems. *Journal of Biomechanics*, *35*(4), 537–542. doi:10.1016/S0021-9290(01)00231-7 PMID:11934425

McAfee, P. C., Garfin, S. R., Rodgers, W. B., Allen, R. T., Phillips, F., & Kim, C. (2011). An attempt at clinically defining and assessing minimally invasive surgery compared with traditional "open" spinal surgery. *SAS Journal*, *5*(4), 125–130. doi:10.1016/j.esas.2011.06.002 PMID:25802679

McDonald, I. (2010). *The Dervish House*. Hachette, UK.

McLaughlin, M., Malone, P., & Jennings, B. (2009). A model for identity in digital ecosystems. In *3rd IEEE International Conference on Digital Ecosystems and Technologies,* 295I, Istanbul, Turkey. 10.1109/DEST.2009.5276727

McRoberts, M. (2010). Beginning Arduino. New York, NY: Apress. doi:10.1007/978-1-4302-3241-4

MEDIA. (2014, Oct. 26). *The Future of Personalized Health Care. Predictive Analytics by @Rock?* Retrieved from https://www.slideshare.net/RockHealth/the-future-of-personalized-health-care-predictive-analytics-press

Mehta, A., Aggrawal, N., & Tiwari, A. (2015). Solar Roadways-The future of roadways. *International Advanced Research Journal in Science, Engineering and Technology (IARJSET), 2.*

Mendis, S., Puska, P., Norrving, B., & World Health Organization. (2011). *Global atlas on cardiovascular disease prevention and control.* Geneva: World Health Organization.

Metcalf, D., Milliard, S. T., Gomez, M., & Schwartz, M. (2016). Wearables and the Internet of Things for Health: Wearable, Interconnected Devices Promise More Efficient and Comprehensive Health Care. *IEEE Pulse*, *7*(5), 35–39. doi:10.1109/MPUL.2016.2592260 PMID:28113167

Meyer, J., & Boll, S. (2014, October). Smart health systems for personal health action plans. In *International Conference on e-Health Networking, Applications and Services*, Natal-RN, Brazil. 10.1109/HealthCom.2014.7001877

Micic, A., Nayac, A., Simplot-Ryl, D., & Stojmenovic, I. (2005). A hybrid randomized protocol for RFID tag identification. In *Proceeding of 1st IEEE International Workshop on Next Generation Wireless Networks* (WoNGeN 2005).

Miler, B. A., & Bisdikian, C. (2001). Bluetooth revealed: the insider's guide to an open specification for global wireless communication. Upper Saddle River, NJ: Prentice Hall.

Miles, S. B., Sarma, S. E., & Williams, J. R. (2008). *RFID Technology and Applications*. New York, NY: Cambridge University Press. doi:10.1017/CBO9780511541155

Miller, H. (1996). The multiple dimensions of information quality. *Information Systems Management*, *13*(2), 79–83. doi:10.1080/10580539608906992

Miller, M. (2018). *Internet Of Things: How smart TVs, smart cars, smart homes, and smart cities are changing the world*. Chennai, India: Pearson.

Mills, G. A., & Nketia, T. A. (2007). Wireless Digital Stethoscope Using Bluetooth Technology. *Proc. IET Seminar Antennas and Propagation for Body-Centric Wireless Communications*, London, U.K., pp.77–83.

Mingjun, W., Zhen, Y., Wei, Z., Xishang, D., Xiaofei, Y., Chenggang, S., . . . Jinghai, H. (2012, October). A research on experimental system for Internet of things major and application project. In *2012 3rd International Conference on System Science, Engineering Design and Manufacturing Informatization (ICSEM),* (Vol. 1, pp. 261-263). IEEE. 10.1109/ICSSEM.2012.6340722

Misfit.com. (n.d.). *Fitness Trackers & Wearable Technology*. Retrieved from http://misfit.com/products/shine

Mishra, A. S., & Deep, V. (2014). Expert System in Agriculture: An Overview. *International Journal of Science Technology & Engineering.*, *1*(5), 45–49.

Mishra, N., Goyal, D., & Sharma, A. D. (2018). Issues in Existing Robotic Service in Restraurants and Hotels. *Proceeding of Third Int.l Conf. on Internet of Things and Connected Technologies (ICIoTCT),* MNIT Jaipur, India. doi:10.2139srn.3166508

Miyazaki, S. (1997). Long-term unrestrained measurement of stride length and walking velocity utilizing a piezoelectric gyroscope. *IEEE Transactions on Biomedical Engineering*, *44*(8), 753–759. doi:10.1109/10.605434 PMID:9254988

Mohan, S., Kumar, E. P., & Paulchamy, B. (2013). *Certain Investigation of Precision Agriculture Robot using Lab view*. IEEE International conference on current trends in Engineering and Technology, India. pp. 319-322. 10.1109/ICCTET.2013.6675975

Moher, D., Liberati, A., Tetzlaff, J., & Altman, D. G. (2009). Preferred reporting items for systematic reviews and meta-analyses: The PRISMA statement. *Annals of Internal Medicine*, *151*(4), 264–269. doi:10.7326/0003-4819-151-4-200908180-00135 PMID:19622511

Mondal, P. (2018). *Five Major Demerits or Problems of Green Revolution in India*. Retrieved from http://www.yourarticlelibrary.com/green-revolution/5-major-demerits-or-problems-of-green-revolution-in-india/20954

Monika, K. A., Rao, N., Prapulla, S. B., & Shobha, G. (2016). *Smart Dustbin-An Efficient Garbage Monitoring System. International Journal of Engineering Science and Computing*, *6*, 7113–7116.

More, S., & Bansode, R. (2015). *Implementation of AES with Time Complexity Measurement for Various Input. Global Journal of Computer Science and Technology, 15(4).*

Mortellaro, M., & DeHennis, A. (2014). Performance characterization of an abiotic and fluorescent-based continuous glucose monitoring system in patients with type 1 diabetes. *Biosensors & Bioelectronics*, *61*, 227–231. doi:10.1016/j.bios.2014.05.022 PMID:24906080

Morton, S., & Redmond, P. (2015). The odd one out: Revisiting and investigating the gender imbalance in ICT study choices. *Journal of Applied Computing and Information Technology*, *19*(1), 2–3.

Moser, E. G., Crew, L. B., & Garg, S. K. (2010). Role of continuous glucose monitoring in diabetes management. *Advances en Diabetologia*, *26*, 73–78.

Mourtzis, D., Papakostas, N., Makris, S., Xanthakis, V., Xanthakis, V., & Chrysolouris, G. (2008). Supply chain modelling and control for producing highly customized products. *CIRP Annals – Manufacturing Technology, 57*(1), 570-586.

Mueller, K. T., Loomis, P. V., Kalafus, R. M., & Sheynblat, L. (1994). U.S. Patent No. 5,323,322. Washington, DC: U.S. Patent and Trademark Office.

Mukesh, P. S., Pandya, M. S., & Pathak, S. (2013). Enhancing AES Algorithm with Arithmetic Coding. *International Conference on Green Computing, Communication and Conservation of Energy (ICGCE)*. 10.1109/ICGCE.2013.6823404

Murakami, Y. (2014). iFarm: Development of Web-based System of Cultivation and Cost Management for Agriculture. In *2014 Eighth International Conference on Complex, Intelligent and Software Intensive Systems, (pp. 624-627)*. 10.1109/CISIS.2014.89

Murphy, M. L. (2009). *The Busy Coder's Guide to Android Development.* USA: CommonsWare; Revised & enlarged edition.

Mynatt, E., Melenhorst, A., Fisk, A., & Rogers, W. (2004). Aware technologies for aging in place: Understanding user needs and attitudes. *IEEE Pervasive Computing*, *3*(2), 36–41. doi:10.1109/MPRV.2004.1316816

Nagalingeswari, K., & Satamraju, K. P. (2017). Efficient Garbage Management System for Smart Cities. *International Journal of Engineering Trends and Technology (IJETT), 50*(5).

Nalawade, S. R., & Devrukhkar, S. (2016). Bus Tracking by Computing Cell Tower Information on Raspberry Pi. (pp. 87–90) *International Conference on Global Trends in Signal Processing, Information Computing and Communication (ICGTSPICC)*. 10.1109/ICGTSPICC.2016.7955275

Nallani, S., & Berlin Hency, V. (2015). *Low power cost effective automatic irrigation system. Optimization and Sciences* (pp. 224–228). EEECOS.

Nasir, M. M., & Mansor, W. (2011). GSM based Motorcycle Security System, (pp. 129–134), *Control and System Graduate Research Colloquium (ICSGRC) IEEE.*

Nassar, W., Al-Qadi, I. L., Flintsch, G. W., & Appea, A. (2000). Evaluation of pavement layer response at the Virginia Smart Road. In *Pavement Subgrade* (pp. 104–118). Unbound Materials, and Nondestructive Testing. doi:10.1061/40509(286)7

Nath, B., Reynolds, F., & Want, R. (2006). RFID technology and applications. *IEEE Pervasive Computing*, *5*(1), 22–24. doi:10.1109/MPRV.2006.13

Naveen, B., Kavya, G. K., Kruthika, S. N., Ranjitha, K. N., & Sahana, C. N. (2018). Automated Waste Segregator Using Arduino. *International Journal of Advance Engineering and Research Development, 5*(05), May -2018.

Navghane, S. S., & Killedar, M. S., & Rohokale, V. M. (2016, May). IoT Based Smart Garbage and Waste Collection Bin. *International Journal of Advanced Research in Electronics and Communication Engineering*, *5*(5), 1577.

Navghane, S. S., Killedar, M. S., & Rohokale, D. V. (2016). *IoT Based Smart Garbage and Waste Collection Bin. International Journal of Advanced Research in Electronics and Communication Engineering*, *5*, 1576–1578.

Nayyar, A., & Puri, V. (2016, November) Smart Farming: IoT Based Smart Sensors Agriculture Stick for Live Temperature and Moisture Monitoring using Arduino, *Conference Paper on Cloud Computing & Solar Technology*. doi:10.1201/9781315364094-121

Nejad, F. M., & Zakeri, H. (2013). The hybrid method and its application to smart pavement management. In *Metaheuristics in Water* (pp. 439–484). Geotechnical and Transport Engineering. doi:10.1016/B978-0-12-398296-4.00019-2

Nentwich, M. M., & Ulbig, M. W. (2015). Diabetic retinopathy - ocular complications of diabetes mellitus. *World Journal of Diabetes*, *6*(3), 489–499. doi:10.4239/wjd.v6.i3.489 PMID:25897358

Nepomilueva, D. (2017). Water scarcity indexes: water availability to satisfy human needs.

Ngai, E. W. T., Cheung, T. C. E., Au, S., & Lai, K. (2007). Mobile commerce integrated with RFID technology in a container depot. *Decision Support Systems*, *43*(1), 62–76. doi:10.1016/j.dss.2005.05.006

Nikzad, N., Chipara, O., & Griswold, W. G. (2014). *APE: An annotation language and middleware for energy-efficient mobile application development*. In Proceedings of the 36th International Conference on Software Engineering, Hyderabad, India. (pp. 515-526). ACM. 10.1145/2568225.2568288

Nikzad, N., Radi, M., Chipara, O., & Griswold, W. G. (2015, November). Managing the energy-delay tradeoff in mobile applications with tempus. In *Proceedings of the 16th Annual Middleware Conference (pp. 259-270). ACM.* 10.1145/2814576.2814803

Nouy, N., Virone, G., Barralon, P., Ye, J., Rialle, V., & Deongeot, J. (2003, June). New trends in health smart homes. In *Proceedings 5th International Workshop on Enterprise Networking and Computing in Healthcare Industry (pp. 118-127). IEEE.*

O'Brien, J. A. (2001). *Management Information Systems: Management Information Technology in the Internetworked Enterprise* (5th ed.). Boston, MA: McGraw-Hill.

Olariu, S., & Weigle, M. C. (2009). *Vehicular Networks: from theory to practice*. CRC Press. doi:10.1201/9781420085891

Olson, P. (2017, Feb. 6). *Report: Jawbone Is Jumping Out Of Consumer Wearables*. Retrieved from http://bit.ly/mpo0617tp5

Ovadia, S. (2014). Automate the internet with "if this then that"(IFTTT). *Behavioral & Social Sciences Librarian*, *33*(4), 208–211. doi:10.1080/01639269.2014.964593

Owayjan, M., Dergham, A., Haber, G., Fakih, N., Hamoush, A., & Abdo, E. (2015). Face recognition security system. In *New Trends in Networking, Computing, E-learning, System Sciences, and Engineering* (pp. 343–348). Cham, Switzerland: Springer.

Pal, K. (2018). A Big Data Framework for Decision Making in Supply Chain. In M. V. Kumar, G. D. Putnik, K. Jayakrishna, V. M. Pillai, & L. Varela (Eds.), Emerging Applications in Supply Chains for Sustainable Business Development, Chapter 1, 1-22. Hershey, PA: IGI Global.

Pal, K. (2018a). Building High Quality Big Data-Based Applications in Supply Chains. In A. Kumar, & S. Saurav (Eds.), Supply Chain Management Strategies and Risk Assessment in Retail Environments (pp. 1–24). Hershey, PA: IGI Global. doi:10.4018/978-1-5225-3056-5.ch001

Pal, K. (2019). Quality Assurance Issues for Big Data Applications in Supply Chain Management. In P. K. Gupta, T. Oren, & M. Singh (Eds.), Predictive Intelligence Using Big Data and the Internet of Things, Chapter 3, 51-76. Hershey, PA: IGI Global.

Pal, K. (2019). Radio Frequency Identification Systems Security Challenges in Supply Chain Management. In J. Rodrigues, A. Gawanmeh, K. Saleem, & S. Parvin (Eds.), Smart Devices, Applications, and Protocols for the IoT. Hershey, PA: IGI Global. doi:10.4018/978-1-5225-7811-6.ch010

Palepu, R. B., & Muley, R. R. (2017). An Analysis of Agricultural Soils by using Data Mining Techniques. *International Journal of Engineering Science and Computing*. October, 15167-15177.

Pal, K. (2017). Supply Chain Coordination Based on Web Services. In H. K. Chan, N. Subraanian, & M. D. Abdulrahman (Eds.), *Supply Chain Management in the Big Data Era* (pp. 137–171). Hershey, PA: IGI Global. doi:10.4018/978-1-5225-0956-1.ch009

Pal, K. (2018b). Ontology-Based Web Service Architecture for Retail Supply Chain Management. In *the Proceedings of 9th International Conference on Ambient Systems, Networks and Technologies,* Porto, Portugal. *Procedia Computer Science*, *130*, 985–990. doi:10.1016/j.procs.2018.04.101

Pal, K., & Karakostas, B. (2014). A multi agent-based service framework for supply chain management, Procedia 53-60. *Computer Science*, *32*, 53–60.

Palvia, P., Leary, D., Mao, E., Midha, V., Pinjani, P., & Salam, A. F. (2004). Research methodologies in MIS: An update. *Communications of the Association for Information Systems*, *14*(1), 526–542.

Panchal, J. R. (2017). Energy Efficient Wireless Sensor Network System for Transportation. International Journal of Engineering Technology. *Management and Applied Sciences, 5*(5), 663–667.

Patel, N. R., Choudhari, P. G., Kale, P. D., Patel, N. R., Raut, G. N., & Bherani, A. (2014). Smart Design of Microcontroller based monitoring system for Agriculture. IEEE International Conference on Circuit, Power and Computer Technology. pp 1710-1713.

Patil, P. R. (2015). Introduction to li-fi (light fidelity) technology. *IJTRE*, *3*(4), 600–603.

Paul, D., Pearlson, K., & McDaniel, R. (1999). Assessing technological barriers to telemedicine: Technology-management implications. *IEEE Transactions on Engineering Management*, *46*(3), 279–288. doi:10.1109/17.775280

Pawar, P., & Vittal, K. P. (2017). Design of smart socket for power optimization in home energy management system, *2017 2nd IEEE International Conference on Recent Trends in Electronics, Information & Communication Technology (RTEICT),* Bangalore, pp. 1739-1744. 10.1109/RTEICT.2017.8256897

Peffers, K., Rothenberger, M., Tuunanen, T., & Vaezi, R. (2012, May). Design science research evaluation. In *International Conference on Design Science Research in Information Systems* (pp. 398-410). Berlin, Germany: Springer.

Peffers, K., Tuunanen, T., Rothenberger, M. A., & Chatterjee, S. (2007). A design science research methodology for information systems research. *Journal of Management Information Systems*, *24*(3), 45–77. doi:10.2753/MIS0742-1222240302

Pelletier-Fleury, N., Fargeon, V., Lanoé, J., & Fardeau, M. (1997). Transaction costs economics as a conceptual framework for the analysis of barriers to the diffusion of telemedicine. *Health Policy (Amsterdam)*, *42*(1), 1–14. doi:10.1016/S0168-8510(97)00038-9 PMID:10173489

Percival, J., & Hanson, J. (2006). Big brother or brave new world? Telecare and its implications for older people's independence and social inclusion. *Critical Social Policy*, *26*(4), 888–909. doi:10.1177/0261018306068480

Pérez-Hoyos, A., Rembold, F., Kerdiles, H., & Gallego, J. (2017). Comparison of Global Land Cover Datasets for Cropland Monitoring. *Remote Sensing*, *9*(11), 1118. doi:10.3390/rs9111118

Petersen, K., Feldt, R., Mujtaba, S., & Mattsson, M. (2008). Systematic mapping studies in software engineering. *EASE*, *8*, 68–77.

Pfützner, A., Strobl, S., Demircik, F., Redert, L., Pfützner, J., Pfützner, A. H., & Lier, A. (2018). Evaluation of a new noninvasive glucose monitoring device by means of standardized meal experiments. *Journal of Diabetes Science and Technology*, *12*(6), 1178–1183. doi:10.1177/1932296818758769 PMID:29451016

Pille, K. Maniyar, R., Bade, N. & Bakliwal, J. M. (2016). Solid Waste Management System, *National Conference on Internet of Things: Towards a Smart Future & Recent Trends in Electronics & Communication (IOTTSF-2016),* in Association with Novateur Publication.

Pipino, L., Lee, Y. W., & Wang, R. Y. (2002). Data quality assessment. *Communications of the ACM*, *45*(4), 211–218. doi:10.1145/505248.506010

Piza, E. L. (2019). *CCTV and Crime Prevention A New Systematic Review and Meta-Analysis.* John Jay College of Criminal Justice, City University of New York.

Porter, M. E. (1985). Technology and Competitive Advantage. *The Journal of Business Strategy*, *5*(3), 60–78. doi:10.1108/eb039075

Pothitos, A. (2016). Mobile Industry Review. Retrieved from http://www.mobileindustryreview.com/2016/10/the-history-of-the-smartphone.html

Poushter, J. (2016). Smartphone ownership and internet usage continues to climb in emerging economies. *Pew Research Center*, *22*, 1–44.

Prater, E., Frazier, G. V., & Reyes, P. M. (2005). Further impact of RFID one-supply chain in grocery retailing, Supply Chain Management. *International Journal (Toronto, Ont.)*, *10*(2), 134–142.

Pushpa, M. K., Gupta, A., Shaikh, S. M., Jha, S., & Suchitra, V. (2015, May). Microcontroller Based Automatic Waste Segregator. *International Journal of Innovative Research in Electrical, Electronics, Instrumentation and Control Engineering*, *3*(5).

Ra, H. K., Salekin, A., Yoon, H. J., Kim, J., Nirjon, S. S., Stone, D. J., . . . Stankovic, J. A. (2016). AsthmaGuide: an asthma monitoring and advice ecosystem. In *IEEE Wireless Health,* Bethesda, MD. 10.1109/WH.2016.7764567

Ragavan, E., Hariharan, C., Aravindraj, N., & Manivannan, S. S. (2016). Real time water quality monitoring system. *International Journal of Pharmacy and Technology*, *8*(4), 26199–26205. doi:10.15680/ijircce.2015.0306016

Rahmani, A. M., Gia, T. N., Negash, B., Anzanpour, A., Azimi, I., Jiang, M., & Liljeberg, P. (2018). (Article in Press). *Future Generation Computer Systems Journal.*

Rai, P., & Varadan, V. K. (2012, October). *Wireless glucose monitoring watch enabled by an implantable self-sustaining glucose sensor system.* Paper presented at the Nanosystems in Engineering and Medicine, Incheon, Republic of Korea.

Rajpal, A., Jain, S., Khare, N., & Shukla, A. K. (2011). Microcontroller-based automatic irrigation system with moisture sensors. In *Proceedings of the International Conference on Science and Engineering* (pp. 94-96).

Rajput, M. (2015). *Why Android Studio Is Better For Android Developers Instead Of Eclipse*. Retrieved from https://dzone.com/articles/why-android-studio-better

Rajput, R. S., Pant, A., & Kumar, S. (2018). Development of Forecasting Model for Sugarcane Productivity using Multiple Linear Regression with Genetic Algorithm. *Periodic Research. 7*(2), 124-128.

Ramlee, R. A., Tang, D. H. Z., & Ismail, M. M. (2012). Smart home system for Disabled People via Wireless Bluetooth. *International Conference on System Engineering and Technology*. 10.1109/ICSEngT.2012.6339347

Rani, P. M. N., Rajesh, T., & Saravanan, R. (2011). Expert Systems in Agriculture: A Review. *Journal of Computer Science and Applications, 3,* 59-71.

Rani, M. U., & Kamalesh, S. (2014). Web based service to monitor automatic irrigation system for the agriculture field using sensors. In *2014 International Conference on Advances in Electrical Engineering (ICAEE).*

Rao, N. H. (2018). Big Data and Climate Smart Agriculture - Status and Implications for Agricultural Research and Innovation in India. *Proceedings of the Indian National Science Academy*. 10.16943/ptinsa/2018/49342

Rao, Y. R. (2017). Automatic smart parking system using Internet of Things (IOT). *Int. J. Eng. Tech. Sci. Res*, *4*, 2394–3386.

Raut, S. H., & Ambulgekar, H. P. (2013, March 4). *Proactive and Reactive Routing Protocols in Multihop Mobile Ad hoc Network*. Retrieved from https://pdfs.semanticscholar.org/4704/972232ae2bf64df8e4fd3dcc93b414fbebdb.pdf

Ravisankar, K., Sidhardha, K., & Prabhadevi, B. (2017). Analysis of Agricultural Data Using Big Data Analytics. *Journal of Chemical and Pharmaceutical Sciences*, *10*(3), 1132-1135.

Ray, P. P. (2018). Continuous glucose monitoring: A systematic review of sensor systems and prospects. *Sensor Review*, *38*(4), 430–437. doi:10.1108/SR-12-2017-0268

Reich, J., & Dunne, L. E. (2016, September). Multi-modal wearable ambient display: an investigation of continuous glucose monitoring. Paper presented at the ACM International Symposium on Wearable Computers, Heidelberg, Germany. 10.1145/2971763.2971801

RelaxLine. (2010, Feb. 20). *BellyBio Interactive Breathing*. Retrieved from https://itunes.apple.com/us/app/bellybio-interactive-breathing/id353763955?mt=8

Reportlinker. (2018, Jan. 15). *The global market for wearable medical devices was valued at $4.8 billion in 2015*. Retrieved from https://www.prnewswire.com/news-releases/the-global-market-for-wearable-medical-devices-was-valued-at-48-billion-in-2015-300582717.html

Riaz, S. (2009). Diabetes mellitus. *Scientific Research and Essays*, *4*(5), 367–373. PMID:20015604

Road Accidents in India 2015. (2015). Government of India Ministry of Road Transport & Highways Transport Research Wing.

Road Accidents in India 2016: 17 deaths on roads every hour, Chennai and Delhi most dangerous. (2017, September 11). Retrieved from http://indianexpress.com/article/india/road-accidents-in-india-2016-17-deaths-on-roads-every-hour-chennai-and-delhi-most-dangerous-4837832/

Roco, M. C., & Bainbridge, W. S. (2003). Overview Converging Technologies for Improving Human Performance. *Converging Technologies for Improving Human Performance*, 1-27. doi:10.1007/978-94-017-0359-8_1

Roetenberg, D., Luinge, H., & Veltink, P. (2003, October). Inertial and magnetic sensing of human movement near ferromagnetic materials. *The Second IEEE and ACM International Symposium on Mixed and Augmented Reality, 2003. Proceedings*. doi:10.1109/ismar.2003.1240714

Rohokale, V. M., Prasad, N. R., & Prasad, R. (2011). A cooperative Internet of Things (IoT) for rural healthcare monitoring and control. *2011 2nd International Conference on Wireless Communication, Vehicular Technology, Information Theory and Aerospace & Electronic Systems Technology (Wireless VITAE)*. doi:10.1109/wirelessvitae.2011.5940920

Roussos, G., Koukara, L., Kourouthanasis, P., Tuominen, J., Seppala, O., & Frissaer, J. (2002). A case study in pervasive retail. Proceedings of the 2nd International Workshop on Mobile Commerce, Atlanta, GA. New York, NY: ACM Press, 90-94. 10.1145/570705.570722

Routing Information Protocol. (2019, June 12). Retrieved from https://en.wikipedia.org/wiki/Routing_Information_Protocol

Rumbaugh, J., Jacobson, I., & Booch, G. (2004). *The Unified Modeling Language Reference Manual* (2nd ed.). Pearson Higher Education.

Ryu, J., Lee, H., Seok, Y., Kwon, T., & Choi, Y. (2007, June). A Hybrid Query Tree Protocol for Tag Collision Arbitration in RFID systems. In *Proceeding of IEEE International Conference on Communications*, 5981-5986. IEEE.

S. A. P., Goswami, D., Tadi, S., & Pandey, K. (n.d.). *Optimized Architecture for AES.* Allahabad, India.

Sabbaghi, A., & Ganesh, V. (2008). Effectiveness and Efficiency of RFID Technology in Supply Chain Management: Strategic Values and Challenges. *Journal of Theoretical and Applied Electronic Commerce Research*, *3*(2), 71–81. doi:10.4067/S0718-18762008000100007

SAM battery Monitor. (2018). *Google Play store.* Available at https://play.google.com/store/apps/details?id=com.gsamlabs.bbm&hl=en_US

Sampri, A., Mavragani, A., & Tsagarakis, K. P. (2016). Evaluating Google Trends as a Tool for Integrating the 'Smart Health'Concept in the Smart Cities' Governance in USA. *Procedia Engineering*, *162*, 585–592. doi:10.1016/j.proeng.2016.11.104

Sangasoongsong, A., Kunthong, J., Sarangan, V., Cai, X., & Bukkapatnam, S. T. S. (2007). A Low-Cost, Portable, *High-Throughput Wireless Sensor System for Phonocardiography Applications*, *2*, 345–356.

Santhi, S., Udayakumar, E., & Gowthaman, T. (2019). SOS Emergency Ad Hoc Wireless Network. In *Computational Intelligence and Sustainable Systems* (pp. 227–234). Cham, Switzerland: Springer.

Santoro, P., Calderaro, V., Galdi, V., & Piccolo, A. (2016). Active smart socket design to perform local control of power demand in residential units, *8th IET International Conference on Power Electronics, Machines and Drives (PEMD 2016)*, Glasgow, UK, pp. 1-5. 10.1049/cp.2016.0380

Sarma, S., Brock, D., & Engels, D. (2001). Radio frequency identification and the electronic product code. *IEEE Micro*, *21*(6), 50–54. doi:10.1109/40.977758

Sauter, T., & Lobashov, M. (2011). How to Access Factory Floor Information Using Internet Technologies and Gateways. *IEEE Transactions on Industrial Informatics*, *7*(4), 699–712. doi:10.1109/TII.2011.2166788

Saygin, C., Sarangapani, J., & Grasman, S. E. (2007). A Systems Approach to Viable RFID Implementation in the Supply Chain. *Springer series in advanced manufacturing,* 1.

Schindler, E. (2003). Location, location, location: what effect will tracking technologies like RFID and GPS have on connected businesses? *networker, 7*(2), 11-14.

Schmidhuber, J. (2015). Deep learning in neural networks: An overview. *Neural Networks*, *61*, 85–117. doi:10.1016/j.neunet.2014.09.003 PMID:25462637

Schoute, F. C. (1983, April). Dynamic frame length Aloha. *IEEE Transactions on Communications*, *COM-31*(4), 565–568. doi:10.1109/TCOM.1983.1095854

Schroeder, C. C. (1998). *U.S. Patent No. 5,787,186.* Washington, DC: U.S. Patent and Trademark Office.

Schwartz, F. L., Marling, C. R., & Bunescu, R. C. (2018). The Promise and Perils of Wearable Physiological Sensors for Diabetes Management. *Journal of Diabetes Science and Technology*, *12*(3), 587–591. doi:10.1177/1932296818763228 PMID:29542348

Science Service Dr. Hempel Digital Health Network. (2018, April 27). *Now, sensors can attach to your skin and internal organs | The age of smart patches.* Retrieved from https://www.dr-hempel-network.com/digital-health-technolgy/wearables-development-smart-patches/

Sethi, P., & Sarangi, S. R. (2017). Internet of things: Architectures, protocols, and applications. *Journal of Electrical and Computer Engineering*, 2017.

Seto, E. Y., Giani, A., Shia, V., Wang, C., Yan, P., Yang, A. Y., ... Bajcsy, R. (2009). A wireless body sensor network for the prevention and management of asthma. In *IEEE International Symposium on Industrial Embedded Systems,* Lausanne, Switzerland. 10.1109/SIES.2009.5196203

Severi, S., Sottile, F., Abreu, G., Pastrone, C., Spirito, M., & Berens, F. (2014, June). M2M technologies: Enablers for a pervasive Internet of Things. In *2014 European Conference on Networks and Communications (EuCNC),* (pp. 1-5). IEEE.

Shahin, A. P., & Hate, S. G. (2016). *Automatic irrigation system using wireless sensor network. 2015 International Conference* Signal Processing, and their Applications (ICCSPA).

Shaker, G., Smith, K., Omer, A. E., Liu, S., Csech, C., Wadhwa, U., & Hughson, R. (2018). Noninvasive monitoring of glucose level changes utilizing a mm-wave radar system. *International Journal of Mobile Human Computer Interaction, 10*(3), 10–29. doi:10.4018/IJMHCI.2018070102

Sharvin, R. (2016). Automated Irrigation System Using X-Bee and LabView. In *3rd International Conference on Electrical, Electronics, Engineering Trends, Communication, Optimization and Sciences (EEECOS).*

Shelby, Z. (2014). Foreword. From *Machine-To-Machine to the Internet of Things*, xi. doi:10.1016/b978-0-12-407684-6.00020-6

Shelby, Z., Hartke, K., & Bormann, C. (2014). *The Constrained Application Protocol (CoAP).* doi:10.17487/rfc7252

Sheng, M., Zhang, Y., Yang, J., Li, C., & Xing, C. (2015, September). Fihuo: A Mobile Smart Health Service Platform. In *Web Information System and Application Conference,* Jinan, China. 10.1109/WISA.2015.26

Shie, M., Lin, P., Su, T., Chen, P., & Hutahaean, A. (2014). Intelligent energy monitoring system based on ZigBee-equipped smart sockets, *2014 International Conference on Intelligent Green Building and Smart Grid (IGBSG)*, Taipei, Taiwan, pp. 1-5. 10.1109/IGBSG.2014.6835281

Shih, D., Sun, P. L., Yen, D. C., & Huang, S. M. (2006). Taxonomy and survey of RFID anti-collision protocols. *Computer Communications*, *29*(11), 2150–2166. doi:10.1016/j.comcom.2005.12.011

Shinde, P. A., & Mane, Y. B. (2015). Advanced vehicle monitoring and tracking system based on Raspberry Pi. *9th International Conference on Intelligent Systems and Control.* 10.1109/ISCO.2015.7282250

Shreyas, B. et al. (2017). Real-time monitoring in agricultural warehouse using IOT. *International Research Journal of Engineering and Technology. 4*(4).

Shu, L., Tao, X. M., & Feng, D. D. (2010). A wearable, wireless electronic interface for textile sensors. In *Proc. IEEE Int. Symp. Circuits Syst. (ISCAS),* Paris, France, pp. 3104– 3107.

Siddiquee, J., Roy, A., Datta, A., Sarkar, P., Saha, S., & Biswas, S. S. (2016). Smart asthma attack prediction system using Internet of Things. In *IEEE 7th Annual Information Technology, Electronics and Mobile Communication Conference,* Vancouver, Canada 10.1109/IEMCON.2016.7746252

Silva, B. N., Khan, M., & Han, T. (2018) Towards substainable smart cities: A review of trends, architectures, components and open challenges in smart cities. *Substainable smart cities and society, 38,* 697-713

Silvani, X., Morandini, F., Innocenti, E., & Peres, S. (2014). Evaluation of a Wireless Sensor Network with Low Cost and Low Energy Consumption for Fire Detection and Monitoring. *Fire Technology*, *51*(4), 971–993. doi:10.100710694-014-0439-9

Singh, R., & Singh, G. (2017). *Review: Role of Data Mining in Agriculture Yield Analysis.* Paper presented at International Conference on Soft Computing Applications in Wireless Communication - SCAWC 2017.

Singhal, S., & Pandey, S. (2001). Solid Waste Management India, Status and Future Direction. *TERI Information Monitoring on Environment Science*, *6*(1), 1–4.

Singh, N. (2003). Emerging technologies to support supply chain management. *Communications of the ACM*, *46*(9), 243–247. doi:10.1145/903893.903943

Sinharoy, A., Mitra, S., & Mondal, P. (2018). Socioeconomic and environmental predictors of asthma-related mortality. *Journal of Environmental and Public Health*, 2018. PMID:29853926

Sleep Time by Azumio. (n.d.). Retrieved from http://www.azumio.com/s/sleeptime/index.html

Smaros, J., & Holmstrom, J. (2000). Viewpoint: Reaching the consumer through e-grocery VMT. *International Journal of Retail & Distribution Management*, *28*(2), 55–61. doi:10.1108/09590550010315098

Smart Greenhouse Remote Monitoring Systems. (n.d.). Retrieved from https://www.postscapes.com/smart-greenhouses

Smart Parking. (n.d.). *Happiest Minds Technology*. Retrieved from https://www.happiestminds.com/whitepapers/smart-parking.pdf

Smith, A. D. (2005). Exploring radio frequency identification technology and its impact on business systems. *Information Management & Computer Security*, *13*(1), 16–28. doi:10.1108/09685220510582647

Sobrinho, A., Silva, L. D., Perkusich, A., Pinheiro, M. E., & Cunha, P. (2018). Design and evaluation of a mobile application to assist the self-monitoring of the chronic kidney disease in developing countries. *BMC Medical Informatics and Decision Making*, *18*(7), PMID:29329530

Solanas, A., Patsakis, C., Conti, M., Vlachos, I. S., Ramos, V., Falcone, F., Postolache, O. … Martınez-Ballesté, A. (2014). Smart health: A context-aware health paradigm within smart cities. *IEEE communication society,* 74-81.

Solanas, A., Patsakis, C., Conti, M., Vlachos, I. S., Ramos, V., Falcone, F., ... Martinez-Balleste, A. (2014). Smart health: A context-aware health paradigm within smart cities. *IEEE Communications Magazine*, *52*(8), 74–81. doi:10.1109/MCOM.2014.6871673

Solar Roadways. (2013). *Solar Roadway–A Real Solution.*

Solar Roadways. (2017). *Welcome to Solar Roadways.*

Sommerville, I. (2015). Software Engineering. Pearson, 10th edition.

Spiryakin, D., & Baranov, A. (2016). Uniform Inbuilt Wireless Sensor Node for Working Conditions Monitoring. *Proceedings of the 2016 Federated Conference on Computer Science and Information Systems*. 10.15439/2016F386

SR, Basavaraju. (2015). Automatic Smart Parking System using Internet of Things (IOT). *International Journal of Scientific and Research Publications, 628.*

Srivastava, A. K., Agarwal, A., & Mathur, A. (2015). *Internet of Things and its enhanced data security. International Journal of Engineering and Applied Sciences. IJEAS.*

Stallings, W. (2011). Network security essentials: Applications and standards (4th ed.). London, UK: Pearson Education.

Stallings, W. (2005). *Cryptography and network security principles and practices* (4th ed.). Prentice Hall. Retrieved from http://www.inf.ufsc.br/~bosco.sobral/ensino/ine5680/material-cripto-seg/2014-1/Stallings/Stallings_Cryptography_and_Network_Security.pdf

Statista. (2018). *Number of apps available in leading app stores as of 3rd quarter 2018.* Available at https://www.statista.com/statistics/276623/number-of-apps-available-in-leading-app-stores/

Statista. (2018). *Number of smartphones sold to end users worldwide from 2007 to 2017 (in million units).* Available at https://www.statista.com/statistics/263437/global-smartphone-sales-to-end-users-since-2007/

Statista. (2018). *Smartphones industry: Statistics & Facts.* Available at https://www.statista.com/topics/840/smartphones/

Statista. (n.d.). Global wearable technology market 2012-2018 | Statistic. Retrieved from http://bit.ly/mpo0617tp1

Staugaard, A. C. (1987). *Robotics and AI: an introduction to applied machine intelligence* (p. 320). Englewood Cliffs, NJ: Prentice-Hall.

Steele, R., Lo, A., Secombe, C., & Wong, Y. K. (2009). Elderly persons' perception and acceptance of using wireless sensor networks to assist healthcare. *International Journal of Medical Informatics*, *78*(12), 788–801. doi:10.1016/j.ijmedinf.2009.08.001 PMID:19717335

Suhr, J. K., & Jung, H. G. (2014). *Sensor Fusion- Based Vacant Parking Slot Detection and Tracking. Intelligent Transportation Systems*. IEEE.

Sundaravadivel, P., Kougianos, E., Mohanty, S. P., & Ganapathiraju, M. K. (2018). Everything You Wanted to Know about Smart Health Care: Evaluating the Different Technologies and Components of the Internet of Things for Better Health. *IEEE Consumer Electronics Magazine*, *7*(1), 1–11. doi:10.1109/MCE.2017.2755378

Survey on impact of lifestyle on our health. Retrieved from www.preveu-eu.org

Suzuki, T., Tanaka, H., Minami, S., Yamada, H., & Miyata, T. (2013, March). Wearable wireless vital monitoring technology for smart health care. In *7th International Symposium Medical Information and Communication Technology,* Tokyo, Japan. 10.1109/ISMICT.2013.6521687

Swanson, B. E., & Ramiller, N. C. (2004). Innovating mindfully with information technology. *Management Information Systems Quarterly*, *28*(4), 553–583. doi:10.2307/25148655

Tajima, M. (2007). Strategic value of RFID in supply chain management. *Journal of Purchasing and Supply Management*, *13*(4), 261–273. doi:10.1016/j.pursup.2007.11.001

Takeda, R., Tadano, S., Todoh, M., Morikawa, M., Nakayasu, M., & Yoshinari, S. (2009). Gait analysis using gravitational acceleration measured by wearable sensors. *Journal of Biomechanics*, *42*(3), 223–233. doi:10.1016/j.jbiomech.2008.10.027 PMID:19121522

Talari, S., Shafie-khah, M., Siano, P., Loia, V., Tommasetti, A., & Catalão, J. P. (2017). A review of smart cities based on the internet of things concept. *Energies*, *10*(4), 421. doi:10.3390/en10040421

Tang, V. W., Zheng, Y., & Cao, J. (2006). An intelligent car park management system based on wireless sensor networks. Pervasive Computing and Applications, (pp. 65–70), *1st International Symposium on IEEE.* 10.1109/SPCA.2006.297498

Tarouco, L. M., Bertholdo, L. M., Granville, L. Z., Arbiza, L. M., Carbone, F., Marotta, M., & De Santanna, J. J. (2012). Internet of Things in healthcare: Interoperatibility and security issues. *2012 IEEE International Conference on Communications (ICC).* 10.1109/ICC.2012.6364830

Tech-Styles: Are Consumers Really Interested in Wearing Tech on their Sleeves? (n.d.). Retrieved from http://bit.ly/mpo0617tp2

Teixeira, A. F., & Postolache, O. (2014, May). *Wireless sensor network and web based information system for asthma trigger factors monitoring.* Paper presented at IEEE International Instrumentation and Measurement Technology Conference, Montevideo, Uruguay.

The Global Asthma report. (2014). Available at www.globalasthmareport.org

The Lancet. (n.d.). *Global, regional, and national life expectancy, all-cause mortality, and cause specific mortality for 249 causes of death, 1980-2015: a systematic analysis for the Global Burden of Disease Study 2015, 388*(10053):1459–1544. doi:10.1016/S0140-6736(16)31012-1

The Lancet.(n.d.). *Global, regional, and national age-sex specific all-cause and cause-specific mortality for 240 causes of death, 1990-2013: a systematic analysis for the Global Burden of Disease Study 2013, 385*(9963): 117–71. doi:10.1016/S0140- 6736(14)61682-2.

Thompson, N. B. (2003). The big showdown? Will RFID technology eventually phase out bar code? Paper. *Film and Foil Converter, 77*(8), 38–40.

Thomson, J., Hass, C., Horn, I., Kleine, E., Mitchell, S., Gary, K., & Amresh, A. (2017, April). *Aspira: Employing a serious game in an mHealth app to improve asthma outcomes.* Paper presented at Serious Games and Applications for Health, Perth, Australia.

To, G., & Mahfouz, M. R. (2013). Modular wireless inertial trackers for biomedical applications. *2013 IEEE Topical Conference on Power Amplifiers for Wireless and Radio Applications.* 10.1109/PAWR.2013.6490226

Tolentino, M. (2013, May 30). *4 Security Challenges for Fitbit, Google Glass + Other Wearable Devices.* Available at siliconangle.com/blog/2013/05/30/4-securitychallenges- for-fitbit-google-glass-other-wearable-devices/

Tong, Y., & Li, Z. Design of Intelligent Socket Based on WiFi, *2017 4th International Conference on Information Science and Control Engineering (ICISCE),* Changsha, China, pp. 952-955. 10.1109/ICISCE.2017.201

Torfs, T., Sterken, T., Brebels, S., Santana, J., Van den Hoven, R., Spiering, V., ... Zonta, D. (2013). Low Power Wireless Sensor Network for Building Monitoring. *IEEE Sensors Journal, 13*(3), 909–915. doi:10.1109/JSEN.2012.2218680

Troiano, R. P., McClain, J. J., Brychta, R. J., & Chen, K. Y. (2014). Evolution of accelerometer methods for physical activity research. *British Journal of Sports Medicine, 48*(13), 1019–1023. doi:10.1136/bjsports-2014-093546 PMID:24782483

Tsow, F., Forzani, E., Rai, A., Wang, R., Tsui, R., Mastroianni, S., ... Tao, N. J. (2009). A Wearable and Wireless Sensor System for Real-Time Monitoring of Toxic Environmental Volatile Organic Compounds. *IEEE Sensors Journal, 9*(12), 1734–1740. doi:10.1109/JSEN.2009.2030747

Turban, E., Mclean, E., Wetherbe, J., Bolloju, N., & Davison, R. (2002). *Information Technology Management: Transforming Business in the Digital Economy* (3rd ed.). New York, NY: Wiley.

Udayakumar, E., ... (2016). Automatic Battery Replacement of Robot, *7th International Conference on Advances in Natural and Engineering (ICECE)*. IEEE. Retrieved from https://www.arduino.cc/en/Main/arduinoBoardMega. *Indian Journal of Science and Technology, 8*(23).

Udayakumar, E., Ramesh, C., Tamilselvan, S., Yogeshwaran, K., & Kanagaraj, T. (2016). Foot Pressure Measurement by using ATMEGA 164 Microcontroller. *Advances in Natural and Applied Sciences, 10*(13), 224-229.

Uddin, J., Reza, S. T., Newaz, Q., Islam, T., & Kim, J. M. . (2012). Automated irrigation system using solar power. *Electrical & Computer Applied Sciences, AENSI Publications, 9*(7), 2015, pp. 33-38.

Uddin, M., Gupta, A., Maly, K., Nadeem, T., Godambe, S., & Zaritsky, A. (2013, December). SmartSpaghetti: Use of smart devices to solve health care problems. In *IEEE International Conference on Bioinformatics and Biomedicine,* Shanghai, China. 10.1109/BIBM.2013.6732598

Uswatte, G., Miltner, W. H., Foo, B., Varma, M., Moran, S., & Taub, E. (2000). Objective Measurement of Functional Upper-Extremity Movement Using Accelerometer Recordings Transformed With a Threshold Filter. *Stroke, 31*(3), 662–667. doi:10.1161/01.STR.31.3.662 PMID:10700501

Uwaoma, C., & Mansingh, G. (2015, January). Towards Real-time Monitoring and Detection of Asthma Symptoms on Resource-constraint Mobile Device. In *12th Annual IEEE Consumer Communications and Networking Conference,* Las Vegas, Nevada. 10.1109/CCNC.2015.7157945

Vanishreeprasad, S., & Pushpalatha, M. N. (2015, May-June). Design and implementation of hybrid cryptosystem using AES and hash function. *IOSR Journal of Electronics and Communication Engineering, 10*(3), 18–24. doi:10.9790/2834-10321824

Vankatesh, J., Aksanli, B., Chan, C. S., Akyurek, A. S., & Rosing, T. S. (2018). Modular and personalized smart health application design in a smart city environment. *Internet of Things Journal, 5*(2), 614–623. doi:10.1109/JIOT.2017.2712558

Varaiya, P. (1993). Smart cars on smart roads: Problems of control. *IEEE Transactions on Automatic Control, 38*(2), 195–207. doi:10.1109/9.250509

Veltink, P. H., & Boom, H. B. (1996). 3D Movement Analysis Using Accelerometry — Theoretical Concepts. *Neuroprosthetics: from Basic Research to Clinical Applications*, 317-325. doi:10.1007/978-3-642-80211-9_39

Veronneau, S., & Roy, J. (2009). RFID benefits, costs, and possibilities: The economical analysis of RFID deployment in a cruise corporation global service supply chain. *International Journal of Production Economics, 122*(2), 692–702. doi:10.1016/j.ijpe.2009.06.038

Vettoretti, M., Cappon, G., Acciaroli, G., Facchinetti, A., & Sparacino, G. (2018). Continuous Glucose Monitoring: Current Use in Diabetes Management and Possible Future Applications. *Journal of Diabetes Science and Technology, 12*(5), 1064–1071. doi:10.1177/1932296818774078 PMID:29783897

Vrba, J., & Vrba, D., Diaz, l. & Fiser, O. (2018). *Metamaterial sensor for microwave non-invasive blood glucose monitoring*. Paper presented at World Congress on Medical Physics and Biomedical Engineering, Prague, Czech Republic.

Wang, H., & He, W. (2011, April). A reservation-based smart parking system. In 2011 IEEE Conference on Computer Communications Workshops (INFOCOM WKSHPS), (pp. 690-695). IEEE

Wang, H., Huang, Q., Zhang, C., & Xia, A. (2010, December). A novel approach for the layout of electric vehicle charging station. In *2010 International Conference on Apperceiving Computing and Intelligence Analysis (ICACIA),* (pp. 64-70). IEEE

Wang, F. Y., Zeng, D., & Yang, L. (2006). Smart cars on smart roads: An IEEE intelligent transportation systems society update. *IEEE Pervasive Computing*, *5*(4), 68–69. doi:10.1109/MPRV.2006.84

Wang, G., Poscente, M. D., Park, S. S., Andrews, C. N., Yadid-Pecht, O., & Mintchev, M. P. (2017). Wearable microsystem for minimally invasive, pseudo-continuous blood glucose monitoring: The e-Mosquito. *IEEE Transactions on Biomedical Circuits and Systems*, *11*(5), 979–987. doi:10.1109/TBCAS.2017.2669440 PMID:28574366

Wang, H., & He, W. (2011). A Reservation-based Smart Parking System. *The First International Workshop on Cyber-Physical Networking Systems.*

Want, R. (2004). RFID: A key to automating everything. *Scientific American*, *290*(January), 56–68. doi:10.1038scient ificamerican0104-56 PMID:14682039

Want, R., Schilit, B. N., & Jenson, S. (2015). Enabling the Internet of Things. *Computer*, *48*(1), 28–35. doi:10.1109/MC.2015.12

Water crisis continues while dam levels rise - The Namibian. (n.d.). Retrieved from https://www.namibian.com.na/161330/archive-read/Water-crisis-continues-while-dam-levels-rise

Watts, G. (2012). UK Biobank opens its data vaults to researchers. *BMJ, 344*(mar30 2), e2459-e2459. doi:10.1136/bmj.e2459

Wearables and the Internet of Things for Health - IEEE PULSE. (2015, May 20). As Fitbit Prepares for IPO, New Consumer Research Reveals Areas of Wearables Market Vulnerability for Fitness Band Leader. Retrieved from http://www.argusinsights.com/fitbit-ipo-release/

Wearables security: Do enterprises need a separate WYOD policy? (n.d.). Retrieved from https://searchsecurity.techtarget.com/answer/Wearables-security-Do-enterprises-need-a-separate-WYOD-policy

Weber, S. (2012). Comparing Key Characteristics of Design Science Research as an Approach and Paradigm. In PACIS (p. 180).

Wei, L., Wu, Q., Yang, M., Ding, W., Li, B., & Gao, R. (2012). Design and Implementation of Smart Parking Management System Based on RFID and Internet. *International Conference on Control Engineering and communication Technology (ICCECT).* 10.1109/ICCECT.2012.12

Weiss, A. (2003). Me and my shadow: RFID tags polarize the debate over privacy vs. efficiency, *networker, 7*(3), 24-30.

Werner, K., & Schill, A. (2009). Automatic Monitoring of Logistics Processes Using Distributed RFID based Event Data. *International Workshop on RFID Technology,* Milan, Italy, 101-108.

What is Smart Farming? (n.d.). Retrieved from https://www.smart-akis.com/index.php/network/what-is-smart-farming/

What Is Water Scarcity? | Fluence. (n.d.). Retrieved from https://www.fluencecorp.com/what-is-water-scarcity/

White, S. A., Walley, K. S., Johnston, J. W., Henderson, P. M., Hale, K. H., & Andrews, W. B., Jr. & Siann, J. I. (2003). U.S. Patent No. 6,531,982. Washington, DC: U.S. Patent and Trademark Office.

White, S. A., Walley, K. S., Johnston, J. W., Henderson, P. M., Hale, K. H., & Andrews, W. B., Jr., & Siann, J. I. (2003). U.S. Patent No. 6,531,982. Washington, DC: U.S. Patent and Trademark Office.

Wikipedia. (2018). Mobile phone. Available at https://en.wikipedia.org/wiki/Mobile_phone

Windhoek water savings still crucial - Local News - Namibian Sun. (n.d.). Retrieved from https://www.namibiansun.com/news/windhoek-water-savings-still-crucial2018-01-12

Wlezien, C. (1995). The public as thermostat: Dynamics of preferences for spending. *American Journal of Political Science*, *39*(4), 981–1000. doi:10.2307/2111666

World Bank Group. (2014). *World development indicators 2014*. World Bank Publications.

World Health Organization. (2007). *Global surveillance, prevention and control of chronic respiratory diseases: a comprehensive approach.*

Wu, F., Rüdiger, C., Redouté, J., & Rasit Yuce, M. (2018). A Wearable Multi-sensor IoT Network System for Environmental Monitoring. *Internet of Things*, 29-38. doi:10.1007/978-3-030-02819-0_3

Wu, F., Rüdiger, C., & Yuce, M. (2017). Real-Time Performance of a Self-Powered Environmental IoT Sensor Network System. *Sensors (Basel)*, *17*(2), 282. doi:10.339017020282 PMID:28157148

Wu, T., Wu, F., Redoute, J., & Yuce, M. R. (2017). An Autonomous Wireless Body Area Network Implementation Towards IoT Connected Healthcare Applications. *IEEE Access: Practical Innovations, Open Solutions*, *5*, 11413–11422. doi:10.1109/ACCESS.2017.2716344

www.thingspeak.com

Xiao, M., Liu, Y. H., & Hu, Q. (2016). Design and Implementation of Socket-Based-Network Connections Smart Home System, *2016 Sixth International Conference on Instrumentation & Measurement, Computer, Communication and Control (IMCCC)*, Harbin, China, pp. 756-760. 10.1109/IMCCC.2016.113

Xu, A., & He, S. (2017). The wireless smart socket control system design, *2017 2nd International Conference on Advanced Robotics and Mechatronics (ICARM)*, Hefei, China, pp. 698-703. 10.1109/ICARM.2017.8273247

Xu, G. (2016). Solar roadways. V sbornike: Rasshirennoye vosproizvodstvo innovatsionnoy ekonomiki i intensifikatsiya sprosa na innovatsii v Rossii Sbornik nauchnykh statey. FGBOU VPO «Rossiyskiy ekonomicheskiy universitet imeni GV Plekhanova, 128-130.

Yang, C., Shen, W., & Wang, X. (2018). The internet of things in manufacturing: Key issues and potential applications. *IEEE Systems, Man, and Cybernetics Magazine*, *4*(1), 6–15. doi:10.1109/MSMC.2017.2702391

Yan, J. (2015). *A guide to understanding machinery prognostics and prognosis oriented maintenance management.* Hoboken, NJ: Wiley.

Yao, C. A., & Ho, K. (2017). Mobile Sensors and Wearable Technology. *Handbook Integrated Care*, 113-119. doi:10.1007/978-3-319-56103-5_7

Yao, H., Shum, A. J., Cowan, M., Lahdesmaki, I., & Parviz, B. A. (2011). A contact lens with embedded sensor for monitoring tear glucose level. *Biosensors & Bioelectronics*, *26*(7), 3290–3296. doi:10.1016/j.bios.2010.12.042 PMID:21257302

Yelapure, S. J., & Kulkarni, R. V. (2012). Literature Review on Expert System in Agriculture. *International Journal of Computer Science and Information Technology. 3*(5), 5086-5089.

Yihua, Z. (2010). VIP customer segmentation based on data mining in mobile-communications industry in Computer Science and Education, (pp. 156–159), 5th *International Conference on IEEE.*

Yoon, H., Xuan, X., Jeong, S., & Park, J. Y. (2018). Wearable, robust, non-enzymatic continuous glucose monitoring system and its in vivo investigation. *Biosensors & Bioelectronics*, *117*, 267–275. doi:10.1016/j.bios.2018.06.008 PMID:29909198

You, I., Choo, K. K. R., & Ho, C. L. (2018). A smartphone-based wearable sensors for monitoring real-time physiological data. *Computers & Electrical Engineering*, *65*, 376–392. doi:10.1016/j.compeleceng.2017.06.031

Yuan, Y., & Zhang, J. J. (2003). Towards an appropriate business model for m-commerce. *International Journal of Mobile Communications*, *1*(1-2), 35–56. doi:10.1504/IJMC.2003.002459

Yung-Cheng, M., Chao, Y. P., & Tsai, T. Y. (2013, September). Smart-clothes—Prototyping of a health monitoring platform. Paper presented at IEEE Third International Conference on Consumer Electronics, Berlin, Germany.

Zaman, N., & Almusalli, F. A. (2017, April). Smartphones power consumption & energy saving techniques. In *2017 International Conference on Innovations in Electrical Engineering and Computational Technologies (ICIEECT) (pp. 1-7). IEEE.*

Zanella, A., Bui, N., Castellani, A., Vangelista, L., & Zorzi, M. (2014). *Internet of Things for Smart Cities. IEEE INTERNET OF THINGS JOURNAL.*

Zanon, M., Mueller, M., Zakharov, P., Talary, M. S., Donath, M., Stahel, W. A., & Caduff, A. (2018). First experiences with a wearable multisensor device in a noninvasive continuous glucose monitoring study at home, Part II: The investigators' view. *Journal of Diabetes Science and Technology*, *12*(3), 554–561. doi:10.1177/1932296817740591 PMID:29145749

Zapata, B. C., Fernández-Alemán, J. L., Idri, A., & Toval, A. (2015). Empirical Studies on Usability of mHealth Apps: A Systematic Literature Review. *Journal of Medical Systems*, *39*(2), 1. doi:10.100710916-014-0182-2 PMID:25600193

Zhang, W., Du, Y., & Wang, M. L. (2015). Noninvasive glucose monitoring using saliva nano-biosensor. *Sensing and Bio-Sensing Research*, *4*, 23–29. doi:10.1016/j.sbsr.2015.02.002

Zhang, Y., Zheng, D., & Deng, R. H. (2018). Security and privacy in smart health: Efficient policy-hiding attribute-based access control. *IEEE Internet of Things Journal*, *5*(3), 2130–2145. doi:10.1109/JIOT.2018.2825289

Zheng, H., He, J., Li, P., Guo, M., Jin, H., Shen, J., ... Chi, C. (2018). Glucose Screening Measurements and Noninvasive Glucose Monitor Methods. *Procedia Computer Science*, *139*, 613–621. doi:10.1016/j.procs.2018.10.202

Zhou, F., & Li, Q. (2014). Parking Guidance System Based on ZigBee and Geomagnetic Sensor Technology. In *Proceedings of 13th International Conference on Distributed Computing and Applications to Business, Engineering and Science (DCABES)* (pp. 268-271). IEEE.

Zhou, F., & Li, Q. (2014), Parking Guidance System Based on ZigBee and Geomagnetic Sensor Technology. *13th International Conference on Distributed Computing and Applications to Business, Engineering and Science (DCABES)*, pp. 268-271. 10.1109/DCABES.2014.58

Zion Market Research. (n.d.). Smart Agriculture Market by Agriculture Type. Retrieved from https://www.zionmarketresearch.com/report/smart-agriculture-market

About the Contributors

Dinesh Goyal is a Professor, Department of Computer Science & Engineering, and Director, Poornima Institute of Engineering & Technology. He holds a Ph.D. degree from Suresh Gyan Vihar University in 2014, and also has 16 years of rich experience in teaching with seven years in pure R&D sector of Image Processing, cloud computing/ information security, and over 90 research papers in international publications followed by 21 papers presented in conferences, more than eight of which are SCI/Scopus/Web of Sciences/Thomson Reuter Indexed Journals. Also, in order to have inquisitiveness for the current trends in technology, he has conducted and organized four conferences and five workshops followed by five National paper publications. In addition to this, he has strived to guide mentor over 35 research scholars during his career, followed by eight doctoral guidance with five as supervisor and three as Co-supervisor.

S. Balamurugan is the Head-R&D, QUANTS IS & Consultancy Services, India. Formerly, he was the Director of Research and Development at Mindnotix Technologies, India. He has authored/edited 35 books, 200 international journals/conferences, and six patents to his credit. He received Three Post-Doctoral Degrees- a Doctor of Science (D.Sc.) degree and Two Doctor of Letters (D.Litt) degrees for his significant contribution to research and development in Engineering. His professional activities include roles as Associate Editor, editorial board member and/or reviewer for more than 100 international journals and conferences. He has been invited as Chief Guest/Resource Person/Keynote Plenary Speaker in many reputed Universities and Colleges at National and International Levels. His research interests include Artificial Intelligence, Augmented Reality, Internet of Things, Big Data Analytics, Cloud Computing, and Wearable Computing. He is a life member of the ACM, ISTE, and CSI.*

* * *

N. Brindha is a Masters in Computer Sciences and Engineering, and submitted a Ph.D. thesis. She is working as an Assistant Professor (Selection Grade) in the Department of Applied Mathematics and Computational Sciences of PSG College of Technology, Coimbatore, India. She has over 14 years of experience in teaching. She has guided more than 50 BSc (CSD) and MSc (CT) projects. She has organized workshops and attended various conferences, workshops, and FTP. Her subject expertise is Multimedia Systems, Distributed Systems, and Software Engineering. Her research interests include Intelligent Information Retrieval, Video Classification and Retrieval using Deep learning and Multimedia Streaming. She has around five publications in various journals and International/National conferences.

Brinta Chowdhury is a Ph.D. candidate in Department of Electrical Engineering at North Carolina A & T State University.

Sesinam Dagadu, before returning to Ghana in late 2013, was working with the British Police under the Digital and Electronics Forensic Service (DEFS), where he built cutting-edge electronic tools for use in forensics, and worked on some of the most high-profile cases in modern Met history. He also has a past building mission-critical, life-saving equipment in the automotive and military sectors for Jaguar Land Rover and the US and British Militaries respectively. Today, Sesinam is the founder of tinyDAVID, the creator of SnooCODE, a resilient and reliable addressing system for the developing world to improve service delivery and identification while improving job prospects at the bottom of the socioeconomic pyramid. His stated professional objectives are "to solve problems and to make new things possible."

S. Deepa has over 12 years of academic experience and currently holds the position of Assistant Professor (Senior Grade) in the department of Applied Mathematics and Computational Sciences at PSG College of Technology, Coimbatore, India. She received her M.Phil (Computer Science) and M.C.A degrees from Bharathiar University, Coimbatore, India. She has guided more than 50 B.Sc (CT), M.Sc (CT), and M.Phil projects. Her research interests include Network analysis and Web Mining. Her research publications appeared in peer- reviewed International journals and conference proceedings.

Udayakumar Easwaran was working as Assistant Professor in Department of Electronics and Communication Engineering at KIT-Kalaignarkarunanidhi Institute of Technology, Coimbatore, India. He is accommodating and versatile with three years of teaching in both UG and PG courses, with a solid commitment to the social and academic growth and development. He is pursuing his PhD at Anna University, Chennai. He completed his Master's degree (Communication Systems) programme from Sri Ramakrishna Institute of Technology, Coimbatore (Anna University, Chennai) in 2015. He published 20 papers in Peer-Reviewed International Journals and 15 papers in National & International Conferences. He is a Life Member in various Professional Societies like IETE, ISTE, IEI, SSI, BMESI, IAENG, ISRD, IASTER, ISOC, SAISE, UACEE, MEACSE, ICSES and SDIWC. He is an Editor for GJECR, IJSET, IJERECE Journals and Reviewer for BPJ, IJPPS, IJSET, RSIJ, IJMBOA, IJSPR, AJPCR, BBRA, IJRE Journals. He also received the Young Scientist award 2019 from Vedant Academics Bangkok Awards, KU Home, Kasetsart University, Bangkok, Thailand, I2OR Bright Researcher Award 2018 from International Institute of Organized Research, Chandigarh, and InSc Young Achiever Award 2019 from Institute of Scholars, Bengaluru, India. He had authored two books on Image processing. He had received funds from ISRO, ICMR, BMESI and Anna University for conduct of National Seminar, Workshop and FDTP. He acted as Organizing Committee Member, Outreach Committee Member, and Advisory Committee Member for many International Conferences. He had participated in many Workshops, Seminars, and FDPs.

Adyanth H received a B.E. degree in Electronics and Communication Engineering from JSS Science and Technology University, Mysore and is currently working as a Network Engineer at Cisco Systems Inc. Adyanth's research interests include the Internet of Things and programming.

V Hashiyana is a Namibian national. He obtained his Doctorate in Computer Science in Russia. Currently, he is a lecturer and HoD of Computer Science Department at University of Namibia. A few of his areas of research are Wireless technologies, Computer Networks and Security, and IoT.

Gloria Ejehiohen Iyawa holds a PhD in Information Systems. She is a Senior Lecturer in the Department of Informatics at the Namibia University of Science and Technology. She serves on four Editorial

Committees and acts as reviewer for both national and international journals. Her work is well published in peer-reviewed forums. Her research interests include: Digital Health, Health Informatics, Software Engineering, and E-Government.

Nikita Jain received M. Tech degree in 2014 from IIIT Delhi (a state university under Govt. of NCT Delhi) with specialisation in Information Security. She graduated from Maharaja Agrasen Institute of Technology in Information Technology, GGSIPU University. She is working as a computer science Assistant Professor in Bharati Vidyapeeth College of Engineering (GGSIPU).She is a PhD Research scholar at Guru Gobind Singh Indraprastha University in Computer Science, specialising in Wireless sensor networks since 2016. Her current research interests are Wireless Sensor networks, Remote sensing and applications, and Social Media mining.

E. Golden Julie received her BE degree in Computer Science and Engg in 2005, and ME degree in Computer Science and Engineering in 2008, from Anna University Chennai. She finished PhD in Anna University Chennai in 2016. Presently, she is working as assistant professor in Regional Centre Anna University, Tirunelveli, India. She has published many research papers in various fields. Her research areas include Wireless Sensor Adhoc Networks and Image Processing. She is a member of ISTE.

Asiya Khan is a lecturer in Control Systems Engineering, Athena SWAN academic lead and Associate Head for Teaching and Learning in the School of Engineering at the University of Plymouth (UoP), and a member of the Autonomous Marine Systems (AMS) research group. Previously, she has worked in industry in the field of payphones and electronic payments, and as a research assistant on the EU FP7 project Adamantium. Her research expertise is in the field of Quality of Service/Experience (QoS/E) monitoring, prediction and control of multimedia communication over wireless access networks. She has published over 40 journal and conference papers, including a ‘Best paper award’ in 2009.

Victor Kulula’s research interest is in Wireless technologies and how they can benefit or simplify everyday activities. Modern technology comes at a high cost. This, therefore, brings about the need for systems that will monitor and alert users at a cost-effective price with accurate and dependable results. He is working as Systems Administrator at Standard Bank of Namibia, where he is responsible for monitoring all critical systems of the bank, ATM and POS transaction, Internet banking, mobile banking, and supporting branches with their day-to-day activities.

Ananya M. B. is a final-year student pursuing BE Electronics and Communication, and interested in cyber security and IoT.

Robson Locatelli Macedo is a Professor at the Federal Rural University Semi-Arid e Master in Computer Science.

Kristine Mae Magtubo is a Doctor of Medicine from the University of the Philippines. She is co-founder and Chief Executive Officer of Veris Tech, Inc., a biomedical device start-up in the Philippines. She was university researcher at the National Telehealth Center where she led multiple projects in telemedicine and public health innovation. Her research interests include innovation for low-resource settings, telemedicine, biomedical devices, local health systems, and innovation policy.

Neetu Marwah is an Assistant Professor at the Department of Electronics and Communication Engineering, Manipal University Jaipur, Jaipur, India, and has completed Ph.D. from Manipal University Jaipur, in Antenna Systems. She has more than 16 years of experience in teaching and research. Her skills include design and development of wideband antenna for mobile and satellite communication, curriculum development, and individual and group training. She has published more than 25 research papers in various International and National journals and conferences of repute and is a life member of Institution of Engineers (India).

Brojo Kishore Mishra is working as a Professor in the department of CSE, School of Engineering, at GIET University, Gunupur-765022, India. He received M.Tech and Ph.D. degrees in Computer Science from the Berhampur University in 2008 and 2012, respectively. Currently, he is guiding three Ph.D research scholars under Biju Pattnaik University of Technology, Odisha. His research interests include Data mining and big data analysis, machine learning, Soft computing, and Evolutionary computation. He has published more than 55 research papers in internationally-reputed journals and referred conferences, and 15 book chapters. He has edited three books, authored two books, and is acting as a member of the editorial board/associate editor / guest editor of various International journals. He served in the capacity of Keynote Speaker, Plenary Speaker, Program Chair, Proceeding chair, Publicity chair, Special session chairperson, and also as member of programme committees of many international conferences. He was a Principal Investigator in a CSI-funded research project. He is a life member of ISTE, CSI, and member of IEEE, ACM, IAENG, UACEE, and ACCS.

Sushruta Mishra has completed his M.Tech degree from IIIT Bhubaneswar, India. He has completed his Ph.D (CSE) from KIIT University, Bhubaneswar. He is working as an Assistant Professor in the School of Computer Engineering, KIIT University.

Stephanie Guimarães Morais has a Bachelor's degree in Technology and Science from Federal Rural University of the Semiarid – Brazil (2018). His current research interest is computing engineering.

Helder Fernando de Araujo Oliveira has a doctoral degree in Computer Science from the Federal University of Campina Grande (2015). He is Assistant Professor at the Federal Rural University of the Semi-Arid. His research interests are computer architecture and embedded systems.

Kamalendu Pal is with the Department of Computer Science, School of Mathematics, Computer Science and Engineering, City, University of London. Kamalendu received his BSc (Hons) degree in Physics from Calcutta University, India; Postgraduate Diploma in Computer Science from Pune, India; MSc degree in Software Systems Technology from Sheffield University; Postgraduate Diploma in Artificial Intelligence from Kingston University; MPhil degree in Computer Science from University College London; and MBA degree from University of Hull, UK. He has published dozens of research papers in international journals and conferences. His research interests include knowledge-based systems, decision-support systems, computer integrated design, software engineering, and service-oriented computing. He is on the editorial board of international computer science journals. He is a member of the British Computer Society, the Institution of Engineering and Technology, and the IEEE Computer Society.

Selvakumar Palanisamy received his B.Sc (2006), M.Sc (2009), and M. Phil (2010) degrees in Chemistry from Madurai Kamaraj University, Tamil Nadu, India. He obtained his PhD degree (2014) in Chemical Engineering from National Taipei University of Technology (NTUT), Taiwan. He was a post-doctoral fellow at Chemical Engineering & Biotechnology (2014-2018), NTUT, Taiwan. Currently, he is working as a Research Associate in the School of Engineering at Manchester Metropolitan University, UK. He is a member of the Royal Society of Chemistry (RSC), UK. His current research interest includes the synthesis of nanomaterials and the development of electrochemical sensors and biosensors for environmental and healthcare applications. He has published over 85 research articles in SCI journals, and he is a consulting editor of the Journal of reports in electrochemistry.

Amiya Ranjan Panda has seven years of research experience in DRDO and one year of teaching experience. He received the B.Tech. degree in information technology from Biju Patnaik University of Technology, Rourkela, India, in 2009, and the M.Tech. degree in computer science and engineering from the Kalinga Institute of Industrial Technology (KIIT), Bhubaneshwar, India, in 2012. He received a Ph.D. degree from Siksha "O" Anusandhan University, Bhubaneshwar, India, in 2017, working in a real-time project "Design, development and implementation of Software Defined Radio based Flight Termination System" of DRDO, ITR, Chandipur, India. He has worked as JRF, SRF, and RA in DRDO, ITR, Chandipur, for more than seven years. Then, he worked as Assistant Professor in Siksha 'O' Anusandhan University for four months, and currently he is working as Assistant Professor in KIIT Deemed to be University, Bhubaneswar. His research interests are Machine learning, IOT, data acquisition system, and Software Defined Radio. He has published 15 international journals. Two of the journals are published in IEEE Transaction of Industrial Informatics. One edited book has been accepted for publication in 2019. He is expert in teaching Data Structure and algorithm, Database Management System (DBMS), Artificial Intelligence, Operating System, Compiler Design, Computer Network, Machine Learning, etc.

Shikha Patel is a fourth-year B.Tech student pursuing computer science and engineering from Kalinga Institute of Industrial Technology. She is interested in researching Iot, Artificial Intelligence, and data mining.

Ravindra Singh Rajput is an Assistant Professor in Computer science with the Department of Mathematics, Statistics and Computer Science, G. B. Pant University of Agriculture & Technology, Pantnagar, Uttarakhand, India, for the last 10 years, and is also pursuing Ph.D. in Computer Science from Suresh Gyan Vihar University, Jaipur. Before joining G. B. Pant University, he has worked with Modi Olivetti Limited, Spice Mobile, National Informatics Center, and Hindustan Aeronautics Limited. His research interests include Cloud Computing, Big Data Analytics, System Reliability, AI, and Data Mining. He has attended six international conferences and published more than 10 research papers, and two book chapters with prestigious publishers. He has also developed more than 20 software applications for government and business organizations.

Jana Shafi obtained her B.Tech (2010), and M.Tech (2012) in Computer Science & Engineering from Maharshi Dayanand University, Rohtak, Haryana, India, and pursuing Ph.D from Sri Padmavati Mahila Visvavidayalam,Tirupati, India. She has 8 years of teaching experience .She is a member of various computer societies. She has 20 papers, one book, and book chapters published. She attended many conferences, including IEEE conferences and workshops. Her Research interests are social networks,

IoT, and soft computing. She has been serving as member of review committee for the Asian Journal of Computer Science Engineering (AJCSE).

T. Shreekanth received his Bachelor of Engineering in Electronics and Communication from PES College of Engineering, Mandya, Karnataka, India, his Master's degree in the area of Bio-Medical Signal Processing from Sri Jayachamarajendra College of Engineering, Mysore, Karnataka, and PhD in the area of Image and Speech signal processing from University of Mysore, India. His areas of interest include Digital Image Processing, Computer Vision and Speech Signal Processing. He has to his credit more than 50 Journal and Conference papers at national and international levels. Since 2011, he has been working as an Assistant Professor in the Department of Electronics and Communication at Sri Jayachamarajendra College of Engineering, Mysore, Karnataka, India.

Rupert Calvin Sievert graduated from the University of the Philippines with a degree in Bachelor of Science in Electronics and Communications Engineering. He worked as software engineer in government-funded projects (Philippines) to develop biomedical devices for maternal & fetal healthcare monitoring and newborn hearing screening.

Ishpal Singh received his B.Tech. degree in Electronics and Communications from the Jawahar Lal Nehru University, New Delhi, India, in 2001, and Master of Technology in Computer Science & Engineering from the Indian Institute of Technology (IIT) Chennai, Tamil Nadu, India, in 2006. He is presently posted as the Military Advisor to the Ministry of Defence, Government of Namibia, and is pursuing his PhD in Computer Science from the Namibia University of Science & Technology. He has published papers in the area of Data Security in Network Management Systems. His current research interests include network security, symmetric encryption algorithms, and cyber security.

Alvaro Sobrinho has a B.Sc. degree in Systems Analysis from the CESMAC University Center (2010), an M.Sc. degree in Computational Modeling of Knowledge from the Federal University of Alagoas (2013), and a Ph.D. degree in Computer Science from the Federal University of Campina Grande, Brazil (2016). He was a researcher and professor at the Federal Rural University of the Semi-Arid from 2017 to 2019. He is currently a researcher and professor at the Federal Rural University of Pernambuco, Academic Unit of Garanhuns. His research interests are formal methods, pervasive computing, and embedded medical systems.

Evelyn R. Sowells-Boone is an Assistant Professor in the Computer Systems Technology department at North Carolina A&T State University's College of Science and Technology. Prior to joining the School of Technology faculty, she held positions at U.S. Department of Energy, and North Carolina A&T State University's Division of Research and College of Engineering. Dr. Sowells-Boone earned a Ph.D. in Electrical Engineering from N. C. A&T University's College of Engineering. She also holds a M.S. and B.S in Computer Science with a concentration in software engineering. Her primary research interests are in the areas of low-power digital systems design, self-timed digital system design, and STEM education. As a result of her work, she has numerous peer-reviewed journal and conference publications. She recently authored a book entitled "Low Power Self-Timed Size Optimization for an Input Data Distribution," which explores innovative techniques to reduce power consumption for portable electronic devices. She was also recently awarded the 2016 Chair's award for Outstanding Junior Researcher and

Teaching Excellence in the Computer System Technology department, and is an NSF ADVANCE IT faculty Scholar. Dr. Sowells-Boone is the lead investigator of the Females in Technology (FiT) summer boot camp grant project for academically-gifted, low income-rising senior and junior high girls for recruitment into the technology degree areas. Evelyn is not only outstanding in teaching and research, but also in service. She received the 2013 Chair's Award for Outstanding Service in the Department of Computer System Technology, and is a member of Upsilon Phi Epsilon, Computer Science Honor Society, American Society of Engineering Education's Electronic Technology and Women in Engineering Divisions, and American Association of University Women.

Nalina Suresh was born in India in 1966. She studied her primary, secondary, and pre-university education in Tumakuru, Karnataka State, India, with excellence performance throughout her pre university education. She obtained B.Tech Degree in Electrical and Electronics Engineering with Cum-Laude from University of Mysore in 1988. She then very briefly worked as visiting teaching faculty at S J College of Engineering in the department of Electrical & Electronics. She obtained her Master's degree in Industrial Electronics (M.Tech) with Cum-Laude from University of Mysore in 1991. She was appointed as Lecturer at Siddaganga Institute of Technology (S.I.T) in the department of Electronics. She served the University of Bangalore, India, for seven years and taught various spectrums of engineering areas. She contributed many technical talks, workshops, and presentations during her professional career at S.I.T. Dr. Nalina Suresh got her first overseas assignment as a Lecturer at Defence University in Ethiopia, East Africa, in the department of Electronics from 1998 to 2000. She taught various subjects for both Diploma and Degree students with defense-specialization subjects. She contributed technical writings for lecturer notes and student journals during her services at Defence University. She got the best teacher award from the University in 1999. She further got an opportunity at National University of Science & Technology (NUST) at Bulawayo, Zimbabwe, as a Lecturer in the Department of Computer Science from June 2000 to February 2002. She taught both Undergraduate and Post Graduate students in the field of Computer Science & Engineering. Then she got an opportunity at the Namibia University of Science and Technology (NUST), Windhoek, as a Lecturer in Electrical and Electronics Engineering Department from March 2002 to March 2005. She taught Diploma, Degree, and B.Tech programmes. She was instrumental in designing many subjects for Degree programmes during her services at the NUST, and contributed to starting a degree programme in Computer Science engineering at NUST. She got the best teacher award from the University in 2004. Dr. Nalina Suresh moved to University of Namibia (UNAM) in April 2005 as a Lecturer at the Department of Computer Science and still serves in that capacity. She taught broad spectrum of modules for more than a decade. She even got a Long Service Award at UNAM. During her tenure at UNAM, she got her PhD in Computer Science from UNAM in 2016. She is very actively involved in conducting departmental activities, leadership duties, science exhibitions, and revising new curriculum. She has served as a project coordinator, mentor, guide, exam coordinator, and student registration worker to mention a few. She got the full faith and confidence from the entire teaching and non-teaching faculty and students from the department of computer science due to her simplicity, humility, and great academic contribution and dedication to the department and to the university in general. She even got best lecturer award in the year 2013 from the department during the annual retreat. Dr. Nalina Suresh is a life member of All India Council for Technical Education (AICTE), Indian Society for Technical Education (ISTE), and Computer Society of India (CSI). She has participated in more than 50 technical workshops, talks, discussions, and presentations. She is also a member of academic, examination, and reviewing committees. She has published a few peer-reviewed scientific papers in international journals, and presented a few conference papers. Her research interests include Cloud Computing, Remote sensor networks, simulation and modelling of Digital techniques, Computational Theory, and Machine Learning.

Shreekanth Thotappa is an Indian national. He obtained his Doctorate in Electronics and communication from JSS Science and Technology University, India. He is working as Associate Professor in the Department of Electronics and Communication, JSS Science and Technology University. His areas of research are Wireless communications and technologies, Digital Signal Processing, Image Processing, and IoT to mention a few.

Bharath V. N. is a final-year student pursuing Electronics and Communication in SJCE MYSURU.

Vijayalakshmi Velusamy is a Senior Lecturer in Electrical and Electronic Engineering at Manchester Metropolitan University, UK. Her multidisciplinary research focuses on the development of Smart Sensors and implementation of integrated biotechnological solutions for a wide range of applications including for environment, agriculture, and healthcare monitoring. She has developed portable biosensors and chemical sensors for environmental monitoring and medical diagnostics. Dr. Velusamy runs an active, multidisciplinary, and multicultural research team, which includes Electrical & Electronic Engineers, Microbiologists, Environmental scientists, Chemists, and Computer Scientists. She has co-authored over 50 publications. Her Sensors Technology research is recognised nationally and internationally for excellence. The research focuses on a sustainable environment, and improves quality of life and socio-economic empowerment.

P. Vetrivelan completed his doctoral degree from PSG College of Technology, Coimbatore, Tamil Nadu, India. He has more than two decades of teaching and research experience and one year of industrial experience. He has published more than 20 research papers in the field of Bio medical Image Processing in Springer and Elsevier journals. He has published more than 40 papers in refereed and indexed international conferences. He is the reviewer for numerous Elsevier-listed journals. He is currently serving as the Associate Professor to the Department of Electronics and Communication Engineering at PSG Institute of Technology and Applied Research, Neelambur, Coimbatore District, Tamil Nadu.

Amtul Waheed received a Bachelor of Computer Application from Osmania University, India, in 2003, a Master of Computer Application from Osmania University in 2006, and a Master of Technology from Jawaharlal Nehru Technological University. She has been working as lecturer in the Computer Science Department in College of Arts and Science, Prince Sattam Bin Abdul Aziz University, Saudi Arabia, since 2011. Her research interests include Web Services, Social Networks, and Internet of things (IoT). She has published many papers in international journals and has presented papers at international conferences. She has been serving as a member of a review committee for the Asian Journal of Computer Science Engineering (AJCSE).

Index

www.ingramcontent.com/pod-product-compliance
Lightning Source LLC
LaVergne TN
LVHW080636011026
849160LV00014B/1232
* 9 7 8 1 5 2 2 5 9 2 4 7 1 *